SpringerWienNewYork

CISM COURSES AND LECTURES

The series presents lecture notes, monographs, edited works and proceedings in the field of Mechanics, Engineering, Computer Science and Applied Mathematics.
Purpose of the series is to make known in the international scientific and technical community results obtained in some of the activities organized by CISM, the International Centre for Mechanical Sciences.

INTERNATIONAL CENTRE FOR MECHANICAL SCIENCES

COURSES AND LECTURES - No. 496

FLUID DYNAMICS OF CAVITATION AND CAVITATING TURBOPUMPS

EDITED BY

LUCA d'AGOSTINO
UNIVERSITY OF PISA, ITALY

MARIA VITTORIA SALVETTI
UNIVERSITY OF PISA, ITALY

SpringerWienNewYork

This volume contains 261 illustrations

Printed in Italy
SPIN 12182640

All contributions have been typeset by the authors.

ISBN 978-3-211-76668-2 SpringerWienNewYork

PREFACE

Cavitation is frequently encountered in hydraulic machinery, where it represents the major source of performance and life degradation and often provides the necessary excitation and compliance for triggering dangerous fluid dynamic instabilities. From the fundamental standpoint cavitation is a complex phenomenon, which poses formidable obstacles in terms of both physical and numerical modeling.

In July 2005 the Advanced School of the Centre International des Sciences Mecaniques (CISM), Udine, Italy, hosted a Course on the "Fluid Dynamics of Cavitation and Cavitating Turbopumps", held by some of the world leading experts in the field: Prof. C.E. Brennen, California Institute of Technology, Pasadena, California, USA, Prof. Y. Tsujimoto, Osaka University, Japan, Dr. J-P. Franc, LEGI, Grenoble, France, Prof. R. Saurel, Institut Universitaire de France and Polytech., Marseille, IUSTI, France, under the joint coordination by Prof. L. d'Agostino and M.V. Salvetti, Università di Pisa, Italy, who also contributed to the lectures.

The lectures of the speakers have been summarized here in a series of papers with the aim of providing a detailed introduction to the physics, fluid dynamics, modeling and numerical simulation of cavitation phenomena in engineering applications, with special emphasis on high performance turbopumps and their cavitation-induced instabilities. In particular, the first paper covers the more fundamental aspects of cavitation (nucleation, bubble dynamics, thermodynamic effects, cavitation erosion, stability of parallel bubbly flows) and the main kinds of cavitating flows (attached cavitation, cloud cavitation, supercavitation, ventilated supercavities, vortex cavitation, shear cavitation). The second paper provides an overview of the hydrodynamics and cavitation phenomena in turbopumps. The third series of papers focuses on the instabilities of cavitating turbopumps (cavitation surge, rotating cavitation, higher order cavitation surge, rotordynamic whirl forces). Finally, the last two papers describe two different approaches for the numerical simulation of cavitating flows. It is hoped that access to the above contributions is useful to students, researchers, scholars and professionals interested in perfecting their knowledge and understanding of cavitating flow phenomena and research in a wide range of engineering applications.

The coordinators wish to acknowledge the constant support of CISM's Advanced School in the organization of the Course, and would like to express their special gratitude to Ms. Paola Agnola for her kind assistance, to Professor Alfredo Soldati for his friendly encouragement and finally to Prof. C.E. Brennen, Prof.

Y. Tsujimoto, Dr. J-P. Franc and Prof. R. Saurel for their prompt collaboration and outstanding contributions to the Course and this publication.

Maria Vittora Salvetti and Luca d'Agostino

CONTENTS

The Rayleigh-Plesset equation: a simple and powerful tool to understand various aspects of cavitation.

Jean-Pierre FRANC

University of Grenoble, France

Abstract. This chapter is a general introduction to cavitation. Various features of cavitating flows are analyzed on the basis of the Rayleigh-Plesset equation. They concern not only the simple configuration of a single spherical bubble but also complex cavitating flows as those observed in cavitating turbopumps. Scaling rules, erosive potential, thermodynamic effect, supercavitation, traveling bubble cavitation, cavitation modeling are some of the topics addressed here. They are examined through this simple, basic equation which proves to be a quite useful tool for a first approach of real cavitation problems.

1 Introduction

Cavitation is the development of vapor structures in an originally liquid flow. Contrary to boiling, the phase change takes place at almost constant temperature and is due to a local drop in pressure generated by the flow itself.

The occurrence of low pressure regions in flows is a well-known phenomenon. For example, in the case of a Venturi, i.e. a converging duct followed by a diverging one, the velocity is maximum at the throat where the cross section is minimum. Then, according to Bernoulli equation, the pressure is minimum there and the risk of cavitation is maximum.

Another example is the flow around a foil at a given angle of attack which is representative of that around the blades of a hydraulic machine. From classical hydrodynamics, it is well-known that the foil is subject to a lift because of a lower pressure on the suction side in comparison to the pressure side. Hence, the suction side is expected to be the place where cavitation will first develop.

A final example is that of vortices which are very common structures in many flows. Because of the rotation and the associated centrifugal forces, the pressure in the core of such structures is lower than outside. Hence vortices are likely to cavitate in their core. There are actually many situations in which cavitating vortices can be observed as tip vortices or coherent vortical structures in turbulent flows like wakes or shear layers.

As known from basic thermodynamics, phase change from liquid to vapor occurs at the vapor pressure p_v which depends only upon the temperature. It is usually a good approximation to consider that the critical pressure for the onset of cavitation is the vapor pressure p_v, although some deviations discussed later may occur.

1.1 The cavitation number

The degree of development of cavitation is characterized by a non dimensional parameter, the cavitation number σ , defined by:

$$\sigma = \frac{p_{ref} - p_v}{\frac{1}{2}\rho V^2} \tag{1.1}$$

In this expression, p_{ref} is a reference pressure taken at a given point in the liquid flow and V is a characteristic flow velocity. Both parameters need to be precisely specified for each practical situation. As an example, in the case of a cavitating flow past a single foil in a hydrodynamic tunnel (see e.g. Figure 3), the reference pressure and velocity are usually chosen as the pressure and velocity in the undisturbed liquid flow, far from the foil.

A non cavitating flow corresponds to large values of the cavitation number. This is easy to understand since large values of the cavitation number usually correspond to large values of the reference pressure. Then, it can be expected that the pressure will be everywhere above the vapor pressure and the flow will remain free of cavitation. It is clear that the cavitation number has no influence on the fully wetted flow which will remain the same whatever the cavitation number may be, provided it is large enough for the flow to remain actually non cavitating. This number is a pertinent parameter only for cavitating flows for which it can be considered as a scaling parameter which measures the global extent of cavitation.

The onset of cavitation generally appears for a critical value of the cavitation number known as the incipient cavitation number σ_i . Starting from the fully wetted flow, cavitation inception can be reached either by decreasing the reference pressure or increasing the flow velocity, both leading to a decrease in cavitation number. Any further decrease will lead to an additional development of cavitation. In the case of Figure 3 for instance, the cavity will grow and its length will increase with a decrease in cavitation number leading to a longer cavity comparable to the supercavity shown in Figure 5. If the reference pressure is now increased, it is generally observed that cavitation disappears for a critical cavitation number somewhat higher than σ_i . Incipient and desinent cavitation numbers are often different and an hysteresis effect is often observed.

1.2 Main types of cavitation

Looking at real cavitating flows as that in a cavitating turbopump (Figure 1) or around a propeller (Figure 2), it appears that the liquid vapor interfaces have generally complicated shapes. There is a wide variety of types of cavitation and basically we can identify the following ones:

- ✓ *attached cavities* as that shown in Figures 3 to 5. Cavitation appears here in the form of a cavity attached to the suction side of the foil. The type of cavitation shown in Figure 3 is known as partial cavitation since the cavity covers only partially the upper side. On the contrary, a supercavity as shown in Figure 5 fully covers the suction side and closes downstream the foil trailing edge.
- ✓ *traveling bubble cavitation* with more or less isolated bubbles according mainly to the nuclei density in the free stream (Figures 6 to 8).
- ✓ *cavitation clouds* which can take various forms. Figure 9 gives an example of two clouds shed by an unsteady partial cavity. This is an illustration of the partial cavitation instability

which is triggered by a re-entrant jet developing upward from the closure region of the cavity.

✓ *cavitating vortices* which can be more or less structured. They are observed in particular at the tip of three-dimensional foils (Figure 10) or in the turbulent wake of bluff bodies where they are less organized because of turbulence (Figure 11).

Secondary effects as interactions between bubbles or with solid walls, fission, coalescence, interface instabilities, re-entrant jet, turbulence… can dramatically complicate previous basic shapes of liquid / vapor interfaces at both large and small scales. The analysis of cavitation can then be particularly difficult because of the geometric complexity of the liquid / vapor interfaces.

1.3 Overview of chapter

Despite this complexity, many basic results can be rather easily derived from the Rayleigh-Plesset equation. This equation, presented in section 2, applies to an isolated spherical bubble which is assumed to remain spherical all along its life. It gives how its radius changes because of the change in pressure it might go through during its life. This is the case, for instance, of an initial microbubble or cavitation nucleus carried by a liquid flow which undergoes pressure changes as it goes along the blades of a hydraulic machine. It grows in low pressure regions, becomes a macroscopic cavitation bubble and finally collapses downstream where the pressure recovers.

Section 3 is devoted to the presentation of a few basic results on single bubbles. First equilibrium is considered and it is shown that the critical pressure for the explosive growth of a nucleus may be significantly smaller than vapor pressure because of surface tension. The two main stages in the typical evolution of a cavitation bubble, i.e. growth and collapse, are then addressed with emphasis on the collapse time, which is a characteristic time scale of great importance in cavitation. Finally, it is shown that a bubble in a liquid is an oscillator because of the elastic behavior of the non condensable gas generally enclosed; the period of oscillation, which is another characteristic time scale, is introduced.

Section 4 is devoted to the presentation of non dimensional forms of the Rayleigh-Plesset equation from which a few conclusions on cavitation scaling are deduced. A first form based on the introduction of characteristic times – pressure, viscous and surface tension times – allows the estimation of the relative importance of each of these phenomena on the dynamics of a single bubble. A second form appropriate to the case of a bubble traveling on the suction side of a blade allows the derivation of scaling laws for traveling bubble cavitation.

Section 5 addresses thermal effects in cavitation. An extended form of the Rayleigh-Plesset equation including thermal effects is derived. Once more it is made non dimensional in order to develop a practical criterion for the estimation of the so called thermodynamic effect in cavitation.

Section 6 is relative to supercavitation, a field in which the Rayleigh-Plesset equation is surprisingly applicable. According to the Logvinovich independence principle, the dynamics of any cross section of a supercavity is independent of the neighboring ones and can be modeled by a Rayleigh type equation. This section shows that the Rayleigh equation, originally derived for spherical bubbles, may also be useful for other cavities whose geometry is actually far from being spherical.

Section 7 is devoted to an analysis of cavitation erosion using once more the Rayleigh equation. Firstly, it is shown that the spherical collapse of a bubble generates a pressure pulse of high amplitude that can largely exceed the yield strength of usual materials and hence cause damage. The flow aggressiveness of a single bubble and consequently of a whole cavitating flow is then analyzed with a special emphasis on the influence of velocity on erosive potential. The section closes with a few general remarks on the erosive potential of various cavitating flows, still based on a discussion of the Rayleigh equation.

In section 8, it is shown that the dynamics of other types of cavities, as ring bubbles, can be modeled by a Rayleigh-Plesset type equation, with some changes and additional terms which take into account the specificities of such cavitating structures, as vorticity.

The chapter ends with a brief presentation of a cavitation model based on the Rayleigh-Plesset equation and often used for simulation. The liquid is assumed to carry cavitation nuclei and the Rayleigh-Plesset equation, which models the evolution of individual bubbles in the cluster, is coupled to Navier-Stokes equations. Such a technique is appropriate to the modeling of complex real cavitating flows, as for instance cloud cavitation generated by a pulsating leading edge cavity.

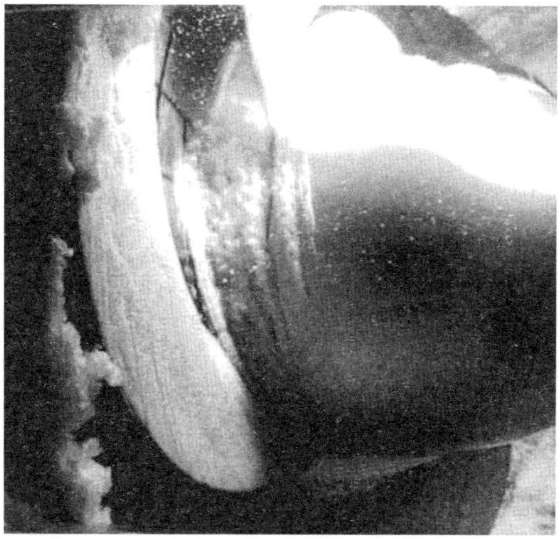

Figure 1. Cavitating flow in the inducer of a rocket engine turbopump (Courtesy of SNECMA Moteurs and CNES)

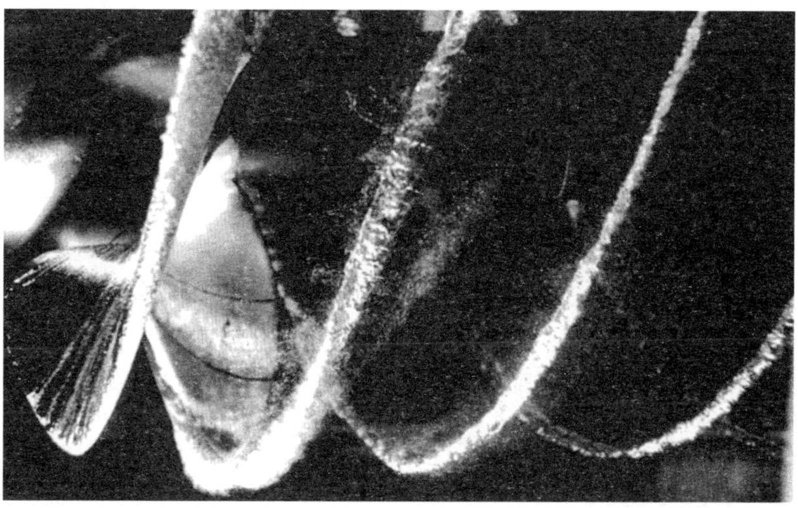

Figure 2. Cavitating flow in a marine propeller (Courtesy of DGA/BEC)

Figure 3. Partial cavity flow on a hydrofoil

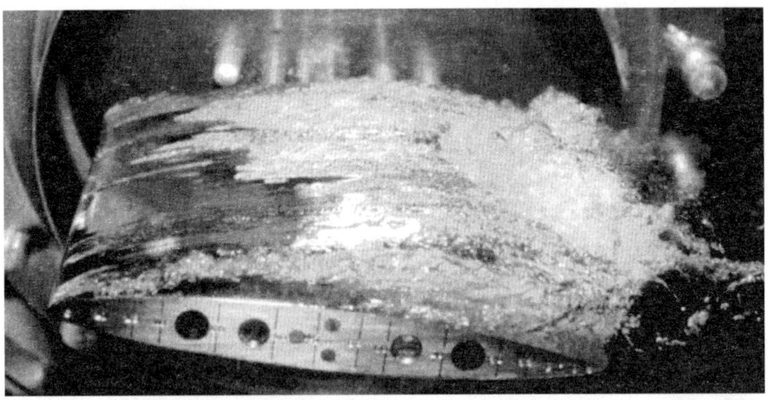

Figure 4. Unstable partial cavitation on a hydrofoil

Figure 5. Supercavity flow around a hydrofoil in a cavitation tunnel

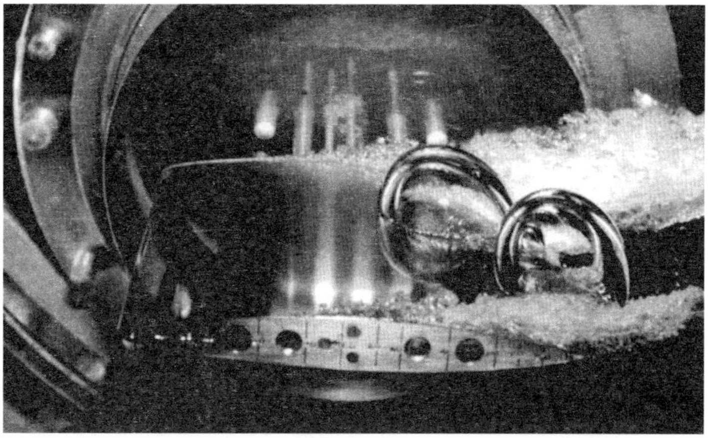

Figure 6. Two cavitation bubbles on the suction side of a hydrofoil

Figure 7. Traveling bubble cavitation at medium angle of attack

Figure 8. Traveling bubble cavitation at large angle of attack

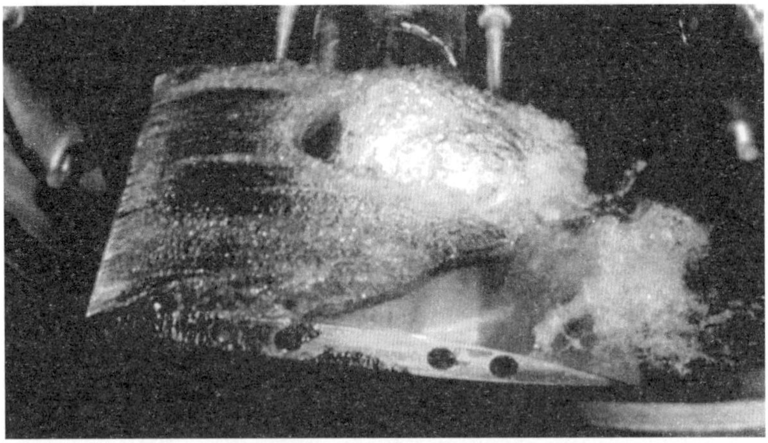

Figure 9. Unsteady leading edge cavity shedding cavitation clouds

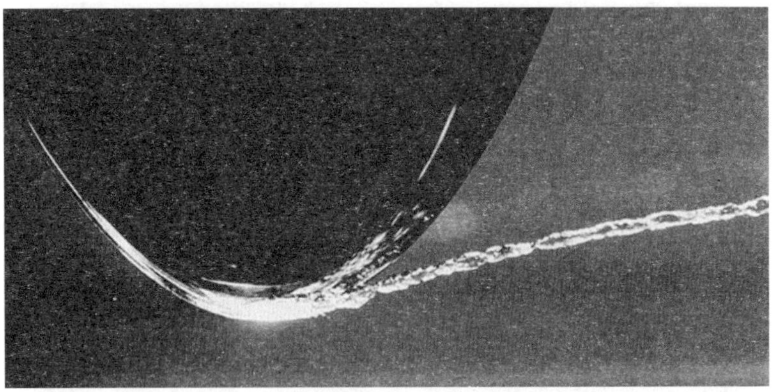

Figure 10. Vortex cavitation at the tip of a three dimensional foil

Figure 11. Cavitating vortices in the turbulent wake of a bluff body

2 The Rayleigh-Plesset equation

2.1 Cavitation nuclei

Cavitation is generally initiated from microscopic nuclei carried by the flow. Such nuclei are points of weakness for the liquid from which macroscopic cavities are generated and grow in low pressure regions.

The simplest and most widely used model of nucleus is that of a microbubble. Such a microbubble, typically of a few microns in diameter, is assumed to be spherical and to contain a gaseous mixture made of the vapor of the liquid and possibly of non condensable gas. The presence of non condensable gas is quite general in practice. In the most common case of water, it is well-known that ordinary water contains dissolved air (essentially oxygen and nitrogen) at least if no special degassing procedure is applied to it. The presence of non condensable gas inside the bubbles is then due to the migration of gas by molecular diffusion through the bubble interface.

Figure 12 illustrates how a microbubble can give birth to a macroscopic cavitation bubble when moving along the suction side of a foil.

Figure 12. Growth of a nucleus on the suction side of a hydrofoil

2.2 The dynamics of a spherical bubble

It is then of major importance for the understanding of cavitation to be able to understand and predict the evolution of such a bubble. This is possible from the Rayleigh-Plesset equation. The driving parameter for the bubble dynamics is obviously the pressure. In the case of Figure 12 for example, the pressure which is applied to the bubble changes during its movement along the suction side. It is usually assumed to be the pressure of the original fully wetted flow at each successive location of the bubble center. This time dependent pressure that the bubble experiences is the driving factor of the bubble dynamics.

For simplicity, the bubble is assumed to evolve in a infinite medium at rest at infinity. The basic input for bubble dynamics is the instantaneous pressure law $p_\infty(t)$ applied to it. Any other flow than the pure radial one induced by the growth or collapse of the bubble is ignored. All the information on the original flow is then supposed to be included in the only $p_\infty(t)$ law.

The expected output is the time evolution of the bubble radius *R(t)*. It satisfies the following second-order differential equation known as the Rayleigh-Plesset equation:

$$\rho\left[R\ddot{R}+\frac{3}{2}\dot{R}^2\right]=\left[p_v-p_\infty(t)\right]+p_{g0}\left(\frac{R_0}{R}\right)^{3k}-\frac{2S}{R}-4\mu\frac{\dot{R}}{R} \tag{2.1}$$

The derivation of this equation together with the main required assumptions can be found in any textbook on cavitation (see e.g. Brennen (1995), Franc and Michel (2004)). In this equation, $\dot{R}$ and $\ddot{R}$ are the first and second order derivatives of the bubble radius with respect to time and R_0 is the initial bubble radius.

On the right hand side, four different terms appear. The first one $p_v - p_\infty(t)$, which measures the closeness of the applied pressure to the vapor pressure, is the driving term for the bubble evolution. It is the most fundamental one since the evolution of the bubble (growth, collapse, oscillations…) will depend essentially upon it.

The second one is the contribution of non condensable gas. Its derivation is based on several assumptions. First it is assumed that the mass of non condensable gas inside the bubble remains constant during its evolution. This is a simplifying assumption that could be evaluated by solving the mass diffusion equation, which is however far beyond the scope of the present introductory chapter.

Secondly, this constant mass of gas is assumed to follow a polytropic thermodynamic behavior characterized by a given polytropic coefficient *k*. If the behavior is isothermal, $k=1$. If it is adiabatic, *k* is the ratio γ of the heat capacities of the enclosed gas. To resolve the ambiguity, it would be necessary to solve an energy equation, which is not essential at this step. The gas transformation can often be assumed as isothermal since the characteristic time for the evolution of a nucleus in real cavitating flows is usually much larger than that required for heat transfer so that temperature equilibrium is continuously achieved. However, for big bubbles resulting of the explosion of a nucleus, the behavior tends to be adiabatic (cf. Franc and Michel (2004)). Nevertheless, we will keep both possibilities by introducing a polytropic coefficient k.

The polytropic behavior is described by the following law between the partial pressure p_g of the gas inside the bubble and its radius *R*:

$$p_g R^{3k} = p_{g0} R_0^{3k} \tag{2.2}$$

where subscript 0 refers to initial conditions. In this equation, R^3 is actually representative of the bubble volume. From the previous equation, the second term on the right hand side of the Rayleigh-Plesset equation appears to be simply the instantaneous partial pressure p_g of the gas inside the bubble.

The third term is the contribution of surface tension. *S* is the usual surface tension coefficient expressed in N/m or J/m^2. Since *R* appears at the denominator, this term is important only for small radii.

The last term accounts for the effect of dynamic viscosity μ of the liquid. Dissipation due to viscosity appears to be proportional to bubble deformation rate $\dot{R}$ and inversely proportional to bubble radius so that it is expected to be significant only for small radii as surface tension.

When the effects of non condensable gas, surface tension and viscosity are negligible, as it is the case for big enough bubbles, the Rayleigh-Plesset equation reduces to the simple Rayleigh equation :

$$\rho\left[R\ddot{R}+\frac{3}{2}\dot{R}^2\right] = p_v - p_\infty(t) \tag{2.3}$$

Furthermore, if the applied pressure p_∞ is constant, the Rayleigh equation can be integrated once to give thc bubble interface velocity:

$$\dot{R}^2 = \frac{2}{3}\frac{p_v - p_\infty}{\rho}\left[1-\left(\frac{R_0}{R}\right)^3\right] \tag{2.4}$$

3 A few basic results

3.1 Bubble equilibrium

Although it is not necessary to know the Rayleigh-Plesset equation to study the equilibrium of a bubble, the condition for equilibrium can easily be deduced from the Rayleigh-Plesset equation by setting all time derivatives to zero and assuming constant the external pressure p_∞. For the analysis of equilibrium, it is common to assume the gas transformation as isothermal ($k=1$) since the temperature can be considered as continuously fixed by that of the liquid. If so, we get the following equilibrium condition:

$$p_\infty = p_{g0}\left[\frac{R_0}{R}\right]^3 + p_v - \frac{2S}{R} \tag{3.1}$$

This equation expresses that the difference between the pressure inside and outside the bubble is due to surface tension. By solving it with respect to R, we can obtain the radius of equilibrium of a bubble at any external pressure p_∞. The corresponding relationship between pressure at infinity and equilibrium radius is shown in Figure 13.

It is important to observe that the equilibrium is not always stable. This point is connected to the existence of a minimum for the equilibrium curve. The corresponding pressure and radius, called critical (for reasons which will become clear at the end of this section) and labeled here by subscript c, are given by:

$$\begin{cases} R_c = \sqrt{\dfrac{3p_{g0}\,R_0^{\,3}}{2S}} \\ p_c = p_v - \dfrac{4S}{3R_c} \end{cases} \tag{3.2}$$

Critical radius and pressure depend on surface tension S and on the group of parameters $p_{g0}R_0^{\ 3}$. Since $R_0^{\ 3}$ is proportional to the volume of the bubble, $p_{g0}R_0^{\ 3}$ is actually proportional to the mass of non condensable gas in the bubble. As a nucleus is completely defined by the quantity of non condensable gas it contains (which is assumed constant), it can also be characterized by either its critical radius or, what is most common, by its critical pressure.

To come back to the problem of stability, consider three different nuclei in equilibrium under conditions 1, 2 and 3 as shown in Figure 13 and, in order to analyze the stability, let us assume that the pressure is, for example, slightly lowered. The equilibrium condition 3.1 is obviously no longer satisfied and the unbalance due to a lower pressure p_∞ is such that the right hand side of the Rayleigh-Plesset equation becomes positive. Hence the bubble will grow as it can be reasonably expected following a pressure drop.

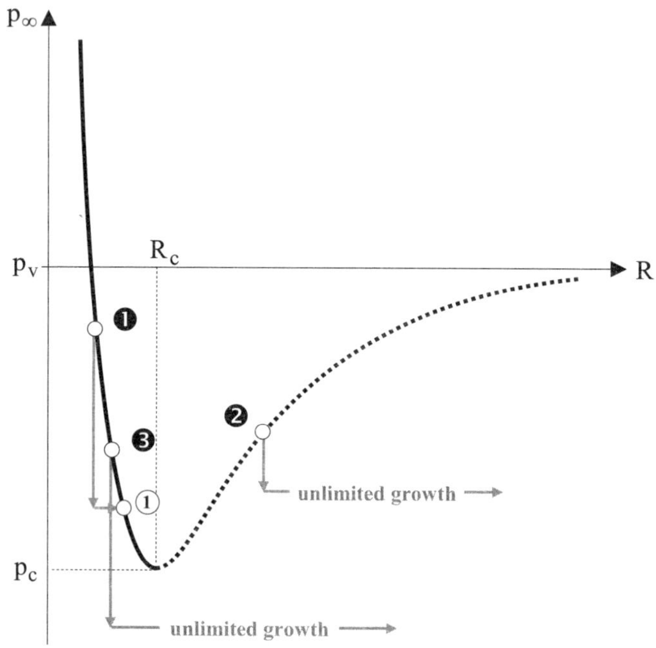

Figure 13. Equilibrium radius of a cavitation nucleus versus pressure at infinity

In the case of nucleus 1, it is clear that the bubble will reach a new equilibrium since the representative point of the bubble follows a path which crosses the equilibrium curve at a new point 1'. This descending part of the equilibrium curve is then stable.

On the contrary, for nucleus 2, the bubble will indefinitely grow without crossing the equilibrium curve, so that the equilibrium is unstable.

Finally, in situation 3 for which the pressure is lowered below the critical pressure which corresponds to the minimum of the equilibrium curve, the bubble will grow indefinitely as well without reaching a new equilibrium. Hence the critical pressure can actually be considered as a threshold for the microbubble to explode and become a macroscopic cavitation bubble.

According to equation 3.2, the critical pressure is smaller than the vapor pressure and the difference is due to surface tension. It is negligible for large nuclei but can become important for small ones. As an example, a nucleus of radius $R_0 = 3\ \mu m$ in equilibrium in water ($p_v = 2300\ Pa$, $S = 0.072\ N/m$) at atmospheric pressure $p_\infty = 10^5\ Pa$ will have a partial pressure of air inside equal to $p_{g0} = p_\infty - p_v + 2S/R \cong 146\ 000\ Pa$, a critical radius $R_c \cong 9\ \mu m$ and a critical pressure $p_c = p_v - 10\ 600\ Pa$ which appears to be significantly smaller than the vapor pressure and even negative. A tension is then necessary to activate this nucleus and make it grow indefinitely.

The difference $p_v - p_c$ is known as the nucleus static delay. In the introduction of this chapter, we mentioned that the threshold pressure for cavitation is usually considered as the vapor pressure. The present model shows that the threshold pressure is actually the nucleus critical pressure which is smaller than the vapor pressure.

In ordinary water, there are generally many nuclei with a wide range of diameters. The weakest points are the biggest nuclei since their critical pressure is the largest. They will then cavitate first. The critical pressure of the biggest nuclei is known as the susceptibility pressure of the liquid sample. It is the critical pressure for cavitation. If no special treatment is applied to the liquid by removing big nuclei, the susceptibility pressure is expected to remain close to the vapor pressure and the assumption of a threshold for cavitation equal to the vapor pressure is quite appropriate.

3.2 Bubble growth

As already observed, the effects of non condensable gas, surface tension and viscosity become negligible when the cavitation bubble is much bigger than the original nucleus. If so, the simplified Rayleigh equation is applicable. If the applied pressure p_∞ equals the vapor pressure p_v, the bubble is in equilibrium. If $p_\infty < p_v$, the bubble will grow ($R > R_0$) and, according to Equation 2.4, the asymptotic growth rate for large radii is:

$$\dot{R} \cong \sqrt{\frac{2}{3}\frac{p_v - p_\infty}{\rho}} \tag{3.3}$$

This equation is valid quite early since, as soon as the cavitation bubble is three times bigger than the initial nucleus, the error on the estimation of the interface velocity using Equation 3.3 instead of Equation 2.4 is smaller than 2%. Previous equation is important in cavitation as it gives the order of magnitude of the growth rate of a cavitation bubble when submitted to a given pressure $p_\infty < p_v$. Let us observe that simple dimensional arguments would have lead to the same formula, except for the numerical coefficient $\sqrt{2/3}$ which cannot be predicted from a pure dimensional analysis.

3.3 Collapse of a pure vapor bubble

If the applied pressure is higher than the vapor pressure, the bubble radius decreases ($R < R_0$). This is the collapse phase. If we continue to ignore the effects of viscosity, non condensable gas and surface tension, the interface velocity during collapse is given by:

$$\dot{R} \cong -\sqrt{\frac{2}{3}\frac{p_\infty - p_v}{\rho}\left[\left(\frac{R_0}{R}\right)^3 - 1\right]} \tag{3.4}$$

Integration of Equation 3.4 allows the computation of the collapse time or bubble lifetime i.e. the time necessary for the bubble to completely disappear until R vanishes ($R=0$). This time is called the Rayleigh time and is given by:

$$\tau_p \cong 0.915\, R_0 \sqrt{\frac{\rho}{p_\infty - p_v}} \tag{3.5}$$

It is interesting to observe that, once more, a simple dimensional analysis would have lead to the same formula except for the coefficient 0.915 which could not be guessed. The subscript p just recalls that this characteristic time scale is directly connected to the pressure difference $p_\infty - p_v$ (see also section 4.1).

It is of crucial importance in cavitation to observe that during the collapse phase of a pure vapor bubble without non condensable gas inside as assumed here, the interface velocity $\dot{R}$ is always increasing and becomes infinite at the very end of the collapse ($R=0$). This final behavior is of course not realistic providing evidence that some original assumptions are no longer physically valid at the ultimate stage of collapse. This is the case when assuming the liquid incompressible, which is obviously unacceptable when the interface velocity comes near to the speed of sound. This is also the case when ignoring the effect of non condensable gas since they are continuously compressed and their pressure drastically increases during collapse. This is another limiting factor which contributes to a reduction in the interface velocity during the final stage of collapse and possibly an inversion of the movement causing bubble rebound.

3.4 Bubble resonance frequency

A bubble in a liquid is a possible oscillator because of the elastic behavior of the non condensable gas that the bubble contains and the inertia of the liquid. Then, a natural resonance frequency is associated to any cavitation bubble in a liquid. However, dissipative losses as those due to viscosity or heat conduction for instance tend to damp out bubble oscillations. From the Rayleigh-Plesset equation, it is easy to predict the pulsating behavior of a bubble and in particular to compute its resonance frequency.

Consider a bubble of radius R_0 in equilibrium at pressure $p_{\infty 0}$. The partial pressure p_{g0} of non condensable gas in the bubble is given by the equilibrium condition (cf. Equation 3.1):

$$p_{\infty 0} = p_v + p_{g0} - \frac{2S}{R_0} \tag{3.6}$$

Suppose that the pressure p_∞ oscillates around equilibrium $p_{\infty 0}$ with pulsation ω and a small amplitude δp according to the relation $p_\infty = p_{\infty 0} + \delta p \sin \omega t$. The variations in bubble radius can be computed using the Rayleigh-Plesset equation which can be written as follows after making use of the equilibrium condition:

$$\rho\left[R\ddot{R}+\frac{3}{2}\dot{R}^2\right]=-\delta p\, sin\, \omega t+p_{g0}\left[\left(\frac{R_0}{R}\right)^{3k}-1\right]+2S\left(\frac{1}{R_0}-\frac{1}{R}\right)-4\mu\frac{\dot{R}}{R} \tag{3.7}$$

By introducing the displacement $\delta R = R - R_0$ instead of R and assuming that δR remains small which is normally the case if the driving pressure fluctuation δp is small enough, the previous equation can be linearized as follows:

$$\rho R_0\, \ddot{\delta R} + 4\mu\frac{\dot{\delta R}}{R_0} + \left(3kp_{g0}-\frac{2S}{R_0}\right)\frac{\delta R}{R_0} \cong -\delta p\, sin\, \omega t \tag{3.8}$$

This is the usual equation of a forced harmonic oscillator whose natural frequency is:

$$f_0=\frac{1}{2\pi R_0}\sqrt{\frac{1}{\rho}\left[3kp_{g0}-\frac{2S}{R_0}\right]} \tag{3.9}$$

Making use of the equilibrium condition 3.6 to estimate the partial pressure of non condensable gas, the resonance frequency is more straightly given by the following equation:

$$f_0=\frac{1}{2\pi R_0}\sqrt{\frac{1}{\rho}\left[3k\left(p_{\infty 0}-p_v+\frac{2S}{R_0}\right)-\frac{2S}{R_0}\right]} \tag{3.10}$$

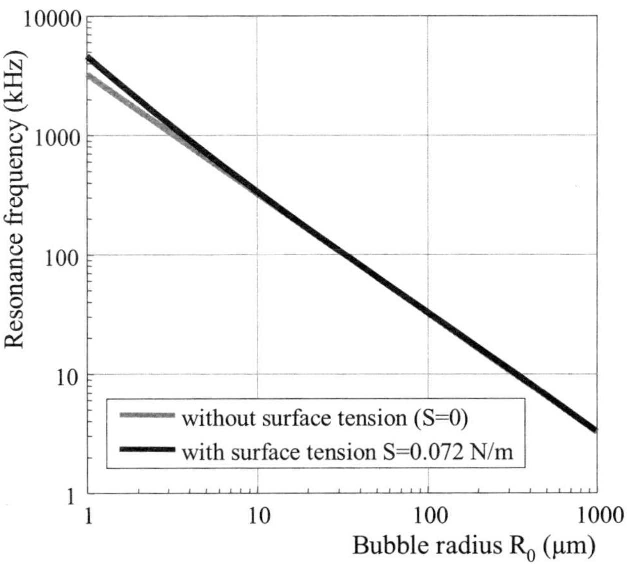

Figure 14. Resonance frequency for cavitation bubbles in water oscillating adiabatically at atmospheric pressure ($p_{\infty 0}=10^5\ Pa,\ p_v=2300\ Pa,\ k=\gamma=1.4,\ S=0.072\ N/m,\ \rho=1000\ kg/m^3$). Without surface tension ($S = 0$), the frequency is inversely proportional to bubble radius.

The resonance frequency of cavitation bubbles of various radii in water oscillating adiabatically ($k = \gamma = 1.4$) at atmospheric pressure is shown in Figure 14. The influence of surface tension appears negligible except for small radii. The resonance frequency can then be considered as inversely proportional to the bubble radius.

The period of natural oscillation is another important characteristic time for a cavitation bubble. In the present case of a bubble submitted to an oscillating pressure field, the amplitude of the harmonic response of the bubble depends greatly on the ratio of the frequency of the applied pressure to the bubble resonance frequency, as for any forced oscillator. More generally, the period of natural oscillations has to be compared to the characteristic time of pressure variations. If they are of the same order of magnitude, bubble oscillations are expected.

Furthermore, let us observe that the second term on the left hand side of Equation 3.8, which involves the first derivative of the bubble radius, demonstrates the damping effect of viscosity on bubble oscillations.

4 Cavitation scaling

4.1 Characteristic time scales for an isolated pure vapor bubble

The Rayleigh-Plesset equation is a rather convenient basis to point out characteristic time scales of the cavitation phenomenon. In the previous sections, we already introduced two of them. Firstly, the Rayleigh time, which depends upon the bubble initial radius and the difference between the applied pressure and the vapor pressure. It is typical of the bubble lifetime when collapsing. Secondly, the resonance frequency, which is mainly connected to the elastic behavior of the non condensable gas, and which characterizes the period of natural oscillations. In the present section, two new time scales are introduced. They are connected to two other physical phenomena, namely viscosity and surface tension. By comparing these various characteristic time scales, it is possible to estimate the importance of each associated physical phenomenon on the dynamics of a cavitation bubble.

We consider here a pure vapor bubble without any non condensable gas inside and we suppose it is submitted to a varying pressure at infinity $p_\infty(t)$. Let a be a characteristic length scale for the bubble radius. If we are interested in the first stage of the bubble evolution, it is natural to choose the initial radius R_0 for a. From purely dimensional arguments, we can then define a viscous time τ_v and a surface tension time τ_S on the basis of length scale a and either viscosity ν or surface tension S. It is easy to check that the two following variables are actually measured in time units:

$$\begin{cases} \tau_v = \dfrac{\rho a^2}{4\,\mu} \\ \tau_S = a\sqrt{\dfrac{\rho a}{2S}} \end{cases} \tag{4.1}$$

In addition, we consider a pressure time defined by:

$$\tau'_p = a\sqrt{\frac{\rho}{p_{\infty ref} - p_v}} \tag{4.2}$$

In this equation, $p_{\infty ref}$ is a reference pressure which is supposed to be a suitable order of magnitude for the applied pressure $p_\infty(t)$. This time scale is similar to the Rayleigh time τ_p introduced in section 3.3 devoted to the analysis of bubble collapse, except that the constant *0.915* has been ignored and the initial bubble radius R_0 has been replaced by *a*, to be more general.

The discussion on the relative importance of these various time scales is conducted on the basis of a non dimensional form of the Rayleigh-Plesset equation (2.1). The bubble radius is made non dimensional using length scale *a*:

$$\overline{R} = \frac{R}{a} \tag{4.3}$$

As for the characteristic time scale, it is a priori unknown. We will then introduce a new constant τ which is assumed to be a characteristic time for the bubble evolution. One of the conclusions of the present approach will be the actual estimation of this time scale. A non dimensional time $\bar{t}$ is then defined by:

$$\bar{t} = \frac{t}{\tau} \tag{4.4}$$

It is of major importance to carefully choose the reference length scale *a* and time scale τ. Actually, they must be pertinent orders of magnitude of the bubble radius and evolution time. In other words, the corresponding non dimensional variables $\overline{R}$ and $\bar{t}$ must be of the order of unity. At the beginning of bubble evolution, the initial radius is obviously a suitable length scale as already mentioned. However, since the bubble size may change by several orders of magnitude with time, it may be necessary to adapt the characteristic length scale during bubble evolution. Furthermore, if the bubble evolution is sufficiently regular, such a well thought non dimensional procedure ensures that the derivatives $\dot{\overline{R}}$ and $\ddot{\overline{R}}$ are also of the order of unity, as usually assumed in classical dimensional analysis.

With this change of variables, the Rayleigh-Plesset equation takes the following non dimensional form (the effect of non condensable gas is not included here):

$$\overline{R}\ddot{\overline{R}} + \frac{3}{2}\dot{\overline{R}}^2 = -\left(\frac{\tau}{\tau'_p}\right)^2 \frac{p_\infty(t) - p_v}{p_{\infty ref} - p_v} - \left(\frac{\tau}{\tau_S}\right)^2 \frac{1}{\overline{R}} - \frac{\tau}{\tau_v}\frac{\dot{\overline{R}}}{\overline{R}} \tag{4.5}$$

From previous equation, it is easy to estimate the relative importance of the three terms driven respectively by pressure, surface tension and viscosity, which appear on the right hand side. Since all variables including the non dimensional radius $\overline{R}$ and its derivatives are of the order of unity, as mentioned above, the dominating term is the one whose characteristic time is the smallest. The two others can then be neglected. This procedure allows the identification of the physical mechanism which controls the evolution of the bubble.

To illustrate the approach, the characteristic pressure, surface tension and viscous times are plotted in Figure 15 against the bubble radius *a*. The pressure time depends upon the reference pressure $p_{\infty ref}$ which has been chosen equal to 1 bar in Figure 15 as an example. From this figure, it appears that the pressure controls the dynamics of the cavitation bubble in a wide range of radii. Surface tension becomes dominating for bubbles smaller than typically 1 µm in radius whereas viscous effects are predominant only for very small bubbles, smaller than about 0.1 µm.

In addition, such an approach can give the order of magnitude of the characteristic time τ for the evolution of the bubble, which remains unknown so far. As an example, consider the case of the collapse of a vapor bubble under a constant pressure $p_\infty(t) = p_{\infty ref}$. If the collapse is pressure dominated, the two last terms in Equation (4.5) are negligible. Since the left hand side is of the order of unity (still because of the suitable choice in reference length and time scales), it is concluded that time τ must be of the same order of magnitude that pressure time τ'_p. This shows that the collapse time is actually given by the pressure time as already concluded in section 3.3 from a detailed computation of the bubble lifetime. Similarly, considering the case of a collapse dominated by surface tension, it can be concluded that the surface tension time τ_S is a relevant order of magnitude for the collapse time under the only effect of surface tension.

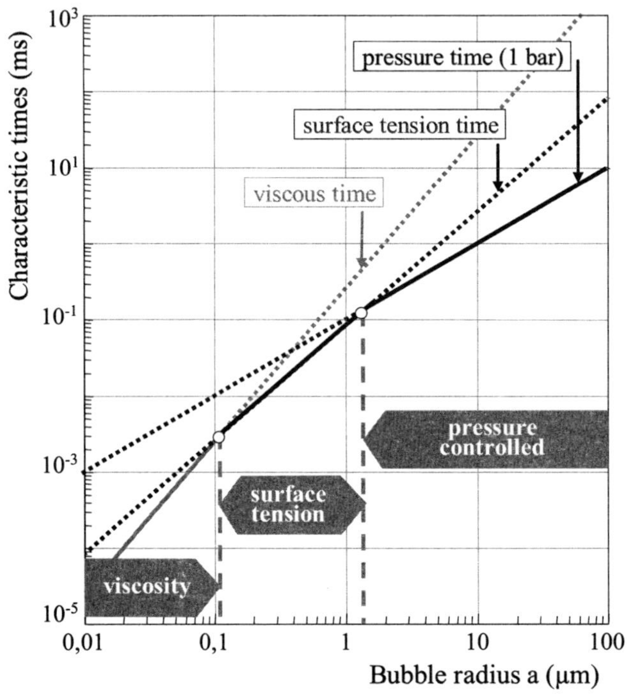

Figure 15. Characteristic time scales versus bubble radius.
Pressure time is estimated at atmospheric pressure ($p_{\infty ref} = 1\ bar$).

4.2 Scaling law for traveling bubble cavitation

We consider here the case of traveling bubble cavitation as shown in Figures 6 to 8. In such a cavitating flow, the cavitation bubbles are generated from nuclei carried by the oncoming liquid. They grow on the suction side of the foil before collapsing in the pressure recovery region.

The problem which is addressed here is the influence of foil size on the dynamics of cavitation bubbles. In particular, we will discuss the scaling law to be satisfied to ensure similarity between a prototype at full scale and a model at smaller scale.

This problem can be approached by using the Rayleigh equation and computing the evolution of a bubble moving along the suction side. The pressure law $p_\infty(t)$ which has to be introduced in the Rayleigh equation is deduced from the computation of the fully-wetted flow around the foil. It is assumed to be given by the successive pressures that the bubble center encounters as it moves along the foil. Clearly, this method ignores possible interactions between bubbles and also the change in the original fully-wetted flow pressure distribution due to the development of cavitation.

In addition, the effects of surface tension, viscosity and gas content are neglected, which is valid as soon as the microbubble becomes a macroscopic cavitation bubble. It is then quite good to consider the Rayleigh equation (2.3):

$$R\frac{d^2R}{dt^2}+\frac{3}{2}\left(\frac{dR}{dt}\right)^2 = \frac{p_v - p_\infty(t)}{\rho} \tag{4.6}$$

In this problem, the time variable t is not really a relevant parameter and it is more appropriate to consider the distance x along the foil. Time derivatives are then changed into space derivatives using the mean flow velocity V of the bubble according to the common transformation $x = Vt$. The previous equation is then transformed as follows:

$$R\frac{d^2R}{dx^2}+\frac{3}{2}\left(\frac{dR}{dx}\right)^2 = \frac{p_v - p_\infty(x)}{\rho V^2} \tag{4.7}$$

where $p_\infty(x)$ is the pressure to which the bubble is submitted at any station x along the foil.

The approach is still based upon a non dimensional form of the Rayleigh equation. The difference in comparison with previous section 4.1 is that the bubble radius R and the distance x are made non dimensional using here the chord length c, so that the new non dimensional variables are:

$$\begin{cases} \overline{R} = \dfrac{R}{c} \\ \overline{x} = \dfrac{x}{c} \end{cases} \tag{4.8}$$

In addition, we introduce the usual non dimensional pressure coefficient C_p defined by:

$$C_p(x) = \frac{p_\infty(x) - p_{ref}}{\frac{1}{2}\rho V^2} \tag{4.9}$$

where p_{ref} is a reference pressure already introduced in section 1.1. Using definition 1.1 of the cavitation parameter, equation (4.7) takes the following non dimensional form:

$$\overline{R}\frac{d^2\overline{R}}{d\overline{x}^2}+\frac{3}{2}\left(\frac{d\overline{R}}{d\overline{x}}\right)^2 = -\frac{1}{2}(C_p + \sigma) \tag{4.10}$$

Let us observe that, in the present approach, the bubble is assumed to have a constant velocity V. If, at any time, the bubble is supposed to take the local flow velocity on the foil, which is more realistic, an additional term appears in equation (4.10) (see e.g. Franc & Michel (2004)). The general conclusions remain however unchanged.

Consider two similar flows around geometrically similar foils at exactly the same angle of attack, the only difference being the chord length which is respectively c and λc. The pressure coefficient distribution is then exactly the same. Provided the cavitation number is the same for both flows, the non dimensional solution $\overline{R}(\overline{x})$ of equation (4.10) is the same too. In other words, at the same relative position $\overline{x}$ on the foil, the radius of the cavitation bubble is proportional to the chord length c. The bigger the foil, the bigger the bubble, as schematically shown in Figure 16.

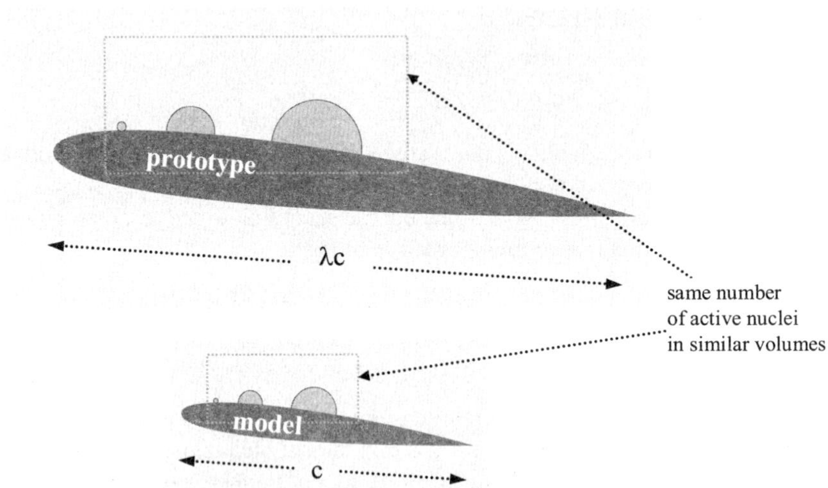

Figure 16. Scaling rules between model and prototype for traveling bubble cavitation

If we now consider the case of traveling bubble cavitation with several bubbles growing together on the suction side, both cavitating flows, model and prototype, will be similar if, in addition, similar volumes of liquid contain exactly the same number of active nuclei. If so, there will be statistically the same number of cavitation bubbles on the foils. The cavitation pattern and then the hydrodynamic performance of the foils will be statistically identical. In terms of nuclei density i.e. of number of microbubbles per unit volume, this requires that the nuclei density N_c for model tests be λ^3 times larger than the nuclei density $N_{\lambda c}$ for the prototype. The following scaling law must then be satisfied between model and prototype:

$$\frac{N_c}{N_{\lambda c}} = \lambda^3 \tag{4.11}$$

This λ^3 law suggests that the density of active nuclei must be much larger for the model at small scale than for the prototype at full scale.

This scaling law does not need to be satisfied when the foil suction side is saturated with cavitation bubbles for large nuclei densities. When saturation occurs, the foil is fully covered with bubbles which merge and form a kind of continuous cavity where the pressure is set to the vapor pressure. Beyond saturation, foil performance becomes independent of nuclei content and there is no need to satisfy the previous λ^3 scaling law. It is the reason why model tests are sometimes conducted with maximum nuclei seeding to get rid of this constraint and also be sure not to overestimate cavitation performances.

5 Thermodynamic effect

As mentioned in the introduction of this chapter, it is generally assumed that temperature is uniform and equal to the liquid bulk temperature all over a cavitating flow. This is the case for water at normal temperature. Strictly speaking, the temperature variation induced by cavitation is so small in cold water that it can actually be neglected.

However, this is not the case for all fluids. For cryogenic liquids used in rocket propulsion for instance, the temperature in the cavitating region T_c may be significantly lower than that of the liquid bulk T_∞. The physical mechanism for this drop in temperature is the following.

Cavitation is basically a vaporization of the liquid which involves a latent heat as for any phase change. Since the cavities are growing inside the liquid itself, the latent heat of vaporization can only be supplied by the liquid surrounding the cavities. Hence, the liquid close to the two phase region, and consequently this region itself, is cooled down.

Although the phenomenon is reversed in the collapse region, the growth region is generally prevailing. As an example, consider the case of a partial cavity attached to the leading edge of a foil, as shown in figure 17. Vaporization takes place on a large and steady upstream part of the cavity whereas condensation occurs around cavity closure, where vapor structures are shed and entrained by the liquid flow. The temperature drop is maximum close to the cavity leading edge and the temperature progressively increases along the cavity and reaches the liquid bulk temperature around cavity closure. There is no overshoot in mean temperature connected to condensation at closure mainly because of an important dispersion of vapor structures and a high turbulence level which both contribute to make the temperature practically uniform in this zone. The following approach will then be focused on the case of a growing bubble which is representative of most real situations.

In the absence of thermodynamic effect, the pressure inside the bubble would be $p_v(T_\infty)$. Because of the drop in temperature, the actual pressure is somewhat smaller. It is equal to $p_v(T_c)$, the vapor pressure at the actual bubble temperature T_c. Note that it is quite a general result to assume thermodynamic equilibrium in cavitating flows. It is only in very special situations, as during the ultimate stage of collapse for instance, that this condition might not be satisfied. The difference in pressure $p_v - p_\infty$ responsible for bubble growth is then smaller and bubble growth is slowed down. Therefore, the thermodynamic effect tends to reduce the development of cavitation. This is a general conclusion valid for any type of cavitation. The length of leading edge cavities for instance may be significantly reduced by thermodynamic effect.

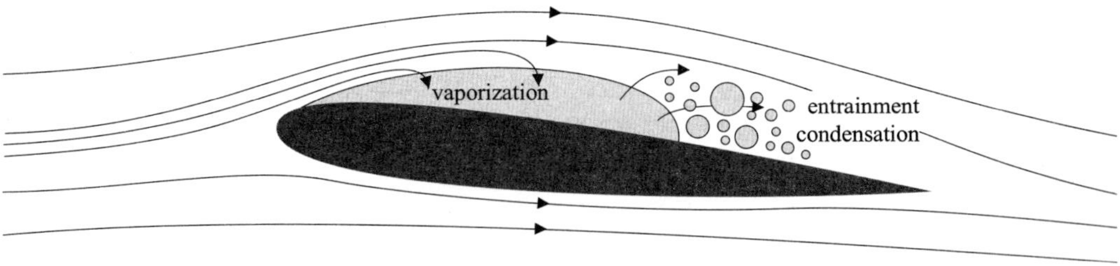

Figure 17. Schematic view of leading edge cavitation

In practice, the problem is to evaluate the order of magnitude of the temperature drop as a function of the thermodynamic properties of the liquid and its vapor and also of the cavitating flow conditions. This requires to write an energy balance which has been ignored so far.

Consider a cavitation nucleus which is growing in a liquid medium whose temperature far from the bubble is T_∞. Let $R(t)$ be the bubble radius at time t and $T_c(t)$ the temperature inside. Initially, it is assumed that the radius of the original microbubble is negligible and that the internal temperature is that of the liquid T_∞.

During bubble growth, the heat necessary for vaporization is supposed to be supplied to the interface by conduction through the liquid. Hence, a thermal boundary layer develops on the bubble wall. The liquid temperature drops from T_∞ to T_c through this boundary layer (see figure 18). As for any diffusive process, the order of magnitude of the boundary layer thickness is $\sqrt{\alpha_\ell t}$ where $\alpha_\ell = \lambda_\ell / (\rho_\ell c_{p\ell})$ is the thermal diffusivity of the liquid (λ_ℓ, ρ_ℓ and $c_{p\ell}$ are the conductivity, density and heat capacity of the liquid). The typical temperature gradient within the boundary layer is $\Delta T / \sqrt{\alpha_\ell t}$ where $\Delta T = T_\infty - T_c$. According to Fourier's law, the conductive heat flux towards the interface is then of the order of $\lambda_\ell \Delta T / \sqrt{\alpha_\ell t}$.

The energy balance expresses that, at any time, the heat supplied by conduction to the interface of area $4\pi R^2$ is used for vaporization and causes the increase of the mass of vapor $\frac{4}{3}\pi R^3 \rho_v$ inside the bubble (ρ_v is the vapor density). Hence, the energy balance writes:

$$\lambda_\ell \frac{\Delta T}{\sqrt{\alpha_\ell t}} 4\pi R^2 = \frac{d}{dt}\left(\frac{4}{3}\pi R^3\right) \rho_v L \tag{5.1}$$

where L is the latent heat of vaporization.

Previous equation leads to the following estimate of the temperature drop at any time:

$$\Delta T \cong \frac{\dot{R}\sqrt{t}}{\sqrt{\alpha_\ell}} \frac{\rho_v L}{\rho_\ell c_{p\ell}} \tag{5.2}$$

Let us observe that the quantity

$$\Delta T^* = \frac{\rho_v L}{\rho_\ell c_{p\ell}} \tag{5.3}$$

has temperature units. This group of parameters appears commonly when thermodynamic effect is concerned. Experiments show that ΔT^* is a relevant order of magnitude of the temperature drop due to thermal effects. Figure 19 gives some typical values of ΔT^* for different fluids at different temperatures. From this figure, the thermodynamic effect appears negligible for water at room temperature whereas it becomes significant for hot water. It is important for liquid hydrogen around 20 K as used in rocket engines.

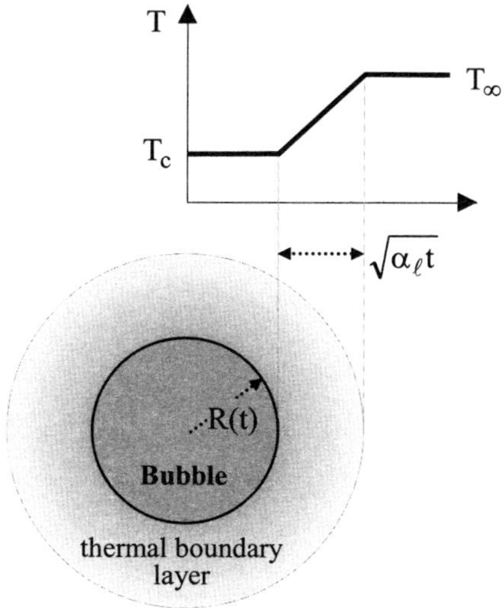

Figure 18. The thermal boundary layer on the bubble interface

The B factor of Stepanov is the non dimensional temperature drop defined by:

$$B = \frac{\Delta T}{\Delta T^*} \tag{5.4}$$

This parameter has a simple physical interpretation. Consider that a volume $\mathcal{V}_v$ of vapor is produced and that the heat required for vaporization is taken from a volume $\mathcal{V}_\ell$ of liquid which is supposed to be cooled of $\Delta T = T_\infty - T_c$. If so, the heat balance writes:

$$\rho_v \, \mathcal{V}_v \, L = \rho_\ell \, \mathcal{V}_\ell \, c_{p\ell} \, \Delta T \tag{5.5}$$

Hence, the B factor appears to be the ratio $\mathcal{V}_v / \mathcal{V}_\ell$ of the volume of vapor produced to the volume of liquid that supplies the latent heat of vaporization.

The drop in temperature ΔT is accompanied by a drop in vapor pressure Δp_v given by:

$$\Delta p_v = p_v(T_\infty) - p_v(T_c) \cong \frac{dp_v}{dT} \Delta T \tag{5.6}$$

The slope of the vapor pressure curve dp_v / dT can be estimated using the famous thermodynamic equation of Clapeyron in which we can usually assume that vapor density is negligible with respect to liquid density:

$$L = T\left[\frac{1}{\rho_v} - \frac{1}{\rho_\ell}\right]\frac{dp_v}{dT} \cong \frac{T}{\rho_v}\frac{dp_v}{dT} \tag{5.7}$$

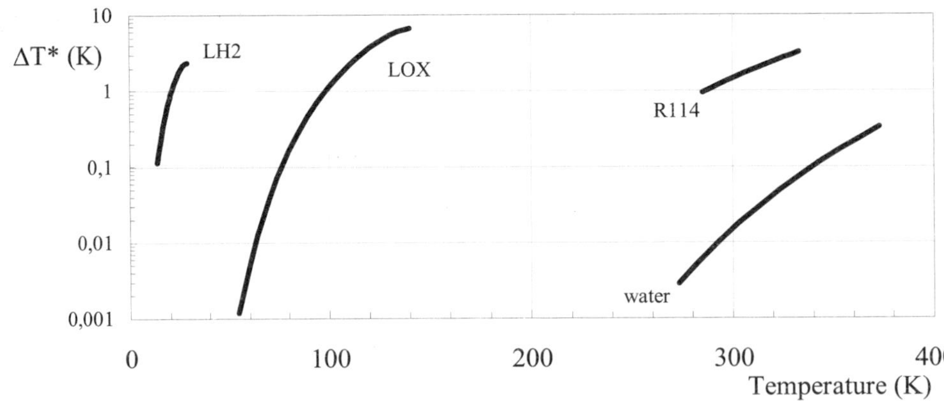

Figure 19. Values of ΔT^* for liquid hydrogen, liquid oxygen, refrigerant 114 and water at different temperatures

By combining equations (5.2), (5.6) and (5.7), the drop in vapor pressure for the growing bubble is then:

$$\Delta p_v \cong \rho_\ell \, \Sigma \, \dot{R}\sqrt{t} \tag{5.8}$$

where parameter Σ, originally introduced by Brennen, is defined by:

$$\Sigma = \frac{(\rho_v L)^2}{\rho_\ell^{\,2} c_{p_\ell} T_\infty \sqrt{\alpha_\ell}} \tag{5.9}$$

Like ΔT^*, the Σ parameter, which has units of m/s$^{3/2}$, increases significantly with thermal effects. It can then be considered as another indicator of the magnitude of the thermodynamic effect. Table 1 gives typical values for cold and hot water.

WATER	20°C	100°C
ΔT^* (K)	0.0102	0.324
Σ (m/s3/2)	3.89	2944

Table 1. Values of ΔT^* and Σ for water at 20°C and 100°C

To come back to the growth of a cavitation bubble in case thermal effects are not negligible, consider the Rayleigh equation (2.3):

$$R\ddot{R} + \frac{3}{2}\dot{R}^2 = \frac{p_v(T_c) - p_\infty}{\rho_\ell} \tag{5.10}$$

In this equation, the vapor pressure which holds for the pressure inside the bubble is taken at temperature T_c. Introducing the vapor pressure at the liquid bulk temperature T_∞, we obtain:

$$\left(R\ddot{R} + \frac{3}{2}\dot{R}^2\right) + \frac{\Delta p_v}{\rho_\ell} = \frac{p_v(T_\infty) - p_\infty}{\rho_\ell} \tag{5.11}$$

This equation is similar to the usual Rayleigh equation except concerning the second term on the left hand side which is new and accounts for thermal effects. Using equation (5.8), this term is transformed as follows:

$$\left(R\ddot{R} + \frac{3}{2}\dot{R}^2 \right) + \Sigma \dot{R}\sqrt{t} \cong \frac{p_v(T_\infty) - p_\infty}{\rho_\ell} \tag{5.12}$$

The thermal term, which is initially zero, takes increasing importance with time. Hence, thermal effects remain negligible as long as:

$$\Sigma \dot{R}\sqrt{t} << \frac{p_v(T_\infty) - p_\infty}{\rho_\ell} \tag{5.13}$$

Since the bubble growth rate $\dot{R}$ is of the order of $\sqrt{(p_v(T_\infty) - p_\infty)/\rho_\ell}$, the previous condition becomes:

$$t << \frac{p_v(T_\infty) - p_\infty}{\rho_\ell \Sigma^2} \tag{5.14}$$

In the practical case of a cavitating flow of characteristic velocity V and characteristic length scale L as that around a blade of chord length L, the typical time t available for bubble growth is the transit time L/V. It has to be compared to the characteristic time $(p_v(T_\infty) - p_\infty)/(\rho_\ell \Sigma^2)$ to estimate whether thermal effects significantly affect cavitation or not. In this equation, p_∞ is the pressure applied to the bubble and responsible for its growth whose typical value is the minimum pressure on the foil. For a given fluid and given cavitating conditions, it is then quite easy from equation (5.14) to evaluate the importance of the thermodynamic effect. In particular, it is clear that the larger Σ, the more important the thermodynamic effect.

6 Supercavitation

Supercavity flows as the one shown in figure 5 are characterized by a long cavity whose length may be much larger than that of the cavitator which generates it. The closure region of a supercavity takes place in the liquid bulk, usually far downstream the cavitator. The smaller the cavitation number, the longer the cavity.

We will consider here the case of an axisymmetric supercavity flow past a cavitator as shown in figure 20. For long cavities, the slender body approximation can be applied. It consists in assuming that the body and its cavity bring only a slight perturbation to the basic flow and then that the axial flow velocity is everywhere close to the velocity at infinity V_∞. Although V_∞ is assumed constant, the present approach applies to unsteady supercavities when unsteadiness is due to a time dependent pressure at infinity p_∞ or a time dependent pressure inside the cavity p_c.

The latter occurs especially in ventilated flows when the cavity is generated by blowing non condensable gas (e.g. air) in the wake of the cavitator. Ventilated flows are similar to natural cavity flows, in spite of some differences connected to the non condensable nature of the injected gas in comparison to vapor, and are then relevant of the same type of approach. A particular feature of ventilated flows is the possibility for the cavity to have a pulsating behavior connected to a periodic release of air in the wake. If so, the cavity length oscillates periodically and the cavity pressure p_c too. Pulsating ventilated supercavities is an example of unsteady supercavity flows which can be modeled by the present approach.

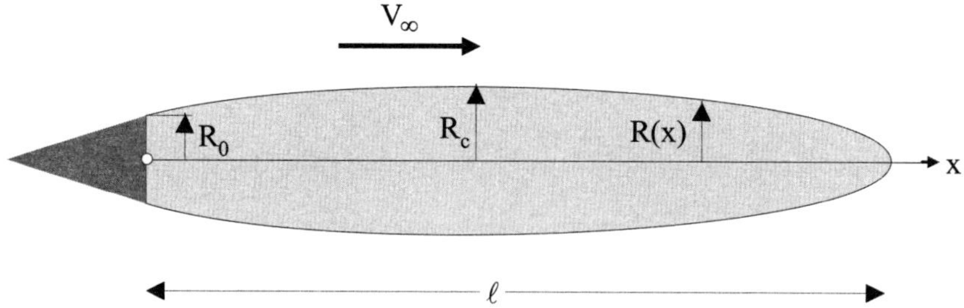

Figure 20. Axisymmetric supercavity flow

Consider a cross section area of the cavity at station x. The corresponding radius and section are respectively R and $S = \pi R^2$. Both depend on the axial position x and in addition on time t in the unsteady case. It can be shown (cf. e.g. Franc & Michel, (2004)) that the evolution of the cross section area S is given by the following equation:

$$\frac{d^2 S}{dt^2} = -\frac{2\pi}{\mu} \frac{p_\infty - p_c(t)}{\rho} \tag{6.1}$$

This equation is derived from Euler equation using the mass conservation equation together with the slender body approximation. *d/dt* is the usual transport derivative given by:

$$\frac{d}{dt} = \frac{\partial}{\partial t} + V_\infty \frac{\partial}{\partial x} \tag{6.2}$$

The μ parameter takes into account the inertia of the liquid surrounding any section of cavity and is equivalent to an added mass coefficient. It is generally assumed constant and can be related to the cavitation number σ. In particular, its asymptotic behavior when σ approaches zero i.e. for long supercavities is the following (see Franc & Michel, (2004)):

$$2\mu \cong ln\frac{1}{\sigma} \tag{6.3}$$

By using $S = \pi R^2$, equation (6.1) becomes:

$$\rho\left[R\ddot{R} + \dot{R}^2\right] = \frac{1}{\mu}(p_c - p_\infty) \tag{6.4}$$

where $\dot{R}$ and $\ddot{R}$ stand respectively for dR/dt and d^2R/dt^2. Equation (6.1) is then very similar to the Rayleigh equation except for the coefficient of $\dot{R}^2$ which is 3/2 in the original Rayleigh equation for the spherical bubble and which is here 1 in axisymmetric configurations. This similarity gives a basis for the physical interpretation of equation (6.1).

Consider a given cross section of the axisymmetric cavity. The use of the transport derivative *d/dt* suggests to follow it as it is advected by the main flow at velocity V_∞. During its movement, the radius of this cut of "cylindrical" bubble changes according to equation (6.4). Like for a spherical bubble, the driving term for the change in radius is the pressure difference between the pressure at infinity and the internal pressure. As shown by equation (6.4), the dynamics of any cross section of the cavity depends only upon this pressure difference and, in particular, it does not depend upon the neighboring cross sections. This result is known as the Logvinovich independence principle of cavity expansion.

It can be considered that, at any time, a new section of cavity is shed by the cavitator basis. It is advected downstream at velocity V_∞ whereas its radius evolves according to the pressure difference $p_\infty - p_c$. The instantaneous shape of the whole cavity is obtained by stacking all sections side by side.

The initial conditions to be added for the resolution of partial differential equation (6.4) are related to the instant when the considered section of the cavity separates from the cavitator basis. Its initial radius R is then the radius of the cavitator basis whereas the initial value of $\dot{R}$ is related to the slope of the cavitator trailing edge. It must be such that continuity of slope is achieved at any time between the cavitator and the cavity at detachment, resulting then in a smooth detachment.

The resolution of partial differential equation (6.1) allows the computation of the shape of the supercavity and its time evolution if the flow is unsteady. As an example, consider the simple case of a steady supercavity flow for which *d/dt* reduces to $V_\infty d/dx$. Equation (6.1) becomes then:

$$\frac{d^2R^2}{dx^2} = -\frac{\sigma}{\mu} \tag{6.5}$$

The solution to this equation is obviously:

$$\left(\frac{R}{R_0}\right)^2 = \frac{x}{R_0}\left[2\dot{R}_0 - \frac{\sigma}{2\mu}\frac{x}{R_0}\right] + 1 \tag{6.6}$$

where R_0 is the radius of the cavitator basis at $x=0$ and $\dot{R}_0$ its slope dR/dx at $x=0$. The origin $x=0$ is chosen at the cavitator basis. For the asymptotic case considered here of a long supercavity at small cavitation number, the radius of the cavity is much larger than the radius R_0 of the cavitator, so that the last term on the right hand side can be neglected and the shape of the cavity is then given approximately by:

$$\left(\frac{R}{R_0}\right)^2 \cong \frac{x}{R_0}\left[2\dot{R}_0 - \frac{\sigma}{2\mu}\frac{x}{R_0}\right] \tag{6.7}$$

This equation shows that the cavity can be approximated by an ellipsoid. The cavity length ℓ, which corresponds to the downstream point where R equals 0, is then given by:

$$\frac{\ell}{R_0} \cong \frac{4\mu}{\sigma}\dot{R}_0 \tag{6.8}$$

The maximum radius R_c of the cavity is obtained at $x = \ell/2$ and is:

$$\frac{R_c}{R_0} \cong \sqrt{\frac{2\mu}{\sigma}}\,\dot{R}_0 \tag{6.9}$$

so that cavity slenderness $\delta = 2R_c/\ell$ is:

$$\delta = \sqrt{\frac{\sigma}{2\mu}} \tag{6.10}$$

When the cavitation number approaches 0, the μ parameter tends to infinity according to asymptotic behavior 6.3. The length and maximum radius of the cavity both tend to infinity but the cavity slenderness tends to zero according to the following asymptotic behavior:

$$\delta \cong \sqrt{\frac{\sigma}{ln(1/\sigma)}} \tag{6.11}$$

Previous equations (6.8) and (6.9) are more often written using the drag coefficient C_D instead of $\dot{R}_0$. The drag can be computed from a global momentum balance. It is given by (see e.g. Franc & Michel, (2004)):

$$D \cong (p_\infty - p_c)\, \pi R_c^2 \tag{6.12}$$

so that the drag coefficient is:

$$C_D = \frac{D}{\frac{1}{2}\rho V_\infty^2\, \pi R_0^2} \cong \sigma \left(\frac{R_c}{R_0}\right)^2 \cong 2\mu \dot{R}_0^2 \tag{6.13}$$

Replacing $\dot{R}_0$ by C_D in equations (6.8) and (6.9), the following equations for the length and radius of the axisymmetric supercavity are obtained:

$$\begin{cases} \dfrac{\ell}{R_0} \cong \dfrac{2}{\sigma}\sqrt{2\mu\, C_D} \cong \dfrac{2}{\sigma}\sqrt{C_D\, ln\dfrac{1}{\sigma}} \\ \dfrac{R_c}{R_0} \cong \sqrt{\dfrac{C_D}{\sigma}} \end{cases} \tag{6.14}$$

These equations are well known asymptotic formulae which were originally derived by Garabedian in 1956 (see Franc & Michel, (2004)) on the basis of a rather complicated mathematical approach. The method presented here and based on equation (6.1) is quite simple. However, to be fully predictive, the method requires to model the μ parameter which is unknown a priori. This is done via equation (6.3) for instance which gives the asymptotic behavior of μ when σ approaches zero. In practice, equation (6.3) was obtained by adjusting the value of μ in order to get the correct value of the cavity slenderness as given by the Garabedian solution. It is then quite normal that equations (6.14) be in agreement with the Garabedian model. Other procedures of adjustment of the μ parameter could however be used, by comparison with tests for instance.

In conclusion, it appears that axisymmetric supercavity flows can be modelled by an equation very similar to the Rayleigh equation, providing some simplifying assumptions. It is also the case for 2D supercavity flows (see Pellone et al. (2004)). One of the advantages of such a method is to provide a rather simple basis for the modeling of supercavity flows, especially in unsteady cases which are not so easy to handle. The adjustment of μ obtained from steady considerations is generally kept for unsteady supercavity flows. In particular, this method proved to be quite efficient for the analysis of pulsating ventilated cavities (see Franc & Michel, (2004)).

7 Cavitation erosion

It is well known that cavitation can lead to erosion. Figure 21 presents a typical example of a component of a volumetric pump eroded by cavitation. For low exposure times, the damage remains superficial. It is characterized by small isolated pits, whose diameter ranges typically from a few microns to a fraction of millimeter and whose depth is of the order of a few percents of their diameter, and even less. Each of them is caused by the collapse of a distinct vapor structure. This preliminary stage, without significant mass loss, is the incubation period. The material surface undergoes mainly plastic deformation.

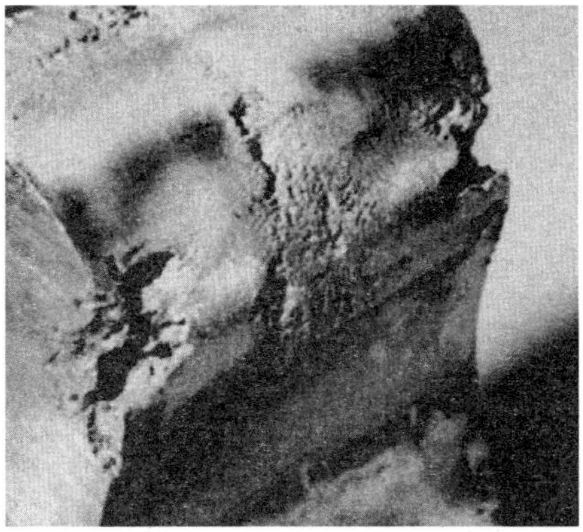

Figure 21. A typical example of cavitation erosion

When the exposure time increases, overlapping becomes more and more important. The wall is then submitted to successive collapses whose number increases with exposure time. Repetitive loading occurs, leading to fatigue followed by rupture and subsequent material removal. During this advanced stage of erosion, the damage rate is often characterized by the mean depth of penetration rate (MDPR) i.e. the thickness of material which is removed per unit exposure time.

The aggressiveness of a cavitating flow depends largely upon flow velocity. There is a threshold velocity below which no damage occurs whatever may be the exposure time. In the case of water and stainless steel 316 L, this threshold normally lies around 15 to 20 m/s. Above, the erosion rate increases very rapidly with velocity. The influence of flow velocity on MDPR is often modeled by a power law with a large exponent, typically between 4 and 9.

As for the evolution of erosion with the development of cavitation, it is generally observed that damage, which is of course zero under non cavitating conditions, first increases when the cavitation number decreases and then decreases for developed cavitation.

The objective of this section is to point out a few basic trends in cavitation erosion from the consideration of the only Rayleigh-Plesset equation. It is shown that trends of practical interest can be obtained concerning for instance scaling rules in cavitation erosion or the influence of cavitation types on the erosive potential.

7.1 Bubble collapse and cavitation erosion

Firstly, the Rayleigh equation will be used to explain why a bubble, when collapsing in the vicinity of a wall, can actually damage it. The existence of damage implies that the wall has been loaded with a stress larger than its yield strength so that plastic deformation actually occurred.

Consider for instance the case of stainless steel 316 L whose yield strength is about 400 MPa. The occurrence of pits proves that the collapse of vapor structures generates impact loads larger than 400 MPa or, in other words, pressure pulses larger than 4000 bar. The point is then to understand how a bubble can generate such a high pressure when collapsing.

The answer partly lies in the distribution of pressure within the fluid during collapse. Consider a pure vapor bubble which is collapsing because of a pressure p_∞ applied far from the bubble greater than the vapor pressure p_v which prevails inside the bubble (cf. section 3.3). If surface tension effects are neglected, pressure on the liquid side at the bubble interface is also p_v. Pressure distribution must then increase from p_v to p_∞ when moving away from the bubble interface.

It is remarkable that, if at the very beginning of the collapse, pressure smoothly increases from p_v to p_∞, it is no longer the case as soon as the bubble radius becomes smaller than a given fraction of its initial radius, namely 63%. From that time, pressure distribution exhibits a maximum very close to the bubble wall as shown in figure 22. This maximum increases all along the collapse phase. Hence, the collapsing bubble gives rise to a kind of pressure wave which propagates inward and steepens during collapse. It must be noticed that this pressure wave is not due to compressibility, which is ignored in the present model, but results only of inertia forces.

The computation of the pressure field inside the liquid during collapse can be found in most textbooks on cavitation (see for example Franc and Michel, (2004)). At any time t, if the bubble radius is R(t), the maximum pressure p_{max} is given by the approximate following non dimensional equation:

$$\Pi = \frac{p_{max} - p_\infty}{p_\infty - p_v} \cong 0.157 \left(\frac{R_0}{R} \right)^3 \tag{7.1}$$

where, as usually R_0 is the initial bubble radius. Furthermore, the analytical computation shows that this maximum takes place close to the bubble wall at a radial distance *r* from the bubble center of the order of:

$$r \cong 1.59\, R \tag{7.2}$$

Both equations (7.1) and (7.2) are asymptotic formulas valid for small enough values of R / R_0 i.e. near the end of the collapse.

As a consequence, a high pressure is generated in the liquid, close to the bubble. Such a high pressure is due to the always increasing velocity of the bubble wall and the focusing effect of the spherical collapse. At the end of the collapse, the interface velocity tends to infinity and the maximum pressure too as shown by equation (7.1). As an example, for a vapor bubble collapsing under pressure at infinity $p_\infty = 1\ bar$, the maximum pressure reaches almost *1 GPa* when the bubble radius is divided by *50* ($R_0 / R = 50$). This value is quite comparable to yield strength and even ultimate strength of usual materials, so that damage can actually be expected.

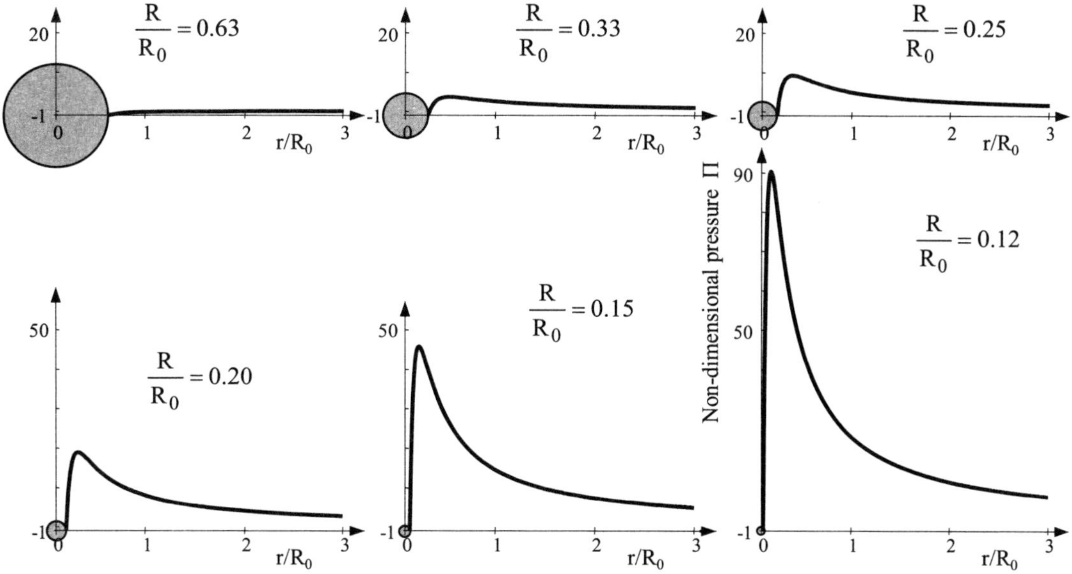

Figure 22. Pressure distribution in the liquid at different times during bubble collapse

As already mentioned in section 3.3, the infinite values given by the present model are not realistic physically and other phenomena (as compressibility and gas content), negligible at the beginning of the collapse, become predominant at the end. Therefore, the actual physics is somewhat more complicated than that given by this simple model.

As an example, the presence of non condensable gas inside the bubble results in a high internal pressure at the end of the collapse which generally forces the bubble to rebound. Computations as well as experiments have shown that an expanding shock wave of high amplitude is generated at the instant of rebound. The impact of this shock wave on the wall is a possible cause of damage.

Another plausible mechanism is the development of a microjet during collapse. As a matter of fact, when a bubble collapses near a wall, spherical symmetry is obviously broken by the wall and the bubble does not remain spherical. A liquid microjet develops towards the wall, goes through the bubble and strikes the wall at high velocity. The resulting impact pressure can also be a reason for damage.

In addition, collective effects, when a cloud of bubbles is collapsing for instance, may also play a role in the damaging process.

Although there is still some controversy on the detailed physical mechanism of cavitation erosion, it is unquestionable that a high pressure is generated during the collapse of a bubble as shown by the Rayleigh equation. Note that the damage potential depends greatly on the distance of the bubble to the wall. The closer to the wall, the higher the impact pressure.

7.2 Scaling law for the aggressiveness of a single bubble

According to the simple previous model (cf. equation (7.1)), the peak of pressure generated by the collapse of a bubble is proportional to the driving pressure $p_\infty - p_v$ responsible for the collapse and to the ratio $(R_0 / R)^3$:

$$p_{max} \approx (p_\infty - p_v)\left(\frac{R_0}{R}\right)^3 \tag{7.3}$$

Besides, equation (3.4) shows that the velocity of the bubble interface during collapse is given by:

$$\dot{R} \approx \sqrt{\frac{p_\infty - p_v}{\rho}\left(\frac{R_0}{R}\right)^3} \tag{7.4}$$

so that, according to this model, the pressure peak scales like:

$$p_{max} \approx \rho \, \dot{R}^2 \tag{7.5}$$

as it could be expected from simple dimensional analysis.

If we refer to the microjet model mentioned in the previous section, assuming that the damage is produced by a liquid jet impacting the wall at high velocity still denoted $\dot{R}$, the impact pressure can be estimated using the classic water hammer formula:

$$p_{max} \approx \rho c \, \dot{R} \tag{7.6}$$

where ρc is the acoustic impedance of the liquid. Note that this equation is valid in the case of a perfectly rigid wall and that a correction has to be applied in case the acoustic impedance of the liquid is not negligible with respect to that of the material.

Although some differences exist between the two models in particular with respect to the exponent of $\dot{R}$, a general conclusion is that the aggressiveness of the collapsing bubble depends primarily upon the interface velocity $\dot{R}$. The point is now to know how the interface velocity scales with the velocity and length scale of the mean flow.

Consider the cavitating flow around a foil as already analyzed in section 4.2 and let us investigate the effect of a change either in length scale (typically the chord length L) or in flow velocity V on the velocity $\dot{R}$ of the bubble interface. As explained in section 4.2, if the bubble radius is made non dimensional using the chord length L of the foil and if time derivatives are replaced by space derivatives, the Rayleigh equation takes the following non dimensional form (cf. equation (4.10)):

$$\overline{R}\frac{d^2\overline{R}}{d\overline{x}^2} + \frac{3}{2}\left(\frac{d\overline{R}}{d\overline{x}}\right)^2 = -\frac{1}{2}(C_p + \sigma) \tag{7.7}$$

For geometrically similar flows – i.e. for the same distribution of pressure coefficient C_p – and for equal cavitation numbers σ – i.e. for equal developments of cavitation –, it was already observed that the evolution of the non dimensional bubble radius $\overline{R}(\overline{x})$ along the foil is unique.

Furthermore, from the non dimensional procedure presented in section 4.2, it can easily be shown that the dimensional value of the velocity of the bubble interface is:

$$\dot{R} = V \frac{d\overline{R}}{d\overline{x}} \tag{7.8}$$

This proves that, for similar flows, $\dot{R}$ is proportional to mean flow velocity. In other words, the dynamics of a bubble traveling on the foil is scaled as the mean flow velocity.

Another consequence of equation (7.8) is that the bubble interface velocity is independent of the length scale which does not appear in equation (7.8). If chord length is doubled for instance, bubble radius is expected to be doubled too. Since the flow velocity is assumed to be the same, the transit time is also doubled, so that finally the bubble interface velocity is unchanged.

The important conclusion of this analysis is that the velocity of the bubble interface depends essentially upon mean flow velocity and that the dependence should be linear. Such a result has to be considered as a first order trend and deviations can be expected due to secondary phenomena as interactions between bubbles, viscous effects.... This result is used in next section to derive scaling laws for the aggressiveness of a cavitating flow.

7.3 Scaling laws for flow aggressiveness

In the previous section, it has been shown that the aggressiveness of a collapsing bubble is proportional to flow velocity. However, as mentioned in the introduction, cavitation damage increases much more rapidly with flow velocity. An analysis of scaling laws for the aggressiveness of a cavitating flow allows to better understand this influence.

A cavitating flow is characterized by a large population of bubbles of various characteristics. If we consider a given surface element of a wall exposed to cavitation, it is quite common to describe flow aggressiveness at this location by a pressure pulse height spectrum (PPHS) $\dot{n}(\Delta p)$. It represents the density $\dot{n}$ per unit time and unit surface area of pressure pulses whose amplitude is larger than Δp.

Considering still similar cavitating flows around blades or hydrofoils as an example, it is expected that PPHS is a function of flow velocity V and chord length L which is chosen as characteristic length scale as previously. In addition, PPHS also depends upon the fluid. A priori, several fluid properties may influence flow aggressiveness as density, viscosity, surface tension... Referring to the simplified approach developed in section 7.2 where inertia is considered as predominant, a unique physical property for the liquid will be considered here as relevant for cavitation erosion. It is either the fluid density ρ if we refer to equation (7.5) or the acoustic impedance ρc if we refer to equation (7.6).

We will then write either:

$$\dot{n} = function(\Delta p, V, L, \rho) \tag{7.9}$$

or:

$$\dot{n} = function(\Delta p, V, L, \rho c) \tag{7.10}$$

Basic dimensional analysis leads to the following non dimensional relations:

$$\frac{\dot{n}\, L^3}{V} = \varphi_1\left(\frac{\Delta p}{\rho V^2}\right) \tag{7.11}$$

or:

$$\frac{\dot{n}\,L^3}{V} = \varphi_2\left(\frac{\Delta p}{\rho c V}\right) \tag{7.12}$$

where φ_1 (respectively φ_2) is a universal function for the considered flow configuration independent of length sale, velocity and fluid properties, provided that geometrical and cavitation scaling are both fulfilled.

To simplify, we will abandon the fluid dependence and consider a unique fluid. Only the influence of length scale and velocity on PPHS will then be examined. The approach could however be extended to account for the effect of other factors on cavitation erosion as a change of material.

To be more general, equations (7.11) and (7.12) are written as follows:

$$\frac{\dot{n}\,L^3}{V} = \varphi\left(\frac{\Delta p}{V^{\alpha}}\right) \tag{7.13}$$

where exponent α is a parameter which ranges a priori between 1 and 2. A common form for function φ is a power law with a negative exponent $-\beta$. A typical order of magnitude for β is 2 to 4 (see e.g. Franc and Michel (2004), Kato et al. (1996)). However, data on β are only few and the β value might significantly depend upon flow configuration. With this choice, PPHS is written as follows:

$$\dot{n} = A\,\frac{V^{\alpha\beta+1}}{L^3}\,\frac{1}{\Delta p^{\beta}} \tag{7.14}$$

It is then possible to estimate pitting rate on a given material. The approach is very basic. The material is characterized by its yield strength σ_Y and it is assumed that a pit is formed if the pulse height Δp is larger than σ_Y and that only elastic behavior occurs otherwise. Pitting rate i.e. the density of pits per unit surface area and unit exposure time is then:

$$\dot{n}_{pits} = A\,\frac{V^{\alpha\beta+1}}{L^3}\,\frac{1}{{\sigma_Y}^{\beta}} \tag{7.15}$$

Hence, it appears that pitting rate increases as the power $\alpha\beta+1$ of flow velocity. Assuming typical values for α and β of 1.5 and 3, we have $\alpha\beta+1 \cong 5.5$. Hence, this approach shows that flow aggressiveness increases with a relatively high power of flow velocity, as shown by experiments.

Two reasons explain this strong influence of flow velocity. Firstly, the aggressiveness of a single bubble increases with flow velocity, as already noticed. Secondly, for a cavitating flow, an increase in flow velocity results in an increase in the production rate of vapor structures. This can be seen on equations (7.11) or (7.12). If we consider vapor structures which follow scaling law (7.5) or (7.6), then, for a constant value of either $\Delta p/\rho V^2$ or $\Delta p/\rho c V$, equations (7.11) and (7.12) show that:

$$\frac{(\dot{n}L^2)\,L}{V} = Constant \tag{7.16}$$

This can be considered as a Strouhal scaling law since $\dot{n}L^2$ is actually a frequency representing the rate of collapse of vapor structures on similar surface areas per unit time. Hence, the collapsing rate increases proportionally to flow velocity. The increase of bubble production rate with flow velocity combined with the increase of the aggressiveness of single bubbles with flow velocity results in a relatively strong increase of the overall aggressiveness of a cavitating flow.

Previous conclusions apply to similar cavitating flows presenting, in particular, the same type of cavitation. Besides, qualitative trends can also be obtained as for the effect of cavitation types on flow aggressiveness. Two main factors have a great influence: the maximum bubble size prior to collapse and the adverse pressure gradient which forces the bubble to collapse. From a qualitative viewpoint, an increase in bubble size has the same effect as an increase in pressure gradient. Both tend to increase the erosive potential.

From an erosion viewpoint, it is then preferable to reduce as much as possible the adverse pressure gradient in order to lengthen bubble lifetime and then minimize the violence of collapse. On the other hand, a small adverse pressure gradient has a destabilizing effect on cavity closure. The length of the cavitating zone and consequently the whole cavitating flow is prone to fluctuations and unsteadiness. In practice, the control of adverse pressure gradient is mainly a matter of design and it may be necessary to find a compromise between erosive potential and flow stability.

Besides pressure gradient, maximum bubble size prior to collapse is another essential parameter for the evaluation of erosive potential. Consider the case of traveling bubble cavitation. For a given flow configuration, bubble maximum size depends greatly upon nuclei density. If it is small, bubbles are rather big. If it is large, bubble growth is limited by neighboring bubbles. This results in smaller bubbles and smaller erosive potential, too. For traveling bubble cavitation, it can then be expected that flow aggressiveness will depend upon nuclei density since it affects maximum bubble size.

As for attached cavitation, its aggressiveness depends upon the type of cavity. An attached closed and stable cavity usually produces small bubbles which are concentrated in the closure region of the cavity. Cavitation damage is then focused at cavity closure where bubbles collapse. This is the case for instance of a thin leading edge cavity attached to a blade at small angle of attack.

An increase in angle of attack leads to an increase in cavity thickness. The cavity becomes more open and sheds more bubbles in its wake. Since the bubble production rate increases, the erosive potential is expected to increase. Moreover, damage generally extends on a larger zone, upstream and downstream of mean cavity closure.

Unsteadiness can significantly enhance the erosive potential of a cavitating flow. It is more especially the case of cloud cavitation which can be quite aggressive from an erosion viewpoint. This is due first to the high density of bubbles contained in a cavitation cloud and also to collective effects during the collapse of a bubble cluster which may enhance the intensity of each elementary collapse.

8 Vortex cavitation

Although the standard Rayleigh-Plesset equation applies to spherical bubbles only, the dynamics of other types of cavitating structures can often be predicted by very similar equations.

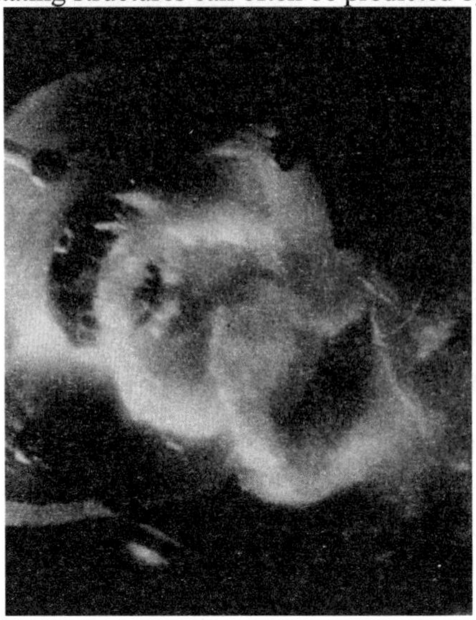

Figure 23. Cavitating vortex rings issuing from a StratoJet® self-resonating cavitating jet. Resonance is achieved through a feedback mechanism between the shear layer at the nozzle lip and the upstream organ pipe tube leading to the orifice. The frequency of emission of the rings is the same as the organ pipe frequency (Courtesy of Dr. Georges Chahine from Dynaflow, Inc.)

It is in particular the case of toroidal cavitating vortices or "ring bubbles" as those periodically produced by a cavitating submerged round jet (figure 23). Chahine and Genoux (1983) have shown that the time evolution of the radius $R(t)$ of a cross section of the ring bubble follows an equation analogous to the Rayleigh-Plesset equation:

$$\rho\left[R\ddot{R}+\dot{R}^2\right]ln\frac{8A}{R}-\frac{1}{2}\rho\dot{R}^2 = \left[p_v-p_\infty(t)\right]+\frac{1}{2}\rho\left[\frac{\Gamma}{2\pi R}\right]^2+p_{g0}\left(\frac{R_0}{R}\right)^{2k}-\frac{S}{R} \qquad (8.1)$$

On the right hand side, we can identify the same terms as in the original Rayleigh-Plesset equation i.e. the driving pressure term together with the non condensable gas and surface tension terms with however slight changes due to differences in geometry. In addition to these conventional terms, an additional one appears which depends upon circulation Γ of the ring bubble. Indeed, such a ring bubble is a toroidal vortex which is provided with a vorticity characterized by the Γ parameter. Physically, circulation tends to resist the collapse in a similar way as non condensable gas. Another parameter which is specific to ring bubbles and which appears on the left hand side of equation (8.1) is the radius A of the ring (see figure 23). Nevertheless, as for the spherical bubble, equation (8.1) allows the analysis of various characteristics of a bubble ring as its collapse time or its stability.

It can also be shown that the dynamics of an axisymmetric cavitating vortex is governed by a Rayleigh-Plesset like equation. Further information on cavitating vortices can be found in Chahine & Genoux (1983) and Franc & Michel (2004).

9 Cavitation modeling

As a first example of the capabilities of the Rayleigh-Plesset equation to model real cavitating flows, let us consider the case of a partial cavity attached to the suction side of a foil as shown in Figure 3. This type of cavity is obviously far from being spherical. However, at least in a preliminary step, it can be approached using the Rayleigh-Plesset equation as follows.

Consider the original fully-wetted flow corresponding to the cavitating case presented in Figure 3. Its characteristics – and particularly the pressure field – are assumed to be known from classical computational techniques. Consider an isolated spherical (or more commonly hemispherical as observed in Figure 6) bubble whose center moves along the wall. This bubble is assumed to be initially a cavitation nucleus i.e. a microbubble of only a few microns in diameter. For small enough values of the cavitation number, this bubble, which is carried by the flow, will first grow in the low pressure region generated by the foil, become a macroscopic cavitation bubble and then collapse downstream where the pressure recovers. To some extent, the envelope of the hemispherical bubble as it moves along the foil gives a rough outline of the original attached cavity. Then, the development of cavitation can be analyzed, at least qualitatively, on the basis of the simple Rayleigh-Plesset equation.

Such a simplified model has obviously many serious limitations. In particular, it ignores any feedback of cavitation on the original fully wetted flow pressure field, which can be considered as valid only near cavitation inception. More realistic, but also more complicated models, can be used according to the accuracy required for prediction. Many of them are still based upon the Rayleigh-Plesset equation as it is the case for the bubble two-phase flow model derived by Kubota et al. (1992) whose principle is presented below.

The framework of the modeling is the standard homogeneous model which provides the simplest technique for analyzing two-phase flows. The liquid / vapor mixture is treated as a pseudofluid which obeys the usual equations of single-phase flow, i.e. continuity:

$$\frac{\partial \rho}{\partial t} + div(\rho \vec{V}) = 0 \tag{9.1}$$

and momentum balance or Navier-Stokes equation :

$$\rho \left[\frac{\partial \vec{V}}{\partial t} + (\vec{V}.\vec{grad})\vec{V} \right] = -\vec{grad}\ p + \mu \left[\vec{\Delta V} + \frac{1}{3}\vec{grad}(div\,\vec{V}) \right] \tag{9.2}$$

In the previous equations, $\vec{V}$ is the velocity of this pseudofluid and ρ its density. Slip between liquid and vapor (bubbles actually) is ignored, so that a unique velocity $\vec{V}$ is considered for the two-phase mixture.

A major feature of the model is the large range of variation of density. Consider the case of water at ambient temperature as a typical example. Density can vary between $\rho_\ell = 1000\ kg/m^3$

in regions occupied by the liquid alone and $\rho_v = 0.017\ kg/m^3$ in pure vapor cavities. From a numerical viewpoint, the large density ratio of the order of *60 000* between liquid and vapor is a real difficulty which generally requires special computational techniques in order to ensure the stability of the numerical scheme.

To be solved, equations (9.1) and (9.2) must be completed by a constitutive equation for this pseudofluid. Some models use an explicit equation. It is the case for instance of the barotropic model which assumes that density is a continuous function of pressure with a drop from the pure liquid density to the pure vapor density at vapor pressure.

The bubble two-phase flow model does not assume any explicit constitutive equation for the two phase mixture. The liquid is supposed to carry nuclei with a bubble number density n. They grow in the low pressure region before collapsing where the pressure recovers. Any cavitating zone is then considered as a kind of bubble cluster.

The density of the two-phase mixture is:

$$\rho = \alpha\rho_v + (1-\alpha)\rho_\ell \tag{9.3}$$

where α is the void fraction. A similar averaging procedure can be used for viscosity. α is computed on the basis of the bubble number density n and bubble radius R by:

$$\alpha = n.\left(\frac{4}{3}\pi R^3\right) \tag{9.4}$$

As for the radius, it is computed using the Rayleigh equation:

$$R\frac{d^2R}{dt^2} + \frac{3}{2}\left(\frac{dR}{dt}\right)^2 = \frac{p_v - p}{\rho_\ell} \tag{9.5}$$

where *d/dt* is the usual transport derivative and p the local pressure.

The model, made of the five equations (9.1) to (9.5), allows in principle the computation of the five unknowns ρ, $\vec{V}$, p, α and R at any time and in any cell of the computational domain.

Several variants have been developed. The original version of Kubota et al. (1992) is based upon an improved version of the Rayleigh-Plesset equation (9.5) which takes into account interactions between bubbles of a cluster, limiting bubble growth in comparison to the case of an isolated bubble. A simplified version has also been used by considering instead of equation (9.5) the asymptotic growth rate 3.3:

$$\dot{R} \cong \sqrt{\frac{2}{3}\frac{p_v - p}{\rho}} \tag{9.6}$$

During the resorption or collapse phase for which $p > p_v$, a similar equation can be used:

$$\dot{R} \cong -\sqrt{\frac{2}{3}\frac{p - p_v}{\rho}} \tag{9.7}$$

Although such a model appears to be appropriate more especially to cloud cavitation, it has been applied successfully to other types of cavitation including even leading edge cavitation. In practice, it may be necessary to limit artificially the void fraction to *1* depending on the computed configuration and model used.

The bubble number density n is a free parameter in the model. It is often assumed constant all over the computational domain. A typical value is $10^6\ nuclei/m^3$ i.e. $1\ nucleus/cm^3$. The initial bubble radius is also a free parameter. Both may have an influence on the cavitating flow structure as noticed by Kubota et al. (1992). More sophisticated models can be imagined using a variable bubble number density. They aim at modeling the effects of coalescence and fragmentation of bubbles which is still a difficult question.

A key parameter of such a model is the vapor production rate $\dot{m}_v$ which measures the mass of vapor that cavitation locally produces (or resorbs) per unit volume and unit time. It appears as a source term in the continuity equation for the vapor phase:

$$\frac{\partial(\alpha\rho_v)}{\partial t} + div(\alpha\rho_v\vec{V}) = \dot{m}_v \tag{9.8}$$

and with the opposite sign in that for the liquid phase:

$$\frac{\partial[(1-\alpha)\rho_\ell]}{\partial t} + div[(1-\alpha)\rho_\ell\vec{V}] = -\dot{m}_v \tag{9.9}$$

Summation of equations (9.8) and (9.9) results in the continuity equation (9.1) for the liquid mixture. By developing equation (9.1) with definition (9.3) of the density for the mixture, a transport equation for void fraction can be obtained (liquid and vapor densities are assumed constant):

$$\frac{\partial\alpha}{\partial t} + \vec{V}.\vec{grad}\,\alpha = \frac{\rho}{\rho_\ell - \rho_v} div\vec{V} \tag{9.10}$$

Comparison of equations (9.8) and (9.10) shows that $div\vec{V}$ is directly related to the vapor production rate:

$$\dot{m}_v = \frac{\rho_\ell\,\rho_v}{\rho_\ell - \rho_v} div\vec{V} \tag{9.11}$$

In conclusion, to close the system made of continuity and Navier-Stokes equations for the two-phase mixture, it is necessary to introduce a cavitation model which specifies the vapor production rate $\dot{m}_v$. The bubble two-phase flow model is one of them. The choice of the cavitation model is crucial for the results of the simulation.

The present section gave only an overview of the principle of simulation of real cavitating flows using the Rayleigh equation. To be effective, such a mathematical model must be associated to an advanced numerical method capable of accounting for large variations in density.

10 Conclusion

Although restrictive assumptions are associated to the Rayleigh-Plesset equation which, strictly speaking, applies to a single spherical bubble in an infinite liquid medium at rest at infinity, various aspects of real cavitating flows can be analyzed, at least qualitatively but sometimes also quantitatively, using this equation. A simple way to obtain general conclusions is to write it in non dimensional form. Depending upon the practical situation and the objectives which are looked for, there are several ways to make it non dimensional, each of them being characterized by a different choice of the relevant time scales and/or length scales.

The use of the bubble radius as characteristic length scale allows the evaluation of the relative importance of pressure, viscosity and surface tension effects and to find out the one which controls bubble evolution. The use of a length scale which characterizes the cavitating flow itself – as the chord length of a blade if, for example, the cavitating flow around a blade is concerned – allows the derivation of scaling laws. As an example, this kind of approach makes it possible to develop guidelines for the analysis of the influence of flow velocity on the erosive potential of a cavitating flow.

The Rayleigh-Plesset equation also applies to other types of cavities, different from the spherical bubble, as cavitating vortices, providing minor adjustments. Moreover, any cross section of a supercavity can also been modeled by a Rayleigh-Plesset type equation, in the framework of the Logvinovich independence principle. Hence, there is quite a variety of cavities whose dynamics can be approached by an equation analogous to the Rayleigh-Plesset equation.

Finally, this equation is widely used for the numerical modeling of complex real cavitating flows. A usual procedure consists in assuming that the liquid carries cavitation nuclei whose growth is controlled by the Rayleigh-Plesset equation. This is a simple way to account for the production of vapor in a cavitating flow by pressure reduction. The corresponding source term is introduced in the continuity equation for the vapor phase which is solved numerically together with the Navier-Stokes equation for the liquid/vapor mixture.

Bibliography

Arndt R.E.A. (2002) Cavitation in vertical flows. *Annu. Rev. Fluid Mech.* **34** 143-175

Blake J.R. and Gibson D.C. (1987) Cavitation bubbles near boundaries. *Annu. Rev. Fluid Mech.* **19** 99-128

Brennen C.E. (1995) Cavitation and bubble dynamics. Oxford University Press

Chahine G.L. and Genoux P.F. (1983) Collapse of a cavitating vortex ring. *J. Fluids Eng.* **105** 400-405

Escaler X., Farhat M., Avellan F. and Egusquiza E. (2003) Cavitation erosion tests on a 2D hydrofoil using surface-mounted obstacles. *Wear* **254** 441-449

Franc J.-P. and Michel J.-M. (2004) Fundamentals of Cavitation. Kluwer

Fruman D.H., Reboud J.L. and Stutz B. (1999) Estimation of the thermal effects in cavitation of thermosensible liquids. *Int. J. Heat and Mass Transfer* **42** 3195-3204

Fujikawa S. and Akamatsu T. (1980) Effects of non-equilibrium condensation of vapour on the pressure wave produced by the collapse of a bubble in a liquid. *J. Fluid Mech.* **97** part 3 481-512

Garabedian P.R. (1956) Calculation of axially symmetric cavities and jets. *Pac. J. Math.* **6** 611-684

Gopalan S. and Katz J. (2000) Flow structure and modeling issues in the closure region of attached cavitation. *Physics of Fluids* **12** No.4 895-911

Kato H., Konno A., Maeda M. and Yamaguchi H. (1996) Possibility of quantitative prediction of cavitation erosion without model test. *J. Fluids Eng.* **118** 582-588.

Knapp R.T., Daily J.W. and Hammitt F.G. (1970) Cavitation. McGraw-Hill

Kubota A., Kato H. and Yamaguchi H. (1992) A new modeling of cavitating flows: a numerical study of unsteady cavitation on a hydrofoil section. *J. Fluid Mech.* **240** 59-96.

Lecoffre Y. (1995) Cavitation erosion, hydrodynamics scaling laws, practical method of long term damage prediction. *Proc. Int. Symp. on Cavitation Deauville (France)* 249-256

Leighton T.G. (1994) The acoustic bubble. Academic Press

Lindau O. and Lauterborn W. (2003) Cinematographic observation of the collapse and rebound of a laser-produced cavitation bubble near a wall. *J. Fluid Mech.* **479** 327-348

Logvinovich G.V. (1969) Hydrodynamics of free surface flows (in Russian). *Nauvoka Dunka* Ed. Kiev

Pellone C., Franc J.-P. and Perrin M. (2004) Modelling of unsteady 2D cavity flows using the Logvinovich independence principle. *C.R. Mécanique* **332** 827-833.

Plesset M.S. (1949) The dynamics of cavitation bubbles. *J. Appl. Mech.* **16** 277 sq.

Prosperetti A. (2004) Bubbles. *Physics of Fluids* **16** No.6 1852-1865

Rayleigh (Lord) (1917) The pressure developed in a liquid during the collapse of a spherical cavity. *Phil. Mag.* **34** 94 sq.

Serebryakov V.V. (1972) The annular model for calculation of axisymmetric cavity flows (in Russian). Hydromechanics *Nauvoka Dunka* Ed. Kiev **27** 25-29

Tomita Y., Robinson P.B., Tong R.P. and Blake J.R. (2002) Growth and collapse of cavitation bubbles near a curved rigid boundary. *J. Fluid Mech.* **466** 259-283

Trevena D.H. (1987) Cavitation and tension in liquids. Adam Hilger

Young F.R. (1989) Cavitation. McGraw-Hill

Hydrodynamics and Cavitation of Pumps

Christopher E. Brennen*,

* California Institute of Technology, Pasadena, CA 91125, USA, e-mail: brennen@its.caltech.edu

Abstract This paper summarizes a set of lectures given on the hydrodynamics and cavitation of pumps presented at CISM in July, 2005. The lectures are based on my book entitled "Hydrodynamics of Pumps" (Brennen 1994) published jointly by Concepts ETI and Oxford University Press and available on the internet at http://caltechbook.library.caltech.edu/22/01/pumps.htm. The author is very grateful to Concepts ETI for permission to utilize large fractions of that book in this summary of the lectures. Readers who wish to explore the subject matter in more detail are encouraged to consult the original book.

1 Introduction

1.1 Cavitation

The word cavitation refers to the formation of vapor bubbles in regions of low pressure within the flow field of a liquid. In some respects, cavitation is similar to boiling, except that the latter is generally considered to occur as a result of an increase of temperature rather than a decrease of pressure. This difference in the direction of the state change in the phase diagram is more significant than might, at first sight, be imagined. It is virtually impossible to cause any rapid uniform change in temperature throughout a finite volume of liquid. Rather, temperature change most often occurs by heat transfer through a solid boundary. Hence, the details of the boiling process generally embrace the detailed interaction of vapor bubbles with a solid surface, and the thermal boundary layer on that surface. On the other hand, a rapid, uniform change in pressure in a liquid is commonplace and, therefore, the details of the cavitation process may differ considerably from those that occur in boiling. Much more detail on the process of cavitation is included in later sections.

It is sufficient at this juncture to observe that cavitation is generally a malevolent process, and that the deleterious consequences can be divided into three categories. First, cavitation can cause damage to the material surfaces close to the area where the bubbles collapse when they are convected into regions of higher pressure. Cavitation damage can be very expensive, and very difficult to eliminate. For most designers of hydraulic machinery, it is the preeminent problem associated with cavitation. Frequently, one begins with the objective of eliminating cavitation completely. However, there are many circumstances in which this proves to be impossible, and the effort must be redirected into minimizing the adverse consequences of the phenomenon.

The second adverse effect of cavitation is that the performance of the pump, or other hydraulic device, may be significantly degraded. In the case of pumps, there is generally a level of inlet pressure at which the performance will decline dramatically, a phenomenon termed cavitation breakdown. This adverse effect has naturally given rise to changes in the design of a pump so as to minimize the degradation of the performance; or, to put it another way, to optimize the performance in the presence of cavitation. One such design modification is the addition of a cavitating inducer upstream of the inlet to a centrifugal or mixed flow pump impeller.

The third adverse effect of cavitation is less well known, and is a consequence of the fact that cavitation affects not only the steady state fluid flow, but also the unsteady or dynamic response of the flow. This change in the dynamic performance leads to instabilities in the flow that do not occur in the absence of cavitation. These instabilities can give rise to oscillating flow rates and pressures that can threaten the structural integrity of the pump or its inlet or discharge ducts. While a complete classification of the various types of unsteady flow arising from cavitation has yet to be constructed, we can, nevertheless, identify a number of specific types of instability, and these are reviewed in later sections of this review.

1.2 Trends in Hydraulic Turbomachinery

Though the constraints on a turbomachine design are as varied as the almost innumerable applications, there are a number of ubiquitous trends which allow us to draw some fairly general conclusions. To do so we make use of the affinity laws that are a consequence of dimensional analysis, and relate performance characteristics to the density of the fluid, ρ, the typical rotational speed, Ω, and the typical diameter, D, of the pump. Thus the volume flow rate through the pump, Q, the total head rise across the pump, H, the torque, T, and the power absorbed by the pump, P, will scale according to

$$Q \quad \alpha \quad \Omega D^3 \tag{1}$$
$$H \quad \alpha \quad \Omega^2 D^2 \tag{2}$$
$$T \quad \alpha \quad \rho D^5 \Omega^2 \tag{3}$$
$$P \quad \alpha \quad \rho D^5 \Omega^3 \tag{4}$$

These simple relations allow basic scaling predictions and initial design estimates. Furthermore, they permit consideration of optimal characteristics, such as the power density which, according to the above, should scale like $\rho D^2 \Omega^3$.

One typical consideration arising out of the affinity laws relates to optimizing the design of a pump for a particular power level, P, and a particular fluid, ρ. This fixes the value of $D^5\Omega^3$. If one wished to make the pump as small as possible (small D) to reduce weight (as is critical in the rocket engine context) or to reduce cost, this would dictate not only a higher rotational speed, Ω, but also a higher impeller tip speed, $\Omega D/2$. However, as we shall see in the next chapter, the propensity for cavitation increases as a parameter called the cavitation number decreases, and the cavitation number is inversely proportional to the square of the tip speed or $\Omega^2 D^2/4$. Consequently, the increase in tip speed suggested above could lead to a cavitation problem. Often, therefore, one designs the smallest pump that will still operate without cavitation, and this implies a particular size and speed for the device.

Furthermore, the typical fluid-induced stresses in the structure will be given by $\rho\Omega^2 D^4/\tau^2$, and, if $D^5\Omega^3$ is fixed and if one maintains the same geometry, D/τ, then the stresses will increase like $D^{-4/3}$ as the size, D, is decreased. Consequently, fluid/structure interaction problems will increase. To counteract this the blades are often made thicker (D/τ is decreased), but this usually leads to a decrease in the hydraulic performance of the turbomachine. Consequently an optimal design often requires a balanced compromise between hydraulic and structural requirements. Rarely does one encounter a design in which this compromise is optimal.

2 Nomenclature

Roman Letters

a	Pipe radius
A	Cross-sectional area
A_{ijk}	Coefficients of pump dynamic characteristics
Ar	Cross-sectional area ratio
B	Breadth of passage or flow
c	Chord of the blade or foil
c	Speed of sound
c_b	Interblade spacing
c_{PL}	Specific heat of liquid
C	Compliance
C_D	Drag coefficient
C_L	Lift coefficient
C_p	Coefficient of pressure
C_{pmin}	Minimum coefficient of pressure
d	Ratio of blade thickness to blade spacing
D	Impeller diameter or typical flow dimension
Df	Diffusion factor
D_T	Determinant of transfer matrix $[T]$
e	Specific internal energy
E	Energy flux
E	Young's modulus
f	Friction coefficient
F	Force
g	Acceleration due to gravity
g_s	Component of g in the s direction
h	Specific enthalpy
h	Blade tip spacing
h_p	Pitch of a helix
h^T	Total specific enthalpy
h^*	Piezometric head
H	Total head rise

I	Acoustic impulse
I_P	Pump impedance
j	Square root of -1
k_L	Thermal conductivity of the liquid
K_G	Gas constant
ℓ	Pipe length or distance to measuring point
L	Lift
L	Inertance
L	Axial length
$\mathcal{L}$	Latent heat
m	Mass flow rate
m_G	Mass of gas in bubble
m_D	Constant related to the drag coefficient
m_L	Constant related to the lift coefficient
M	Mach number, u/c
n	Coordinate measured normal to a surface
N	Specific speed
$NPSP$	Net positive suction pressure
$NPSE$	Net positive suction energy
$NPSH$	Net positive suction head
p	Pressure
p_A	Radiated acoustic pressure
p^T	Total pressure
p_G	Partial pressure of gas
p_S	Sound pressure level
p_V	Vapor pressure
P	Power
$\tilde{q}^n$	Vector of fluctuating quantities
Q	Volume flow rate (or heat)
$\mathcal{Q}$	Rate of heat addition
r	Radial coordinate in turbomachine
R	Radial dimension in turbomachine
R	Bubble radius
R	Resistance
R_N	Cavitation nucleus radius
Re	Reynolds number
s	Coordinate measured in the direction of flow
s	Solidity
$\mathcal{S}$	Surface tension of the saturated vapor/liquid interface
S	Suction specific speed
S_i	Inception suction specific speed
S_a	Fractional head loss suction specific speed
S_b	Breakdown suction specific speed
Sf	Slip factor
t	Time

T	Temperature or torque
T_{ij}	Transfer matrix elements
$[T]$	Transfer matrix based on $\tilde{p}^T, \tilde{m}$
$[T^*]$	Transfer matrix based on $\tilde{p}, \tilde{m}$
$[TP]$	Pump transfer matrix
$[TS]$	System transfer matrix
u	Velocity in the s or x directions
u_i	Velocity vector
U	Fluid velocity
U_∞	Velocity of upstream uniform flow
v	Fluid velocity in non-rotating frame
V	Volume or fluid velocity
w	Fluid velocity in rotating frame
$\dot{W}$	Rate of work done on the fluid
z	Elevation
Z_{CF}	Common factor of Z_R and Z_S
Z_R	Number of rotor blades
Z_S	Number of stator blades

Greek Letters

α	Angle of incidence
α_L	Thermal diffusivity of liquid
β	Angle of relative velocity vector
β_b	Blade angle relative to cross-plane
γ_n	Wave propagation speed
Γ	Geometric constant
δ	Deviation angle at flow discharge
δ	Clearance
ϵ	Eccentricity
ϵ	Angle of turn
η	Efficiency
θ	Angular coordinate
θ_c	Camber angle
θ^*	Momentum thickness of a blade wake
Θ	Thermal term in the Rayleigh-Plesset equation
ϑ	Inclination of discharge flow to the axis of rotation
κ	Bulk modulus of the liquid
μ	Dynamic viscosity
ν	Kinematic viscosity
ρ	Density of fluid
σ	Cavitation number
σ_i	Cavitation inception number

σ_a	Fractional head loss cavitation number
σ_b	Breakdown cavitation number
σ_c	Choked cavitation number
σ_{TH}	Thoma cavitation factor
Σ	Thermal parameter for bubble growth
τ	Blade thickness
ϕ	Flow coefficient
ψ	Head coefficient
ψ_0	Head coefficient at zero flow
ω_P	Bubble resonant frequency
Ω	Radian frequency of shaft rotation

Subscripts

On any variable, Q:

Q_o	Initial value, upstream value or reservoir value
Q_1	Value at inlet
Q_2	Value at discharge
Q_a	Component in the axial direction
Q_b	Pertaining to the blade
Q_∞	Value far from the bubble or in the upstream flow
Q_B	Value in the bubble
Q_C	Critical value
Q_D	Design value
Q_E	Equilibrium value
Q_G	Value for the gas
Q_{H1}	Value at the inlet hub
Q_{H2}	Value at the discharge hub
Q_i	Components of vector Q
Q_i	Pertaining to a section, i, of the hydraulic system
Q_L	Saturated liquid value
Q_m	Meridional component
Q_M	Mean or maximum value
Q_N	Nominal conditions or pertaining to nuclei
Q_P	Pertaining to the pump
Q_r	Component in the radial direction
Q_s	Component in the s direction
Q_{T1}	Value at the inlet tip
Q_{T2}	Value at the discharge tip
Q_V	Saturated vapor value
Q_θ	Component in the circumferential (or θ) direction

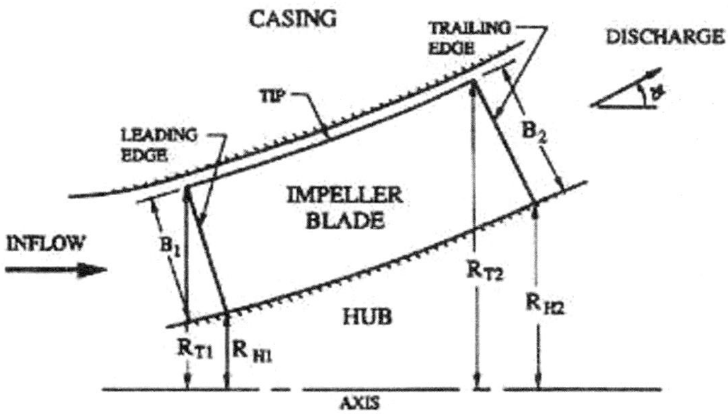

Figure 1. Cross-sectional view through the axis of a pump impeller.

Superscripts and other qualifiers

On any variable, Q:

$\bar{Q}$	Mean value of Q <u>or</u> complex conjugate of Q
$\tilde{Q}$	Complex amplitude of Q
$\dot{Q}$	Time derivative of Q
$\ddot{Q}$	Second time derivative of Q
$Re\{Q\}$	Real part of Q
$Im\{Q\}$	Imaginary part of Q

3 Basic Hydrodynamics

3.1 Geometric Notation

The geometry of a generalized turbomachine rotor is sketched in figure 1, and consists of a set of rotor blades (number $= Z_R$) attached to a hub and operating within a static casing. The radii of the inlet blade tip, inlet blade hub, discharge blade tip, and discharge blade hub are denoted by R_{T1}, R_{H1}, R_{T2}, and R_{H2}, respectively. The discharge blade passage is inclined to the axis of rotation at an angle, ϑ, which would be close to 90° in the case of a centrifugal pump, and much smaller in the case of an axial flow machine. In practice, many pumps and turbines are of the "mixed flow" type , in which the typical or mean discharge flow is at some intermediate angle, $0 < \vartheta < 90°$.

The flow through a general rotor is normally visualized by developing a meridional surface (figure 2), that can either correspond to an axisymmetric stream surface, or be some estimate thereof. On this meridional surface (see figure 2) the fluid velocity in a non-rotating coordinate system is denoted by $v(r)$ (with subscripts 1 and 2 denoting particular values at inlet and discharge) and the corresponding velocity relative to the rotating blades is denoted by $w(r)$. The velocities, v and w, have components v_θ and w_θ in the circumferential direction, and v_m and w_m in the meridional direction. Axial and radial components are denoted by the subscripts a and r. The velocity of the blades is Ωr. As shown in figure 2, the flow angle $\beta(r)$ is defined as the angle between the relative velocity vector in the meridional plane and a plane perpendicular to the axis of rotation. The blade angle $\beta_b(r)$ is defined as the inclination of the tangent to the blade in the meridional plane and the plane perpendicular to the axis of rotation. If the flow is precisely parallel to the blades, $\beta = \beta_b$. Specific values of the blade angle at the leading and trailing edges (1 and 2) and at the hub and tip (H and T) are denoted by the corresponding suffices, so that, for example, β_{bT2} is the blade angle at the discharge tip.

At the leading edge it is important to know the angle $\alpha(r)$ with which the flow meets the blades, and, as defined in figure 3,

$$\alpha(r) = \beta_{b1}(r) - \beta_1(r). \tag{5}$$

This angle, α, is called the incidence angle, and, for simplicity, we shall denote the values of the incidence angle at the tip, $\alpha(R_{T1})$, and at the hub, $\alpha(R_{H1})$, by α_T and α_H, respectively. Since the inlet flow can often be assumed to be purely axial ($v_1(r) = v_{a1}$ and parallel with the axis of rotation), it follows that $\beta_1(r) = \tan^{-1}(v_{a1}/\Omega r)$, and this can be used in conjunction with equation (5) in evaluating the incidence angle for a given flow rate.

The incidence angle should not be confused with the "angle of attack", which is the angle between the incoming relative flow direction and the chord line (the line joining the leading edge to the trailing edge). Note, however, that, in an axial flow pump with straight helicoidal blades, the angle of attack is equal to the incidence angle.

At the trailing edge, the difference between the flow angle and the blade angle is again important. To a first approximation one often assumes that the flow is parallel to the blades, so that $\beta_2(r) = \beta_{b2}(r)$. A departure from this idealistic assumption is denoted by the deviation angle, $\delta(r)$, where, as shown in figure 3:

$$\delta(r) = \beta_{b2}(r) - \beta_2(r) \tag{6}$$

This is normally a function of the ratio of the width of the passage between the blades to the length of the same passage, a geometric parameter known as the solidity which is defined more precisely below. Other angles, that are often used, are the angle through which the flow is turned, known as the *deflection angle*, $\beta_2 - \beta_1$, and the corresponding angle through which the blades have turned, known as the *camber angle* and denoted by $\theta_c = \beta_{b2} - \beta_{b1}$.

Deviation angles in radial machines are traditionally represented by the slip velocity, $v_{\theta s}$, which is the difference between the actual and ideal circumferential velocities of the discharge flow, as shown in figure 4. It follows that

$$v_{\theta s} = \Omega R_2 - v_{\theta 2} - v_{r2} \cot \beta_{b2} \tag{7}$$

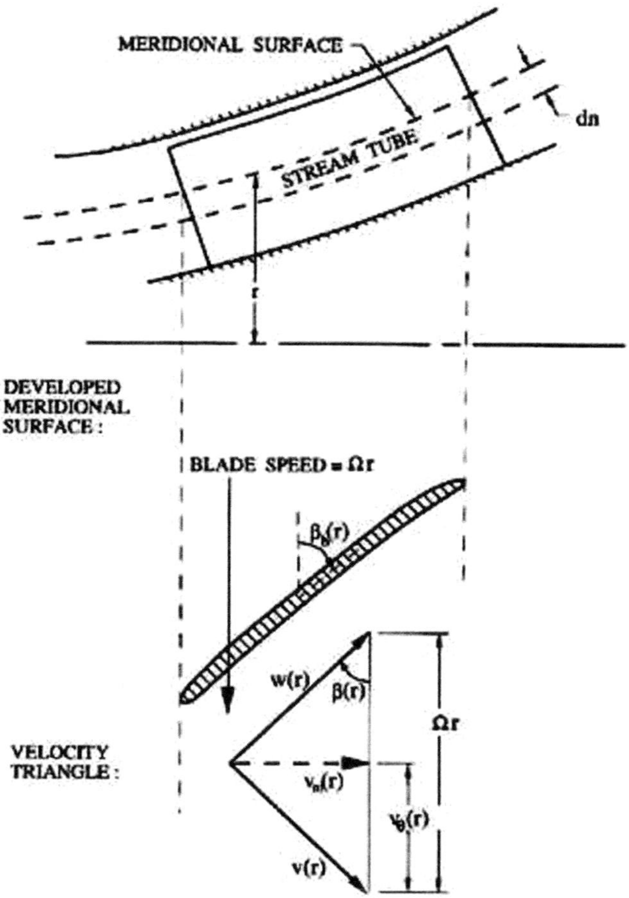

Figure 2. Developed meridional surface and velocity triangle.

This, in turn, is used to define a parameter known as the slip factor, Sf, where

$$Sf = 1 - \frac{v_{\theta s}}{\Omega R_2} = 1 - \phi_2 \left(\cot \beta_2 - \cot \beta_{b2}\right) \tag{8}$$

Other, slightly different "slip factors" have also been used in the literature; for example, Stodola (1927), who originated the concept, defined the slip factor as $1 - v_{\theta s}/\Omega R_2(1 - \phi_2 \cot \beta_{b2})$. However, the definition (8) is now widely used. It follows that the deviation angle, δ, and the slip factor, Sf, are related by

$$\delta = \beta_{b2} - \cot^{-1}\left(\cot \beta_{b2} + \frac{(1 - Sf)}{\phi_2}\right) \tag{9}$$

where the flow coefficient, ϕ_2, is defined later in equation (21).

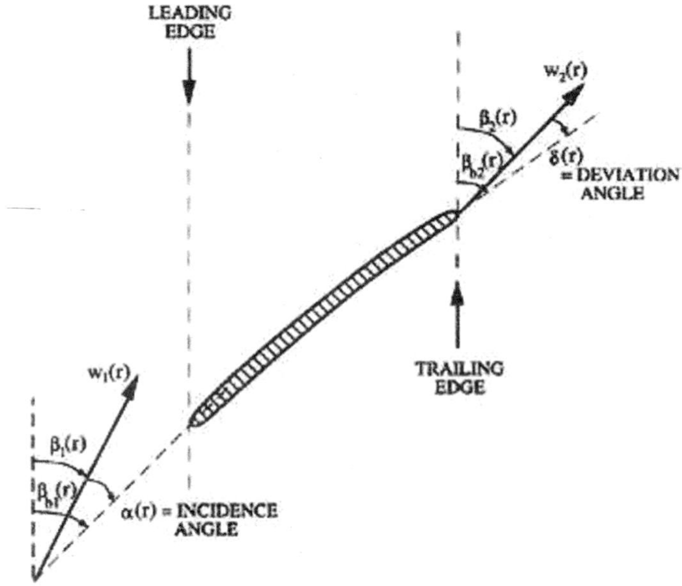

Figure 3. Repeat of figure 2 showing the definitions of the incidence angle at the leading edge and the deviation angle at the trailing edge.

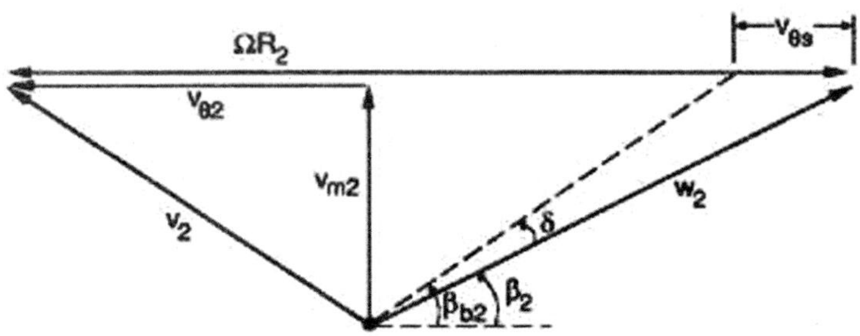

Figure 4. Velocity vectors at discharge indicating the slip velocity, $v_{\theta s}$.

3.2 Cascades

We now turn to some specific geometric features that occur frequently in discussions of pumps and other turbomachines. In a purely axial flow machine, the development of a cylindrical surface within the machine produces a *linear cascade* of the type shown in figure 5(a). The centerplane of the blades can be created using a "generator", say $z = z^*(r)$, which is a line in the $rz-$plane. If this line is rotated through a helical path, it describes a helicoidal surface of the form

$$z = z^*(r) + \frac{h_p \theta}{2\pi} \tag{10}$$

where h_p is the "pitch" of the helix. Of course, in many machines, the pitch is also a function of θ so that the flow is turned by the blades. If, however, the pitch is constant, the development of a cylindrical surface will yield a cascade with straight blades and constant blade angle, β_b. Moreover, the blade thickness is often neglected, and the blades in figure 5(a) then become infinitely thin lines. Such a cascade of infinitely thin, flat blades is referred to as a *flat plate cascade.*

It is convenient to use the term "simple" cascade to refer to those geometries for which the blade angle, β_b, is constant whether in an axial, radial, or mixed flow machine. Clearly, the flat plate cascade is the axial flow version of a simple cascade.

Now compare the geometries of the cascades at different radii within an axial flow machine. Often the pitch at a given axial position is the same at all radii. Then it follows that the radial variation in the blade angle, $\beta_b(r)$, must be given by

$$\beta_b(r) = \tan^{-1}\left[\frac{R_T \tan \beta_{bT}}{r}\right] \tag{11}$$

where β_{bT} is the blade angle at the tip, $r = R_T$.

In a centrifugal machine in which the flow is purely radial, a cross-section of the flow would be as shown in figure 5(b), an array known as a *radial cascade.* In a *s*imple radial

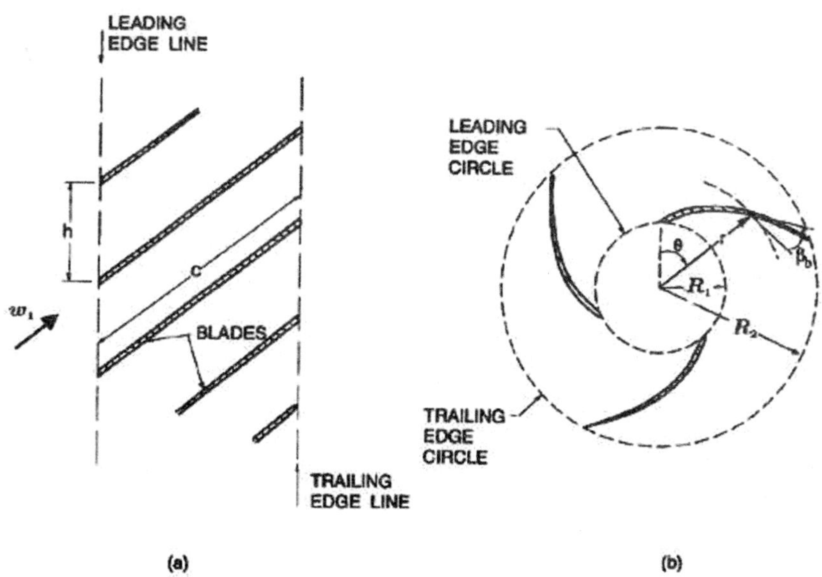

Figure 5. Schematics of (a) a linear cascade and (b) a radial cascade.

cascade, the angle, β_b, is uniform along the length of the blades. The resulting blade geometry is known as a logarithmic spiral, since it follows that the coordinates of the blades are given by the equation

$$\theta - \theta_0 = A \ln r \tag{12}$$

where $A = \cot \beta_b$ and θ_0 are constants. Logarithmic spiral blades are therefore equivalent to straight blades in a linear cascade. Note that a fluid particle in a flow of uniform circulation and constant source strength at the origin will follow a logarithmic spiral since all velocities will be of the form C/r where C is a uniform constant.

In any of type of pump, the ratio of the length of a blade passage to its width is important in determining the degree to which the flow is guided by the blades. The solidity, s, is the geometric parameter that is used as a measure of this geometric characteristic, and s can be defined for any simple cascade as follows. If we identify the difference between the θ coordinates for the same point on adjacent blades (call this $\Delta\theta_A$) and the difference between the θ coordinates for the leading and trailing edges of a blade (call this $\Delta\theta_B$), then the solidity for a simple cascade is defined by

$$s = \frac{\Delta\theta_B}{\Delta\theta_A \cos \beta_b} \tag{13}$$

Applying this to the linear cascade of figure 5(a), we find the familiar

$$s = c/h \tag{14}$$

In an axial flow pump this corresponds to $s = Z_R c / 2\pi R_{T1}$, where c is the chord of the blade measured in the developed meridional plane of the blade tips. On the other hand, for the radial cascade of figure 5(b), equation (13) yields the following expression for the solidity:

$$s = Z_R \ell n \left(R_2/R_1 \right) / 2\pi \sin \beta_b \tag{15}$$

which is, therefore, geometrically equivalent to c/h in the linear cascade.

In practice, there exist many "mixed flow" pumps whose geometries lie between that of an axial flow machine ($\vartheta = 0$, figure 1) and that of a radial machine ($\vartheta = \pi/2$). The most general analysis of such a pump would require a cascade geometry in which figures 5(a) and 5(b) were projections of the geometry of a meridional surface (figure 2) onto a cylindrical surface and onto a plane perpendicular to the axis, respectively. (Note that the β_b marked in figure 5(b) is not appropriate when that diagram is used as a projection). We shall not attempt such generality here; rather, we observe that the meridional surface in many machines is close to conical. Denoting the inclination of the cone to the axis by ϑ, we can use equation (13) to obtain an expression for the solidity of a simple cascade in this conical geometry,

$$s = Z_R \ell n \left(R_2/R_1 \right) / 2\pi \sin \beta_b \sin \vartheta \tag{16}$$

Clearly, this includes the expressions (14) and (15) as special cases.

3.3 Flow Notation

The flow variables that are important are, of course, the static pressure, p, the total pressure, p^T, and the volume flow rate, Q. Often the total pressure is defined by the total head, $p^T/\rho g$. Moreover, in most situations of interest in the context of turbomachinery, the potential energy associated with the earth's gravitational field is negligible relative to the kinetic energy of the flow, so that, by definition

$$p^T = p + \frac{1}{2}\rho v^2 \tag{17}$$

$$p^T = p + \frac{1}{2}\rho\left(v_m^2 + v_\theta^2\right) \tag{18}$$

$$p^T = p + \frac{1}{2}\rho\left(w^2 + 2r\Omega v_\theta - \Omega^2 r^2\right) \tag{19}$$

using the velocity triangle of figure 2. In an incompressible flow, the total pressure represents the total mechanical energy per unit volume of fluid, and, therefore, the change in total pressure across the pump, $p_2^T - p_1^T$, is a fundamental measure of the mechanical energy imparted to the fluid by the pump.

It follows that, in a pump with an incompressible fluid, the overall characteristics that are important are the volume flow rate, Q, and the total pressure rise, $\rho g H$, where $H = (p_2^T - p_1^T)/\rho g$ is the total head rise. These dimensional characteristics are conveniently nondimensionalized by defining a head coefficient, ψ,

$$\psi \quad = \quad (p_2^T - p_1^T)/\rho R_{T2}^2\Omega^2 \quad = \quad gH/R_{T2}^2\Omega^2 \tag{20}$$

and one of two alternative flow coefficients, ϕ_1 and ϕ_2:

$$\phi_1 = Q/A_1 R_{T1}\Omega \qquad or \qquad \phi_2 = Q/A_2 R_{T2}\Omega \tag{21}$$

where A_1 and A_2 are the inlet and discharge areas, respectively. The discharge flow coefficient is the nondimensional parameter most often used to describe the flow rate. However, in discussions of cavitation, which occurs at the inlet to a pump impeller, the inlet flow coefficient is a more sensible parameter. Note that, for a purely axial inflow, the incidence angle is determined by the flow coefficient, ϕ_1:

$$\alpha(r) = \beta_{b1}(r) - \tan^{-1}(\phi_1 r/R_{T1}) \tag{22}$$

Furthermore, for a given deviation angle, specifying ϕ_2 fixes the geometry of the velocity triangle at discharge from the pump.

Frequently, the conditions at inlet and/or discharge are nonuniform and one must subdivide the flow into annular streamtubes, as indicated in figure 2. Each streamtube must then be analysed separately, using the blade geometry pertinent at that radius. The mass flow rate, m, through an individual streamtube is given by

$$m = 2\pi\rho r v_m dn \tag{23}$$

where n is a coordinate measured normal to the meridional surface, and, in the present text, will be useful in describing the discharge geometry.

Conservation of mass requires that m have the same value at inlet and discharge. This yields a relation between the inlet and discharge meridional velocities, that involves the cross-sectional areas of the streamtube at these two locations. The total volume flow rate through the turbomachine, Q, is then related to the velocity distribution at any location by the integral

$$Q = \int 2\pi r v_m(r) dn \tag{24}$$

The total head rise across the machine, H, is given by the integral of the total rate of work done on the flow divided by the total mass flow rate:

$$H = \frac{1}{Q} \int \frac{(p_2^T(r) - p_1^T(r))}{\rho g} 2\pi r v_m(r) dn \tag{25}$$

These integral expressions for the flow rate and head rise will be used in later chapters.

3.4 Specific Speed

At the beginning of any pump design process, neither the size nor the shape of the machine is known. The task the pump is required to perform is to use a shaft rotating at a frequency, Ω (in rad/s), to pump a certain flow rate, Q (in m^3/s) through a head rise, H (in m). As in all fluid mechanical formulations, one should first seek a nondimensional parameter (or parameters) which distinguishes the nature of this task. In this case, there is one and only one nondimensional parametric group that is appropriate and this is known as the "specific speed", denoted by N. The form of the specific speed is readily determined by dimensional analysis:

$$N = \frac{\Omega Q^{\frac{1}{2}}}{(gH)^{\frac{3}{4}}} \tag{26}$$

Though originally constructed to allow evaluation of the shaft speed needed to produce a particular head and flow, the name "specific speed" is slightly misleading, because N is just as much a function of flow rate and head rise as it is of shaft speed. Perhaps a more general name, like "the basic performance parameter", would be more appropriate. Note that the specific speed is a size-independent parameter, since the size of the machine is not known at the beginning of the design process.

The above definition of the specific speed has employed a consistent set of units, so that N is truly dimensionless. With these consistent units, the values of N for most common turbomachines lie in the range between 0.1 and 4.0 (see below). Unfortunately, it has been traditional in industry to use an inconsistent set of units in calculating N. In the USA, the g is dropped from the denominator, and values for the speed, flow rate, and head in rpm, gpm, and ft are used in calculating N. This yields values that are a factor of 2734.6 larger than the values of N obtained using consistent units. The situation is even more confused since the Europeans use another set of inconsistent units (rpm, m^3/s, head in m, and no g) while the British employ a definition similar to the U.S., but with Imperial gallons rather than U.S. gallons. One can only hope that the pump (and turbine) industries would cease the use of these inconsistent measures that would be regarded with derision by any engineer outside of the industry. In this monograph, we shall use the dimensionally consistent and, therefore, universal definition of N.

Note that, since Q and gH were separately nondimensionalized in the definitions (20) and (21), N can be related to the corresponding flow and head coefficients by

$$N = \left[\frac{\pi}{\cos \vartheta} \left(1 - \frac{R_{H2}^2}{R_{T2}^2} \right) \right]^{\frac{1}{2}} \frac{\phi_2^{\frac{1}{2}}}{\psi^{\frac{3}{4}}} \tag{27}$$

In the case of a purely centrifugal discharge ($\vartheta = \pi/2$), the quantity within the square brackets reduces to $2\pi B_2/R_{T2}$.

Since turbomachines are designed for specific tasks, the subscripted N_D will be used to denote the design value of the specific speed for a given machine.

3.5 Pump Geometries

Since the task specifications for a pump (or turbine or compressor or other machine) can be reduced to the single parameter, N_D, it is not surprising that the overall or global geometries of pumps, that have evolved over many decades, can be seen to fit quite neatly into a single parameter family of shapes. This family is depicted in figure 6. These geometries reflect the fact that an axial flow machine, whether a pump, turbine, or compressor, is more efficient at high specific speeds (high flow rate, low head) while a radial machine, that uses the centrifugal effect, is more efficient at low specific speeds (low flow rate, high head). The same basic family of geometries is presented quantitatively in figure 7, where the anticipated head and flow coefficients are also plotted. While the existence of this parametric family of designs has emerged almost exclusively as a result of trial and error, some useful perspectives can be obtained from an approximate analysis of the effects of the pump geometry on the hydraulic performance (Brennen 1994).

Normally, turbomachines are designed to have their maximum efficiency at the design specific speed, N_D. Thus, in any graph of efficiency against specific speed, each pump geometry will trace out a curve with a maximum at its optimum specific speed, as illustrated by the individual curves in figure 8. Furthermore, Balje (1981) has made note of another interesting feature of this family of curves in the graph of efficiency against specific speed. First, he corrects the curves for the different viscous effects which can occur in machines of different size and speed, by comparing the data on efficiency at the same effective Reynolds number. Then, as can be seen in figure 8, the family of curves for the efficiency of different types of machines has an upper envelope with a maximum at a specific speed of unity. Maximum possible efficiencies decline for values of N_D greater or less than unity. Thus the "ideal" pump would seem to be that with a design specific speed of unity, and the maximum obtainable efficiency seems to be greatest at this specific speed. Fortunately, from a design point of view, one of the specifications has some flexibility, namely the shaft speed, Ω. Though the desired flow rate and head rise are usually fixed, it may be possible to choose the drive motor to turn at a speed, Ω, which brings the design specific speed close to the optimum value of unity.

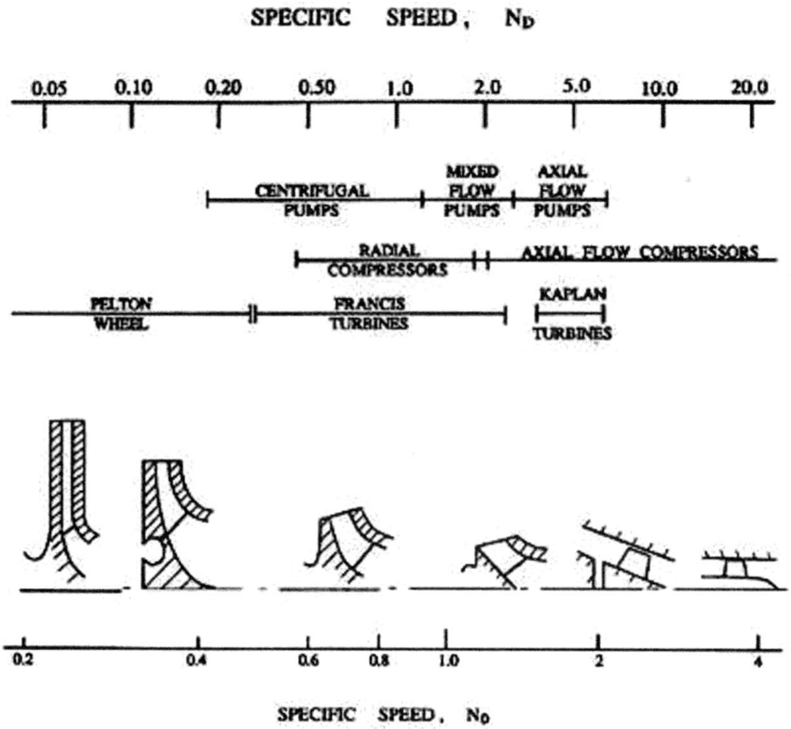

Figure 6. Ranges of specific speeds for typical turbomachines and typical pump geometries for different design speeds (from Sabersky, Acosta and Hauptmann 1989).

3.6 Energy Balance

The next step in the assessment of the performance of a turbomachine is to consider the application of the first and second laws of thermodynamics to such devices. In doing so we shall characterize the inlet and discharge flows by their pressure, velocity, enthalpy, etc., assuming that these are uniform flows. It is understood that when the inlet and discharge flows are non-uniform, the analysis actually applies to a single streamtube and the complete energy balance requires integration over all of the streamtubes.

The basic thermodynamic measure of the energy stored in a unit mass of flowing fluid is the total specific enthalpy (total enthalpy per unit mass) denoted by h^T and defined by

$$h^T = h + \frac{1}{2}|u|^2 + gz = e + \frac{p}{\rho} + \frac{1}{2}|u|^2 + gz \tag{28}$$

where e is the specific internal energy, $|u|$ is the magnitude of the fluid velocity, and z is the vertical elevation. This expression omits any energy associated with additional external forces (for example, those due to a magnetic field), and assumes that the process is chemically inert.

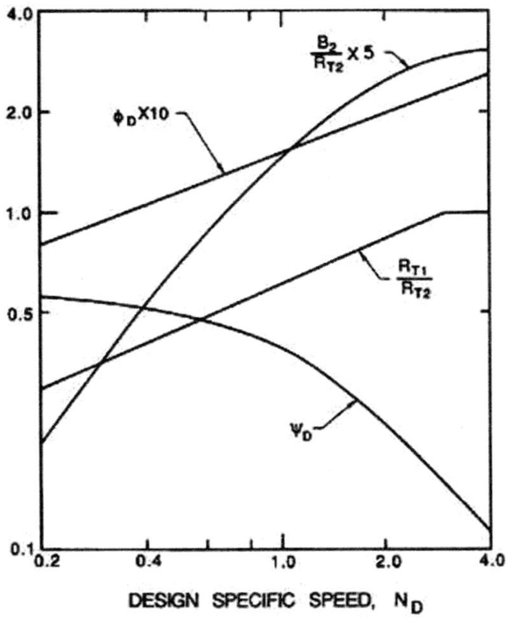

Figure 7. General design guidelines for pumps indicating the optimum ratio of inlet to discharge tip radius, R_{T1}/R_{T2}, and discharge width ratio, B_2/R_{T2}, for various design specific speeds, N_D. Also shown are approximate pump performance parameters, the design flow coefficient, ϕ_D, and the design head coefficient, ψ_D (adapted from Sabersky, Acosta and Hauptmann 1989).

Consider the steady state operation of a fluid machine in which the entering fluid has a total specific enthalpy of h_1^T, the discharging fluid has a total specific enthalpy of h_2^T, the mass flow rate is m, the net rate of heat addition to the machine is $\mathcal{Q}$, and the net rate of work done on the fluid in the machine by external means is $\dot{W}$. It follows from the first law of thermodynamics that

$$m(h_2^T - h_1^T) = \mathcal{Q} + \dot{W} \tag{29}$$

Now consider incompressible, inviscid flow. It is a fundamental property of such a flow that it contains no mechanism for an exchange of thermal and mechanical energy, and, therefore, equation 29 divides into two parts, governing the mechanical and thermal components of the total enthalpy, as follows

$$(p/\rho + \frac{1}{2}|u|^2 + gz)_2 - (p/\rho + \frac{1}{2}|u|^2 + gz)_1 = \frac{(p_2^T - p_1^T)}{\rho} = \frac{\dot{W}}{m} \tag{30}$$

$$e_2 - e_1 = \mathcal{Q}/m \tag{31}$$

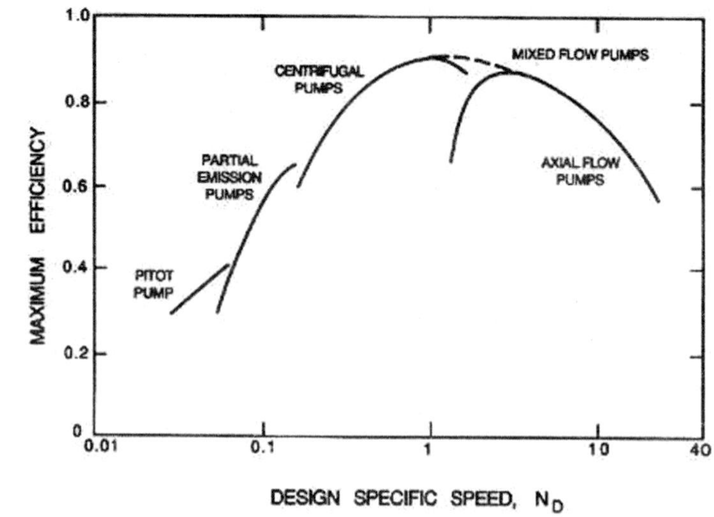

Figure 8. Compilation by Balje (1981) of maximum efficiencies for various kinds of pumps as a function of design specific speed, N_D. Since efficiency is also a function of Reynolds number the data has been corrected to a Reynolds number, $2\Omega R_{T2}^2/\nu$, of 10^8.

Thus, for incompressible inviscid flow, the fluid mechanical problem (for which equation 30 represents the basic energy balance) can be decoupled from the heat transfer problem (for which the heat balance is represented by equation 31).

It follows that, if T is the torque applied by the impeller to the fluid, then the rate of work done on the fluid is $\dot{W} = T\Omega$. Consequently, in the case of an ideal fluid which is incompressible and inviscid, equation 30 yields a relation connecting the total pressure rise across the pump, $p_2^T - p_1^T$, the mass flow rate, m, and the torque:

$$m\frac{(p_2^T - p_1^T)}{\rho} = T\Omega \tag{32}$$

Furthermore, the second law of thermodynamics implies that, in the presence of irreversible effects such as those caused by viscosity, the equality in equation 32 should be replaced by an inequality, namely a "less than" sign. Consequently, in a real pump operating with an incompressible fluid, viscous effects will cause some of the input energy to be converted to heat rather than to an increase in the stored energy in the fluid. It follows that the right hand side of equation 32 is the actual work done on the fluid by the impeller, and the left hand side is the fraction of that work which ends up as mechanical energy stored in the fluid. It is, therefore, appropriate to define a quantity, η_P, known as the pump hydraulic efficiency, to represent that fraction of the work done on the fluid that ends up as an increase in the mechanical energy stored in the fluid:

$$\eta_P = m\left(p_2^T - p_1^T\right)/\rho T\Omega \tag{33}$$

Of course, additional mechanical losses may occur in a pump. These can cause the rate of work transmitted through the external shaft of the pump to be greater than the rate at which the impeller does work on the fluid. For example, losses may occur in the bearings or as a result of the "disk friction" losses caused by the fluid dynamic drag on other, non-active surfaces rotating with the shaft. Consequently, the overall (or shaft) efficiency, η_S, may be significantly smaller than η_P. For approximate evaluations of these additional losses, the reader is referred to the work of Balje (1981).

Despite all these loss mechanisms, pumps can be surprisingly efficient. A well designed centrifugal pump should have an overall efficiency in the neighborhood of 85% and some very large pumps (for example those in the Grand Coulee Dam) can exceed 90%. Even centrifugal pumps with quite simple and crude geometries can often be 60% efficient.

3.7 Noncavitating Pump Performance

It is useful at this point to develop an approximate and idealized evaluation of the hydraulic performance of a pump in the absence of cavitation. This will take the form of an analytical expression for the head rise (or ψ) as a function of the flow rate (or ϕ_2).

To simplify this analysis it is assumed that the flow is incompressible, axisymmetric and steady in the rotating framework of the impeller blades; that the blades are infinitely thin; and that viscous losses can be neglected. Under these conditions the flow in any streamtube, such as depicted in figure 2, will follow the Bernoulli equation for a rotating system (see, for example, Sabersky, Acosta and Hauptmann 1989),

$$\frac{2p_1}{\rho} + w_1^2 - r_1^2\Omega^2 = \frac{2p_2}{\rho} + w_2^2 - r_2^2\Omega^2 \tag{34}$$

This equation can be usefully interpreted as an energy equation as follows. The terms $p + \frac{1}{2}\rho w^2$ on either side are the total pressure or mechanical energy per unit volume of fluid, and this quantity would be the same at inlet and discharge were it not for the fact that "potential" energy is stored in the rotating fluid. The term $\rho(r_1^2 - r_2^2)\Omega^2/2$ represents the difference in this "potential" energy at inlet and discharge. Clearly, when there are losses, equation (34) will no longer be true.

Using the definition of the total pressure (equation (17)) and the relations between the velocities derived from the velocity triangles of figure 2, equation (34) can be manipulated to yield the following expression for the total pressure rise, $(p_2^T - p_1^T)$, for a given streamtube:

$$p_2^T - p_1^T = p_2 - p_1 + \frac{\rho}{2}\left(v_2^2 - v_1^2\right) \tag{35}$$

$$= \rho(\Omega r_2 v_{\theta 2} - \Omega r_1 v_{\theta 1}) \tag{36}$$

In the absence of inlet swirl ($v_{\theta 1} = 0$), this leads to the nondimensional performance characteristic

$$\psi = 1 - \phi_2 \cot \beta_{b2} \tag{37}$$

using the definitions in equations (20) and (21). Here we have assumed that the inlet and discharge conditions are uniform which, in effect, restricts the result to a turbomachine in which the widths, B_1 and B_2 (figure 1), are such that $B_1 \ll R_{T1}, B_2 \ll R_{T2}$, and in which the velocities of the flow and the impeller are uniform across both the inlet and the discharge. Usually this is not the case, and the results given by equations (36) and (37) then become applicable to each individual streamtube. Integration over all the streamtubes is necessary to obtain the performance characteristic for the machine. An example of this integration was given in section 3.3. Even in these nonuniform cases, the simple expression (37) is widely used in combination with some mean or effective discharge blade angle, β_{b2}, to estimate the performance of a pump.

It is important to note that the above results can be connected with those of the preceding section by applying the angular momentum theorem (Newton's second law of motion applied to rotational motion) to relate the torque, T, to the net flux of angular momentum out of the pump:

$$T = m(r_2 v_{\theta 2} - r_1 v_{\theta 1}) \tag{38}$$

where, as before, m is the mass flow rate. Note that this momentum equation (38) holds whether or not there are viscous losses. In the absence of viscous losses, a second expression for the torque, T, follows from equation 32. By equating the two expressions, the result (36) for the performance in the absence of viscous losses is obtained by an alternative method.

3.8 Several Specific Impellers and Pumps

Throughout this review, we shall make reference to experimental data on various phenomena obtained with several specific impellers and pumps. It is appropriate at this

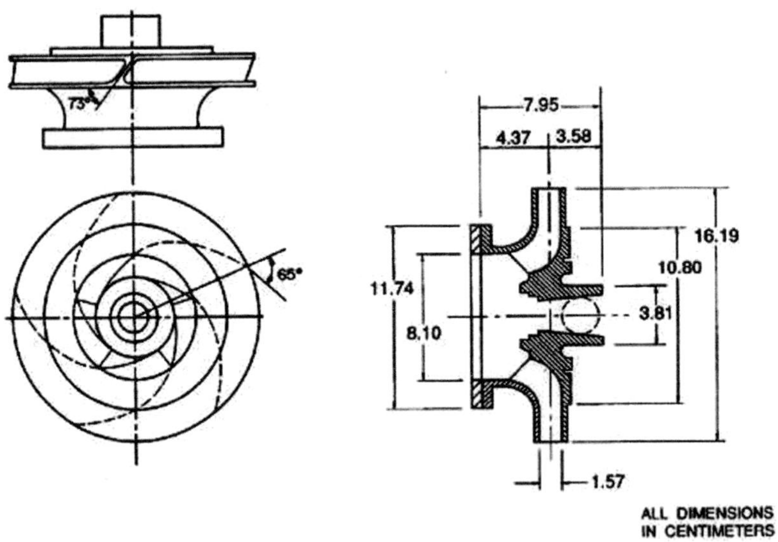

Figure 9. A centrifugal pump impeller designated Impeller X.

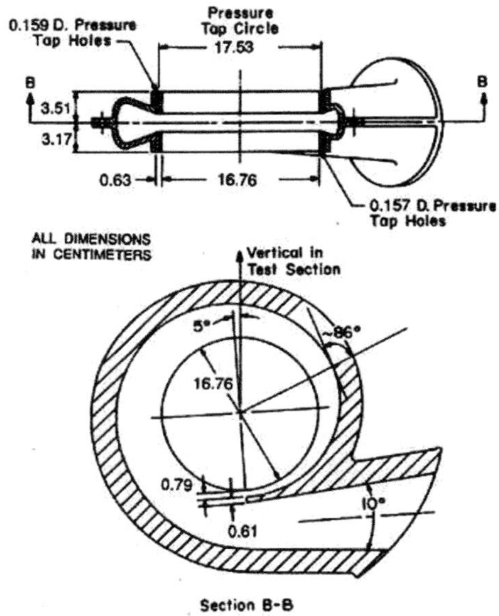

Figure 10. A vaneless spiral volute (designated Volute A) designed to be matched to Impeller X.

point to include a brief description of these components. The descriptions will also serve as convenient examples of pump geometries.

Impeller X, which is shown in figure 9, is a five-bladed centrifugal pump impeller made by Byron Jackson Pump Division of Borg Warner International Products. It has a discharge radius, $R_{T2} = 8.1\ cm$, a discharge blade angle, β_{bT2}, of 23°, and a design specific speed, N_D, of 0.57. Impeller X was often tested in combination with Volute A (figure 10), a single exit, spiral volute with a base circle of 18.3 cm and a spiral angle of 4°. It is designed to match Impeller X at a flow coefficient of $\phi_2 = 0.092$. This implies that the principles of fluid continuity and momentum have been utilized in the design, so that the volute collects a circumferentially uniform discharge from the impeller and channels it to the discharge line in such a way that the pressure in the volute is circumferentially uniform, and in a way that minimizes the viscous losses in the decelerating flow. For given volute and impeller geometries, these objectives can only by met at one "design" flow coefficient, as described in section 5.2. We would therefore expect that the hydraulic losses would increase, and the efficiency decrease, at off-design conditions. It is valuable to emphasize that the performance of a pump depends not only on the separate designs of the impeller and volute but also on the matching of the two components.

Two particular axial flow pumps or inducers, designed to function with cavitation, will also be referred to frequently. These are shown in figure 11. In a number of contexts, data for several simple 9° helical inducers ($\beta_{bT1} = 9°$) will be used for illustrative purposes, and a typical geometry is shown on the left of figure 11. The second inducer geometry is pertinent to a somewhat lower specific speed. Impellers IV (7.58 cm diameter) and VI

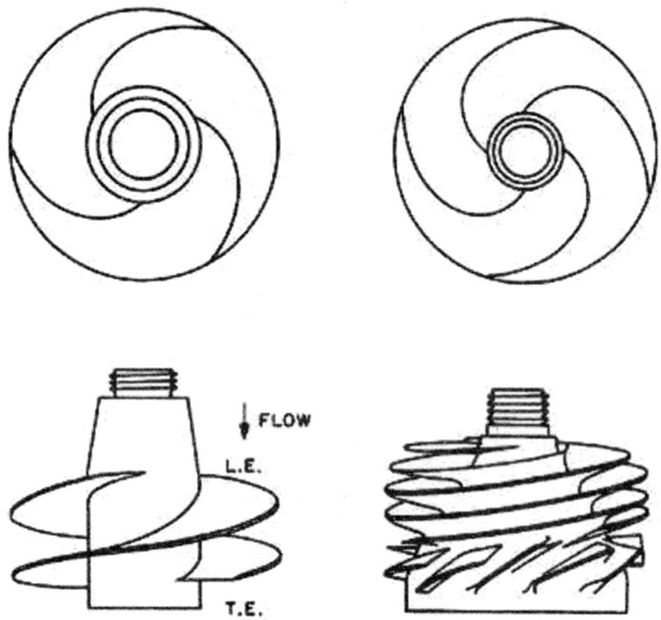

Figure 11. Two cavitating inducers for which performance data is presented. On the left a 7.58 *cm* diameter, 9° helical Impeller V (a 10.2 *cm* version is designated Impeller VII). On the right a 7.58 *cm* diameter scale model of the impeller in the SSME low pressure LOX turbopump, Impeller IV (a 10.2 *cm* version is designated Impeller VI).

(10.2 *cm* diameter) were scale models of the low pressure liquid oxygen impeller in the Space Shuttle Main Engine (SSME). These have a design flow coefficient of about 0.076.

4 Two-dimensional performance analysis

4.1 Introduction

In this and the following section, we briefly survey the more detailed analyses of the flow in axial and centrifugal pumps, and provide a survey of some of the models used to synthesize the noncavitating performance of these turbomachines. The survey begins in this chapter with a summary of some of the results that emerge from a more detailed analysis of the two-dimensional flow in the meridional plane of the turbomachine, while neglecting most of the three-dimensional effects. Subsections 4.2 through 4.4 address the analyses of linear cascades for axial flow machines, while subsection 4.5 summarizes the analyses of radial cascades for centrifugal machines. Three-dimensional effects are addressed later.

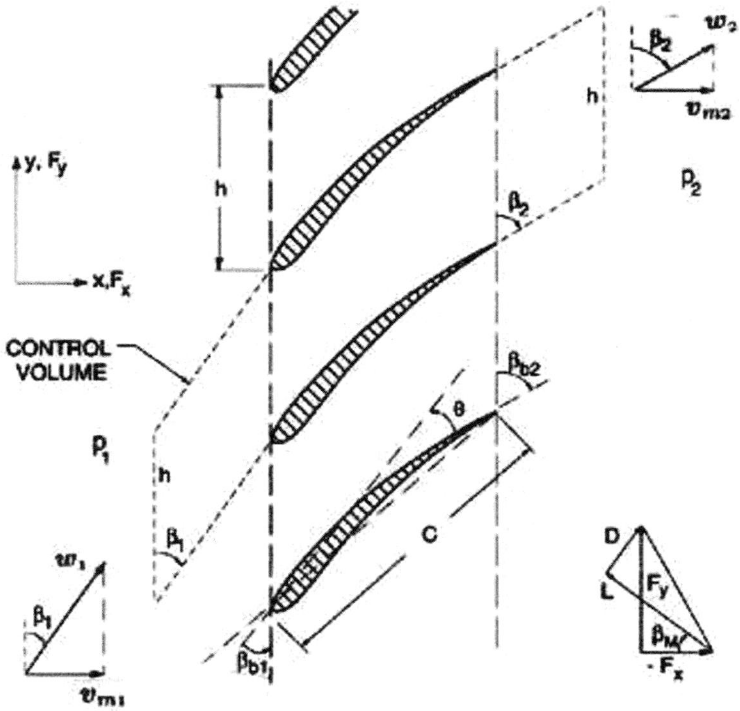

Figure 12. Schematic of a linear cascade showing the blade geometry, the periodic control volume and the definition of the lift, L, and drag, D, forces on a blade.

4.2 Linear Cascade Analyses

The fluid mechanics of a linear cascade will now be examined in more detail, so that the role played by the geometry of the blades and information on the resulting forces on individual blades may be used to supplement the analysis of section 3.7. Referring to the periodic control volume indicated in figure 12, and applying the momentum theorem to this control volume, the forces, F_x and F_y, imposed by the fluid on each blade (per unit depth normal to the sketch), are given by

$$F_x = -(p_2 - p_1)h \tag{39}$$

$$F_y = \rho h v_m (w_1 \cos\beta_1 - w_2 \cos\beta_2) \tag{40}$$

where, as a result of continuity, $v_{m1} = v_{m2} = v_m$. Note that F_y is entirely consistent with the expression (38) for the torque, T.

To proceed, we define the vector mean of the relative velocities, w_1 and w_2, as having a magnitude w_M and a direction β_M, where by simple geometry

$$\cot\beta_M = \frac{1}{2}(\cot\beta_1 + \cot\beta_2) \tag{41}$$

$$w_M = v_m / \sin\beta_M \tag{42}$$

It is conventional and appropriate (as discussed below) to define the lift, L, and the drag, D, components of the total force on a blade, $(F_x^2 + F_y^2)^{\frac{1}{2}}$, as the components normal and tangential to the vector mean velocity, w_M. More specifically, as shown in figure 12,

$$L = -F_x \cos\beta_M + F_y \sin\beta_M \tag{43}$$

$$D = F_x \sin\beta_M + F_y \cos\beta_M \tag{44}$$

where L and D are forces per unit depth normal to the sketch. Nondimensional lift and drag coefficients are defined as

$$C_L = L \Big/ \frac{1}{2}\rho w_M^2 c \ ; \ C_D = D \Big/ \frac{1}{2}\rho w_M^2 c \tag{45}$$

The list of fundamental relations is complete if we write the expression for the pressure difference across the cascade as

$$p_1 - p_2 = \Delta p_L^T + \frac{\rho}{2}\left(w_1^2 - w_2^2\right) \tag{46}$$

where Δp_L^T denotes the total pressure loss across the cascade caused by viscous effects. In frictionless flow, $\Delta p_L^T = 0$, and the relation (46) becomes the Bernoulli equation in rotating coordinates (equation (34) with $r_1 = r_2$ as is appropriate here). A nondimensional loss coefficient, f, is defined as

$$f = \Delta p_L^T \Big/ \frac{1}{2}\rho w_M^2 \tag{47}$$

Equations (39) through (47) can be manipulated to obtain expressions for the lift and drag coefficients as follows

$$C_D = 2f \sin\beta_M / s \tag{48}$$

$$C_L = \frac{2}{s}\left[\frac{\psi}{\phi}\sin\beta_M + \frac{f(\phi - \cos\beta_M \sin\beta_M)}{\sin\beta_M}\right] \tag{49}$$

where $s = c/h$ is the solidity, ψ is the head coefficient, $(p_2^T - p_1^T)/\rho\Omega^2 R^2$, and ϕ is the flow coefficient, $v_m/\Omega R$. Note that in frictionless flow $C_D = 0$ and $C_L = 2\psi \sin\beta_M/\phi s$; then the total force (lift) on the foil is perpendicular to the direction defined by the β_M of equation (41). This provides confirmation that the directions we chose in defining L and D (see figure 12) were appropriate for, in frictionless flow, C_D must indeed be zero.

Also note that equations (39) through (47) yield the head/flow characteristic given by

$$\psi = \phi\left(\cot\beta_1 - \cot\beta_2\right) - f\phi^2\left(1 + \cot^2\beta_M\right) \tag{50}$$

which, when there is no inlet swirl or prerotation so that $\tan\beta_1 = \phi$, becomes

$$\psi = 1 - \phi\cot\beta_2 - f\left[\phi^2 + \frac{1}{4}\left(1 + \phi\cot\beta_2\right)^2\right] \tag{51}$$

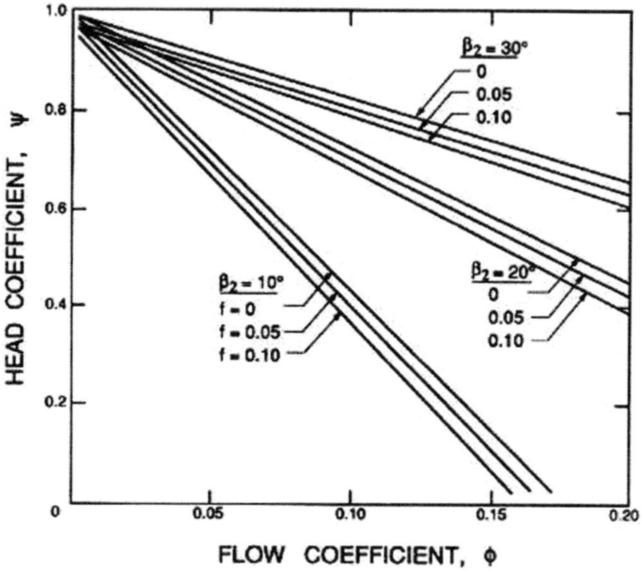

Figure 13. Calculated head/flow characteristics for some linear cascades.

In frictionless flow, when the discharge is parallel with the blades ($\beta_2 = \beta_{b2}$), this, of course, reduces to the characteristic equation (37). Note that the use of the relation (51) allows us to write the expression (49) for the lift coefficient as

$$C_L = \frac{2}{s}\left[2\sin\beta_M(\cot\beta_1 - \cot\beta_M) - f\cos\beta_M\right] \tag{52}$$

Figure 13 presents examples of typical head/flow characteristics resulting from equation (51) for some chosen values of β_2 and the friction coefficient, f. It should be noted that, in any real turbomachine, f will not be constant but will vary substantially with the flow coefficient, ϕ, which determines the angle of incidence and other flow characteristics. More realistic cases are presented a little later in figure 14.

The observant reader will have noted that all of the preceding equations of this section involve only the inclinations of the flow and *not* of the blades, which have existed only as ill-defined objects that achieve the turning of the flow. In order to progress further, it is necessary to obtain a detailed solution of the flow, one result of which will be the connection between the flow angles (β_M, β_2) and the geometry of the blades, including the blade angles (β_b, β_{b1}, β_{b2}). A large literature exists describing methods for the solutions of these flows, but such detail is beyond the scope of this text. As in most high Reynolds number flows, one begins with potential flow solutions, for which the reader should consult a modern text, such as that by Horlock (1973), or the valuable review by Roudebush (1965). König (1922) produced one of the earliest potential flow solutions, namely that for a simple flat plate cascade of infinitely thin blades. This was used

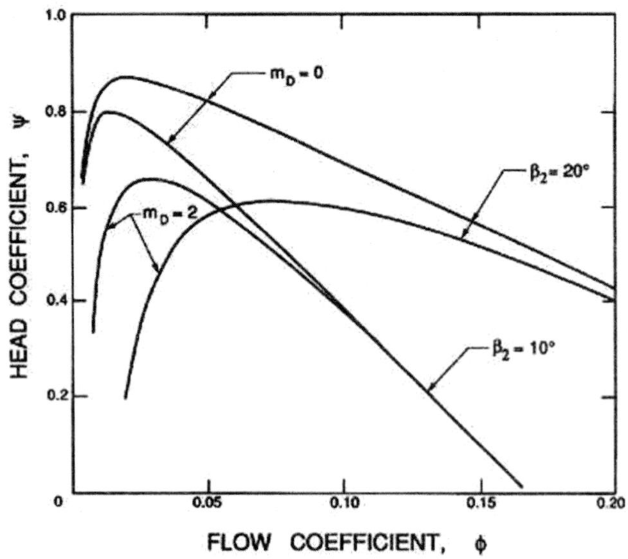

Figure 14. Calculated head/flow characteristics for a linear cascade using blade drag coefficients given by equation (56) with $C_{D0} = 0.02$. The corresponding characteristics with $C_{D0} = m_D = 0$ are shown in figure 13.

to generate figure 15. Such potential flow methods must be supplemented by viscous analyses of the boundary layers on the blades and the associated wakes in the discharge flow. Leiblein (1965) provided an excellent review of these viscous flow methods, and some of his basic methodology will be introduced later.

To begin with, however, one can obtain some useful insights by employing our basic knowledge and understanding of lift and drag coefficients obtained from tests, both those on single blades (airfoils, hydrofoils) and those on cascades of blades. One such observation is that the lift coefficient, C_L, is proportional to the sine of the angle of attack, where the angle of attack is defined as the angle between the mean flow direction, β_M, and a mean blade angle, β_{bM}. Thus

$$C_L = m_L \sin(\beta_{bM} - \beta_M) \tag{53}$$

where m_L is a constant, a property of the blade or cascade geometry. In the case of frictionless flow ($f = 0$), the expression (53) may be substituted into equation (52), resulting in an expression for β_M. When this is used with equation (51), the following head/flow characteristic results:

$$\psi = \frac{2m_L s \sin\beta_{bM}}{4 + m_L s \sin\beta_{bM}} \left[1 - \phi\left(\cot\beta_{bM} + \frac{v_{\theta 1}}{v_{m1}}\right)\right] \tag{54}$$

where, for convenience, the first factor on the right-hand side is denoted by

$$\psi_0 = \frac{2m_L s \sin\beta_{bM}}{4 + m_L s \sin\beta_{bM}} = \left[1 + \frac{\cot\beta_2 - \cot\beta_{b2}}{\cot\beta_1 - \cot\beta_2}\right]^{-1} \tag{55}$$

The factor, ψ_0, is known as the frictionless shut-off head coefficient, since it is equal to the head coefficient at zero flow rate. The second expression for ψ_0 follows from the preceding equations, and will be used later. Note that, unlike equation (51), the head/flow characteristic of equation (54) is given in terms of m_L and practical quantities, such as the blade angle, β_{bM}, and the inlet swirl or prerotation, $v_{\theta 1}/v_{m1}$.

It is also useful to consider the drag coefficient, C_D, for it clearly defines f and the viscous losses in the cascade. Instead of being linear with angle of attack, C_D will be an even function so an appropriate empirical result corresponding to equation (53) would be

$$C_D = C_{D0} + m_D \sin^2\left(\beta_{bM} - \beta_M\right) \tag{56}$$

where C_{D0} and m_D are constants. Some head/flow characteristics resulting from typical values of C_{D0} and m_D are shown in figure 14. Note that these performance curves have a shape that is closer to practical performance curves than the constant friction factor results of figure 13.

4.3 Deviation Angle

While the simple, empirical approach of the last section has practical and educational value, it is also valuable to consider the structure of the flow in more detail, and to examine how higher level solutions to the flow might be used to predict the performance of a cascade of a particular geometry. In doing so, it is important to distinguish between performance characteristics that are the result of idealized inviscid flow and those that are caused by viscous effects. Consider, first, the inviscid flow effects. König (1922) was the first to solve the potential flow through a linear cascade, in particular for a simple cascade of infinitely thin, straight blades. The solution leads to values of the deviation, δ, that, in turn, allow evaluation of the shut-off head coefficient, ψ_0, through equation (55). This is shown as a function of solidity in figure 15. Note that for solidities greater than about unity, the idealized, potential flow exits the blade passages parallel to the blades, and hence $\psi_0 \rightarrow 1$.

Another approach to the same issue of relating the flow angle, β_2, to the blade angles, is to employ an empirical rule for the deviation angle, $\delta = \beta_{b2} - \beta_2$ (equation (6)), in terms of other geometric properties of the cascade. One early empirical relation suggested by Constant (1939) (see Horlock 1973) relates the deviation to the camber angle, θ_c, and the solidity, s, through

$$\delta_N = C\ \theta_c / s^{\frac{1}{2}} \tag{57}$$

where the subscript N refers to nominal conditions, somewhat arbitrarily defined as the operating condition at which the deflection $(\beta_2 - \beta_1)$ has a value that is 80% of that at which stall would occur. Constant suggested a value of 0.26 for the constant, C. Note that β_2 can then be evaluated and the head rise obtained from the characteristic (50). Later investigators explored the variations in the deviation angle with other flow parameters

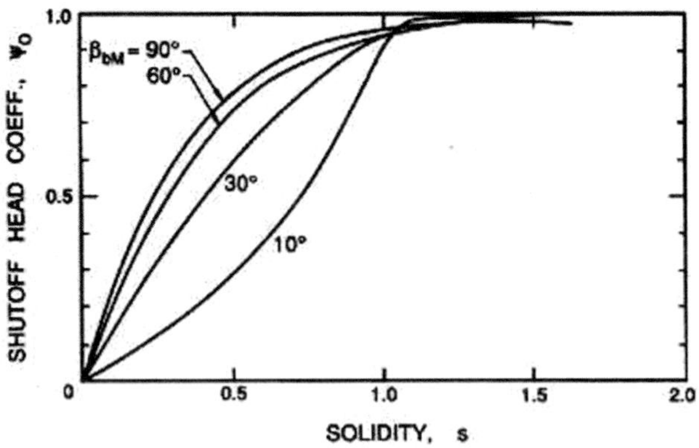

Figure 15. The performance parameter, ψ_0, as a function of solidity, s, for flat plate cascades with different blade angles, β_b. Adapted by Wislicensus (1947) (see also Sabersky, Acosta and Hauptmann 1989) from the potential flow theory of König (1922).

(see, for example, Howell 1942), and devised more complex correlations for use in the design of axial flow rotors (Horlock 1973). However, the basic studies of Leiblein on the boundary layers in linear cascades, and the role which these viscous effects play in determining the deviation and the losses, superceded much of this empirical work.

4.4 Viscous Effects in Linear Cascades

It is also of value to examine in more detail the mechanism of viscous loss in a cascade. Even in two-dimensional cascade flow, the growth of the boundary layers on the pressure and suction surfaces of the blades, and the wakes they form downstream of the blades (see figure 16), are complex, and not amenable to simple analysis. However, as the reviews by Roudebush and Lieblein (1965) and Lieblein (1965) demonstrate, it is nevertheless possible to provide some qualitative guidelines for the resulting viscous effects on cascade performance. In this respect, the diffusion factor, introduced by Lieblein *et* al. (1953), is a useful concept that is based on the following approximations. First, we note that under normal operating conditions, the boundary layer on the suction surface will be much thicker than that on the pressure surface of the foil, so that, to a first approximation, we may neglect the latter. Then, the thickness of the wake (and therefore the total pressure loss) will be primarily determined by that fraction of the suction surface over which the velocity gradient is adverse, since that is where the majority of the boundary layer growth occurs. Therefore, Lieblein *et* al. argued, the momentum thickness of the wake, θ^*, should correlate with a parameter they termed the *d*iffusion factor, given by $(w_{max} - w_2)/w_{max}$, where w_{max} is the maximum velocity on the suction surface. One should visualize deceleration or diffusion of the flow from w_{max} to w_2, and that this diffusion is the primary factor in determining the wake thickness. However, since w_{max} is not easily determined, Lieblein *et* al. suggest an approximation to the diffusion factor that is denoted Df, and given by

$$Df = 1 - \frac{w_2}{w_1} + \frac{v_{\theta 2} - v_{\theta 1}}{2 s w_1} = 1 - \frac{\sin\beta_1}{\sin\beta_2} + \frac{\sin\beta_1(\cot\beta_1 - \cot\beta_2)}{2s} \tag{58}$$

Figure 17 shows the correlation of the momentum thickness of the wake (normalized by the chord) with this diffusion factor, Df, for three foil profiles. Such correlations are now commonly used to determine the viscous loss due to blade boundary layers and wakes. Note that, once θ^*/c has been determined from such a correlation, the drag coefficient, C_D, and the friction or loss coefficient follow from equations (45), (47), and (48) and the fact that $D = \rho w_2^2 \theta^*$:

$$C_D = \frac{2\sin^2\beta_M}{\sin^2\beta_2}\frac{\theta^*}{c} \quad ; \quad f = \frac{s\sin\beta_M}{\sin^2\beta_2}\frac{\theta^*}{c} \tag{59}$$

The data shown in figure 17 were for a specific Reynolds number, Re, and the correlations must, therefore, be supplemented by a statement on the variation of the loss coefficient with Re. A number of correlations of this type exist (Roudebush and Lieblein 1965), and exhibit the expected decrease in the loss coefficient with increasing Re. For more detail on viscous losses in a cascade, the reader should consult the aforementioned papers by Lieblein.

In an actual turbomachine, there are several additional viscous loss mechanisms that were not included in the cascade analyses discussed above. Most obviously, there are additional viscous layers on the inner and outer surfaces that bound the flow, the hub and the shroud (or casing). These often give rise to complex, three-dimensional secondary flows that lead to additional viscous losses (Horlock and Lakshminarayana 1973). Moreover, the rotation of other, "non-active" surfaces of the impeller will lead to viscous

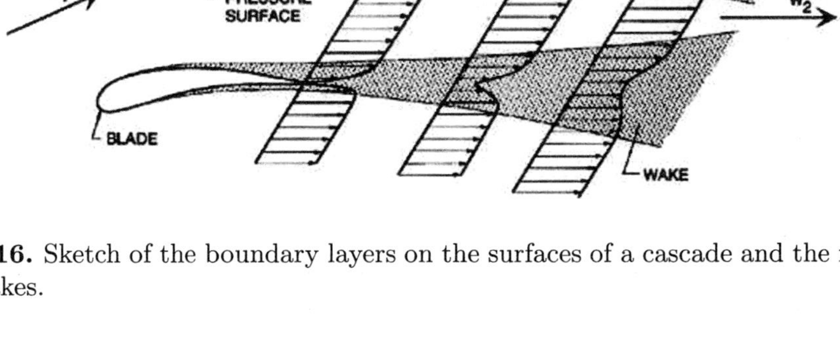

Figure 16. Sketch of the boundary layers on the surfaces of a cascade and the resulting blade wakes.

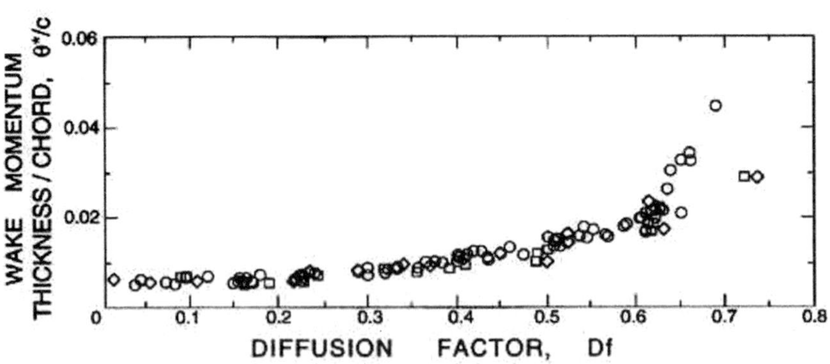

Figure 17. Correlation of the ratio of the momentum thickness of the blade wakes, θ^*, to the chord, c, with the diffusion factor, Df, for cascades of blades with three different profiles: NACA $65-(A_{10})10$ series (○) and two British C.4 parabolic arc profiles (□ and ◇). The maximum thickness of the blades is $0.1c$ and the Reynolds number is 2.5×10^5. Adapted from Lieblein (1965).

shear stresses, and thence to losses known as "disk friction losses" in the terminology of turbomachines. Also, leakage flows from the discharge back to the suction, or from one stage back to a preceding stage in a multistage pump, constitute effective losses that must be included in any realistic evaluation of the losses in an actual turbomachine (Balje 1981).

4.5 Radial Cascade Analyses

Two-dimensional models for centrifugal or radial turbomachines begin with analyses of the flow in a radial cascade (section 3.2 and figure 18), the counterpart of the linear cascade for axial flow machines. More specifically, the counterpart of the linear flat plate cascade is the logarithmic spiral cascade, defined in section 3.2, and shown in more detail in figure 18. There exist simple conformal mappings that allow potential flow solutions for the linear cascade to be converted into solutions for the corresponding radial cascade flow, though the proper interpretation of these solutions requires special care. The resulting head/flow characteristic for frictionless flow in a radial cascade of infinitely thin logarithmic spiral blades is given in a classic paper by Busemann (1928), and takes the form

$$\psi = Sf_B - \psi_0 \phi \left(\cot \beta_b + \frac{v_{\theta 1}}{v_{m1}} \right) \tag{60}$$

The terms Sf_B and ψ_0 result from quite separate and distinct fluid mechanical effects. The term involving ψ_0 is a consequence of the frictionless, potential flow head rise through any simple, nonrotating cascade whether of axial, radial, or mixed flow geometry. Therefore, ψ_0 is identical to the quantity, ψ_0, defined by equation (55) in the context of a linear cascade. The values for ψ_0 for a simple cascade of infinitely thin blades, whether linear, radial or mixed flow, are as given in figure 15. The ψ_0 term can be thought of

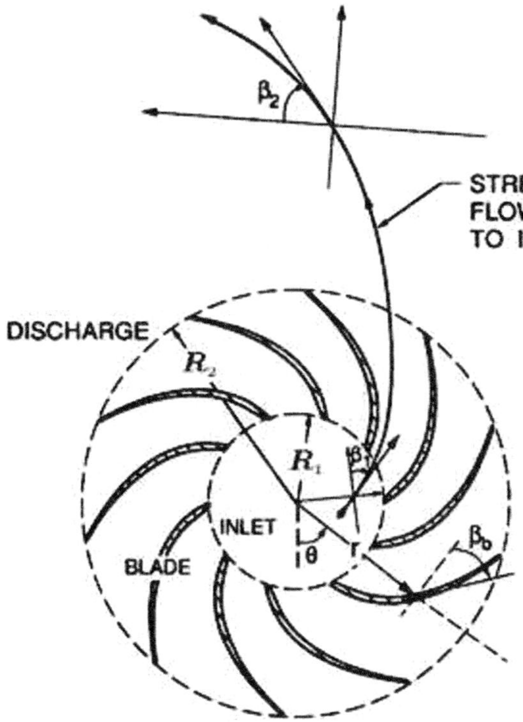

Figure 18. Schematic of the radial cascade corresponding to the linear cascade of figure 12.

as the "through flow" effect, and, as demonstrated by figure 15, the value of ψ_0 rapidly approaches unity when the solidity increases to a value a little greater than one.

However, it is important to recognize that the ψ_0 term is the result of a frictionless, potential flow solution in which the vorticity is zero. This solution would be directly applicable to a static or nonrotating radial cascade in which the flow entering the cacade has no component of the vorticity vector in the axial direction. This would be the case for a nonswirling axial flow that is deflected to enter a nonrotating, radial cascade in which the axial velocity is zero. But, relative to a *r*otating radial cascade (or centrifugal pump impeller), such an inlet flow does have vorticity, specifically a vorticity with magnitude 2Ω and a direction of rotation opposite to the direction of rotation of the impeller. Consequently, the frictionless flow through the impeller is not irrotational, but has a constant and uniform vorticity of -2Ω.

In inviscid fluid mechanics, one frequently obtains solutions for these kinds of rotational flows in the following way. First, one obtains the solution for the irrotational flow, which is represented by ψ_0 in the current problem. Mathematically, this is the complementary solution. Then one adds to this a particular solution that satisfies all

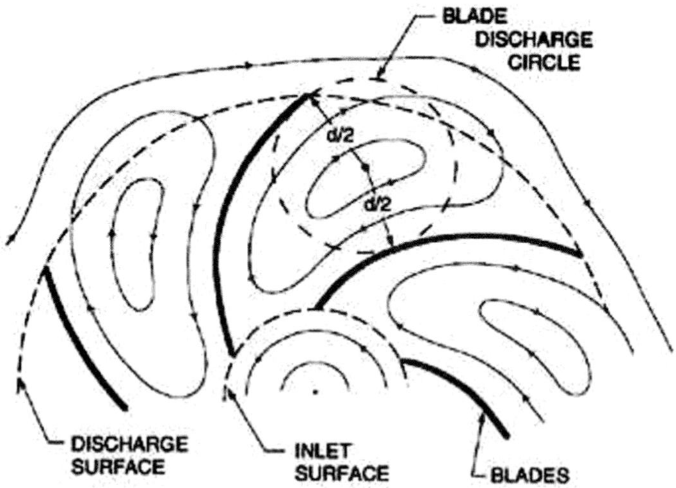

Figure 19. A sketch of the displacement component of the inviscid flow through a rotating radial cascade.

the same boundary conditions, but has a uniform vorticity, -2Ω. In the present context, this particular, or rotational, solution leads to the term, Sf_B, which, therefore, has a quite different origin from the irrotational term, ψ_0. The division into the rotational solution and the irrotational solution is such that all the net volumetric flow through the impeller is included in the irrotational (or ψ_0) component. The rotational solution has no through flow, but simply consists of a rotation of the fluid within each blade passage, as sketched in figure 19. Busemann (1928) called this the *d*isplacement flow; other authors refer to its rotating cells as *r*elative eddies (Balje 1980, Dixon 1978). In his pioneering work on the fluid mechanics of turbomachines, Stodola (1927) was among the first to recognize the importance of this rotational component of the solution. Busemann (1928) first calculated its effect upon the head/flow characteristic for the case of infinitely thin, logarithmic spiral blades, in other words the simple cascade in the radial configuration. For reasons which will become clear shortly, the function, Sf_B, is known as the Busemann slip factor, and Busemann's solutions lead to the values presented in figure 20 when the solidity, $s > 1.1$. Note that the values of Sf_B are invariably less than or equal to unity, and, therefore, the effect of the displacement flow is to cause a decrease in the head. This deficiency can, however, be minimized by using a large number of blades. As the number of blades gets larger, Sf_B tends to unity as the rotational flow within an individual blade passage increasingly weakens. In practice, however, the frictional losses will increase with the number of blades. Consequently, there is an important compromise that must be made in choosing the number of blades. As figure 20 shows, this compromise will depend on the blade angle. Furthermore, the compromise must also take into account the structural requirements for the blades. Thus, radial machines for use with liquids usually have a smaller number of blades than those used for gases. The reason for

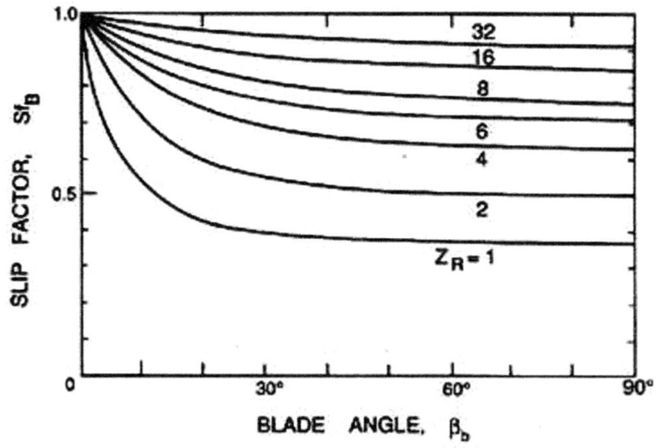

Figure 20. The Busemann slip factor, Sf_B, plotted against the blade angle, β_b, for various numbers of blades, Z_R. The results shown are for radial cascades of infinitely thin logarithmic spiral blades with solidities, $s > 1.1$. Adapted by Sabersky, Acosta and Hauptmann (1989) and Wislicenus (1947) from Busemann's (1928) theory.

this is that a liquid turbomachine requires much thicker blades, and, therefore, each blade creates much more flow blockage than in the case of a gas turbomachine. Consequently, liquid machines tend to have a smaller number of blades, typically eight for the range of specific speeds for which radial machines are designed ($N_D < 1.5$) (Stepanoff 1948, Anderson). Another popular engineering criterion (Stepanoff 1948) is that Z_R should be one third of the discharge blade angle, β_b (in degrees).

The decrease in the head induced by the displacement flow is due to the nonuniformity in the discharge flow; this nonuniformity results in a *m*ean angle of discharge (denoted by β_2) that is different from the discharge blade angle, β_{b2}, and, therefore, implies an effective deviation angle or *s*lip, Sf (see section 3.1). In fact, it is clear that the relations (20), (36), (60), and (8) imply that $Sf = Sf_B$, and, hence, the terminology used above. Stodola (1927) recognized that slip would be a consequence of the displacement flow, and estimated the magnitude of the slip velocity, $v_{\theta s}$, in the following approximate way. He argued that the slip velocity could be roughly estimated as $\Omega d/2$, where $d/2$ is the radius of the *b*lade discharge circle shown in figure 19. He visualized this as representative of the rotating cell of fluid in a blade passage, and that the rotation of this cell at Ω would lead to the aforementioned $v_{\theta s}$. Then, provide Z_R is not too small, $d \approx 2\pi R_2 \sin\beta_{b2}$, and it follows that

$$v_{\theta s} = \pi\Omega R_2 \sin\beta_{b2}/Z_R \tag{61}$$

and, from equation (8), that the estimated slip factor, Sf_S, is

$$Sf_S = 1 - \frac{\pi \sin\beta_{b2}}{Z_R} \tag{62}$$

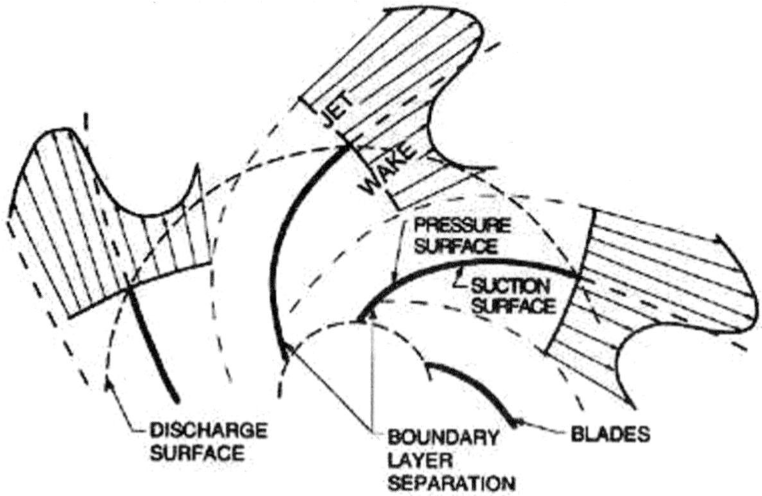

Figure 21. A sketch of actual discharge flow from a centrifugal pump or compressor including the alternating pattern of jets and wakes resulting from flow separation from the suction surfaces.

Numerical comparisons with the more exact results of Busemann presented in figure 20, show that equation (62) gives a reasonable first approximation. For example, an impeller with four blades, a blade angle of 25°, and a solidity greater than unity, has a Stodola slip factor of $Sf_S = 0.668$ compared to the value of $Sf_B = 0.712$ from Busemann's more exact theory.

There is a substantial literature on slip factors for centrifugal pumps. Some of this focuses on the calculation of slip factors for inviscid flow in radial cascades with blades that are more complex than the infinitely thin, logarithmic spiral blades used by Busemann. Useful reviews of some of this work can be found, for example, in the work of Wislicenus (1947), Stanitz (1952), and Ferguson (1963). Other researchers attempt to find slip factors that provide the best fit to experimental data. In doing so, they also attempt to account for viscous effects in addition to the inviscid effect for which the slip factor was originally devised. As an example of this approach, the reader may consult Wiesner (1967), who reviews the existing, empirical slip factors, and suggests one that seems to yield the best comparison with the experimental measurements.

4.6 Viscous Effects in Radial Flows

We now turn to a discussion of the viscous effects in centrifugal pumps. Clearly a radial cascade will experience viscous boundary layers on the blades that are similar to those discussed earlier for axial flow machines (see section 4.4). However, two complicating factors tend to generate loss mechanisms that are considerably more complicated. These two factors are flow separation and secondary flow.

Normally, the flow in a centrifugal pump separates from the suction surface near the leading edge, and produces a substantial wake on the suction surfaces of each of the blades. Fischer and Thoma (1932) first identified this phenomenon, and observed that the wake can occur even at design flow. Normally, it extends all the way to the impeller discharge. Consequently, the discharge flow consists of a low velocity zone or wake next to the suction surface, and, necessarily, a flow of increased velocity in the rest of the blade passage. This "jet-wake structure" of the discharge is sketched in figure 21. Note that this viscous effect tends to counteract the displacement flow of figure 19. Since the work of Fischer and Thoma, many others have studied this aspect of flows in centrifugal pumps and compressors (see, for example, Acosta and Bowerman 1957, Johnston and Dean 1966, Eckardt 1976), and it is now recognized as essential to take these features into account in constructing any model of the flow in radial turbomachines. Modern analyses of the flow in radial turbomachines usually incorporate the basic features of the jet-wake structure in the blade passages (for example, Sturge and Cumpsty 1975, Howard and Osborne 1977). Sturge and Cumpsty have calculated the shape of the wake in a typical, two-dimensional radial cascade, using numerical methods to solve a free streamline problem similar to those discussed in chapter 8.

At design flow, the wake or boundary layer on the suction surface may be quite thin, but as the flow coefficient, ϕ, is decreased, the increased incidence leads to larger wakes (Fischer and Thoma 1932, Johnston and Dean 1966). Clearly, the nonuniformity of the discharge flow implies an "effective" slip due to these viscous effects. This slip will not only depend on the geometry of the blades but will also be a function of the flow coefficient and the Reynolds number. The change with flow coefficient is particularly interesting. As ϕ is decreased below the design value and the wake grows in width, an increasing fraction of the flow is concentrated in the jet. Johnston and Dean (1966) showed that this results in a flow that *more* closely follows the geometry of the pressure surface, and, therefore, to a *decrease* in the slip. This can be a major effect in radial compressors. Johnston and Dean made measurements in an 18-bladed radial compressor impeller with a 90° discharge blade angle (for which $Sf_S = 0.825$), and found that the effective slip factor increased monotonically from a value of about 0.8 at $\phi_2 = 0.5$ to a value of 1.0 at $\phi_2 = 0.15$. However, this increase in the slip factor did not produce an increase in the head rise, because the increase in the viscous losses was greater than the potential gain from the decrease in the slip.

Finally, it is important to recognize that secondary flows can also have a substantial effect on the development of the blade wakes, and, therefore, on the jet-wake structure. Moreover, the geometric differences between the typical radial compressor and the typical centrifugal pump can lead to significant differences in the secondary flows, the loss mechanisms, and the jet-wake structure. The typical centrifugal pump geometry was illustrated in figure 7, to which we should append the typical number of blades, $Z_R = 8$. A typical example is the geometry at $N_D = 0.6$, namely $R_{T1}/R_{T2} \approx 0.5$ and $B_2 \approx 0.2R_{T2}$. Assuming $Z_R = 8$ and a typical blade angle at discharge of 25°, it follows that the blade passage flow at discharge has cross-sectional dimensions normal to the relative velocity vector of $0.2R_{T2} \times 0.3R_{T2}$, while the length of the blade passage is approximately $1.2R_{T2}$. Thus the blade passage is fairly wide relative to its length. In contrast, the typical radial compressor has a much smaller value of B_2/R_{T2}, and a much larger number of blades. As a result, not only is the blade passage much narrower relative to its length, but also the typical cross-section of the discharge flow is far from square, being significantly narrower in the axial direction. The viscous boundary layers on the suction and pressure surfaces of the blades, and on the hub and shroud (or casing), will have a greater effect the smaller the cross-sectional dimensions of the blade passage are relative to its length. Moreover, the secondary flows that occur in the corners of this passage amplify these viscous effects. Consequently, the flow that discharges from a blade passage of a typical radial compressor is more radically altered by these viscous effects than the flow discharging from a typical centrifugal pump.

5 Other flow features

5.1 Introduction

In this section we briefly survey some of the other important features of the flows through turbomachines, namely the secondary flows that can have important bearing on both the non-cavitating and the cavitating performance of the machine.

5.2 Discharge Flow Management

To this point the entire focus has been on the flow within the impeller or rotor of the pump. However, the flow that discharges from the impeller requires careful handling in order to preserve the gains in energy imparted to the fluid. In many machines this requires the conversion of velocity head to pressure by means of a diffuser. This inevitably implies hydraulic losses, and considerable care needs to be taken to minimize these losses. The design of axial and radial diffusers, with and without vanes to recover the swirl velocity, is a major topic, whose details are beyond the scope of this book. The reader is referred to the treatise by Japikse (1984).

Such diffusers are more common in compressors than in pumps. Typical pump configurations are as follows. Axial flow pumps often employ a set of stator vanes before (or in) the axial diffuser in order to recover the swirl velocities. Special care needs to be taken to match the swirl angles of the flow exiting the impeller with the inlet angles of the stator vanes. It is advisable, where possible, to measure the impeller discharge flow directly before finalizing a design. In some designs, the axial diffuser will be followed by a spiral collector or "volute" in order to recover the energy in the remaining swirl and axial velocities.

In the case of centrifugal pumps, a radial flow diffuser with vanes may or may not be used. Often it is not, and the flow discharges directly into the volute. The proper design of this volute is an important component of centrifugal pump design (Anderson 1955, Worster 1963, Stepanoff 1948). The objective is to design a volute in which the flow is carefully matched to the flow exiting the impeller, so that the losses are minimized and so that the pressure is uniform around the impeller discharge. The basic concept is sketched in figure 22. The flow discharges from the impeller with a velocity, $v_{\theta 2}$, in the tangential direction and a velocity, v_{r2}, in the radial direction, given by

$$v_{r2} = Q/2\pi R_2 B_2 \tag{63}$$

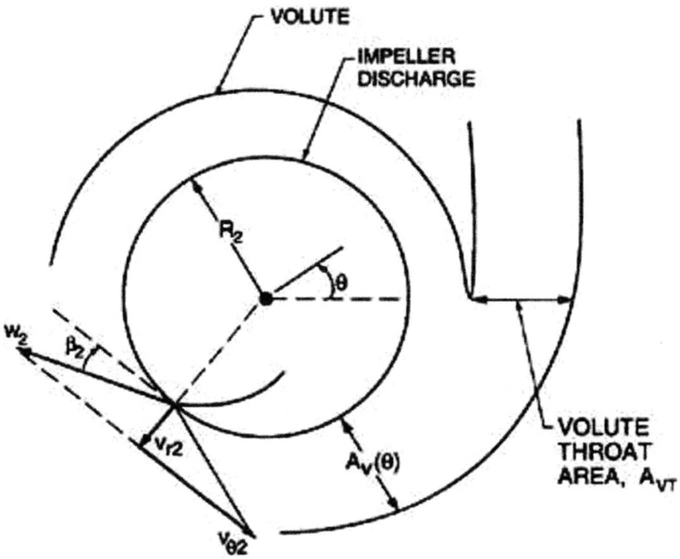

Figure 22. Volute flow notation.

where B_2 is the impeller discharge width. For simplicity, it will be assumed that the discharge from the impeller is circumferentially uniform; in fact, nonuniform volute flow will lead to a nonuniform flow in the impeller that is unsteady in the rotating frame. Though this complication is often important, it is omitted from the present, simple analysis.

If we further assume radially uniform velocity at each angular location in the volute (also an assumption that needs to be modified in a more accurate analysis), then it follows from the application of conservation of mass to an element, $d\theta$, of the volute that the discharge flow will be matched to the flow in the volute if

$$v_{\theta 2}\frac{dA_V}{d\theta} = v_{r2}R_2B_2 \tag{64}$$

This requires a circumferentially uniform rate of increase of the volute area of $dA_V/d\theta = v_{r2}R_2B_2/v_{\theta 2}$ over the entire development of the spiral. If the area of the clearance between the cutwater and the impeller discharge is denoted by A_{VC}, and the volute exit area is denoted by A_{VT}, then A_V should have the following linear behavior:

$$A_V = A_{VC} + \frac{\theta}{2\pi}A_{VT} \tag{65}$$

It follows that $dA_V/d\theta = A_{VT}/2\pi$ and hence

$$\phi^{-1}\tan\beta_2 - 1 = \frac{2\pi R_2B_2\tan\beta_2}{A_{VT}} \tag{66}$$

Consequently, for a given impeller operating at a given design flow coefficient, ϕ_D, there exists a specific area ratio, $2\pi R_2 B_2 \tan\beta_2 / A_{VT}$, for the volute geometry. This parameter is close to the ratio which Anderson (1955) used in his design methodology (see also Worster 1963), namely the ratio of the cross-sectional area of the flow leaving the impeller ($2\pi R_2 B_2 \sin\beta_2$) to the volute throat area (A_{VT}). For more detailed analyses of the flow in a volute, the reader is referred to Pfleiderer (1932), Stepanoff (1948), and Lazarkiewicz and Troskolanski (1965). For example, Pfleiderer explored the radially nonuniform distributions of velocity within the volute and the consequences for the design methodology.

5.3 Prerotation

Perhaps no aspect of turbomachinery flow is more misrepresented and misunderstood than the phenomenon of "prerotation", and, while other secondary flows are important in turbomachines we choose to comment on this particular feature because of its importance for hydraulic performance and its interaction with cavitation.

It is first essential to distinguish between two separate phenomena both of which lead to a swirling flow entering the pump. These two phenomena have very different fluid mechanical origins. Here we shall distinguish them by the separate terms, "backflow-induced swirl" and "inlet prerotation". Both imply a swirl component of the flow entering the pump. In fluid mechanical terms, the flow has axial vorticity (if the axis of rotation is parallel with the axis of the inlet duct) with a magnitude equal to twice the rate of angular rotation of the swirl motion. Moreover, there are some basic properties of such swirling flows that are important to the understanding of prerotation. These are derived from the vorticity transport theorem (see, for example, Batchelor 1967). In the context of the steady flow in an inlet duct, this theorem tells us that the vorticity will only change with axial location for two reasons: (a) because vorticity is diffused into the flow by the action of viscosity, or (b) because the flow is accelerated or decelerated as a result of a change in the cross-sectional area of the flow. The second mechanism results in an increase in the swirl velocity due to the stretching of the vortex line, and is similar to the increase in rotation experienced by figure skaters when they draw their arms in closer to their body. When the moment of inertia is decreased, conservation of angular momentum results in an increase in the rotation rate. Thus, for example, a nozzle in the inlet line would increase the magnitude of any preexisting swirl.

For simplicity, however, we shall first consider inlet ducting of uniform and symmetric cross-sectional area, so that only the first mechanism exists. In inviscid flow, it follows that, if there is a location far upstream at which the swirl (or axial vorticity) is zero, then, in the absence of viscous effects, the swirl will be everywhere zero. This important result, which is a version of Kelvin's theorem (Batchelor 1967), is not widely recognized in discussions of prerotation. Moreover, the result is not altered by the existence of viscous effects, since purely axial motion cannot generate axial vorticity. However, there are two common circumstances in which prerotation can be generated without violation of the above theorem, and these give rise to the two phenomena named earlier.

The first of these common circumstances arises because of one of the most important secondary flows that can occur in pumps, namely the phenomenon of "backflow". This

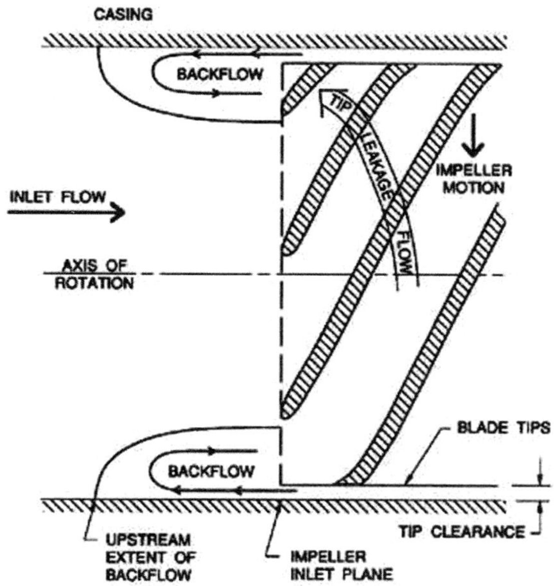

Figure 23. Lateral view of impeller inlet flow showing tip leakage flow leading to backflow.

is caused by the leakage flow between the tip of the blades of an impeller (we consider first an unshrouded impeller) and the pump casing. The circumstances are depicted in figure 23. Below a certain critical flow coefficient, the pressure difference driving the leakage flow becomes sufficiently large that the tip leakage jet penetrates upstream of the inlet plane of the impeller, and thus forms an annular region of "backflow" in the inlet duct. After penetrating upstream a certain distance, the fluid of this jet is then entrained back into the main inlet flow. The upstream penetration distance increases with decreasing flow coefficient, and can reach many diameters upstream of the inlet plane. In some pump development programs (such as the Rocketdyne J-2 liquid oxygen pump) efforts have been made to insert a "backflow deflector" in order to improve pump performance (Jakobsen 1971). The intention of such a device is to prevent the backflow from penetrating too far upstream, to reduce the distortion of the inlet flow field, and to recover, as far as is possible, the swirl energy in the backflow. More recently, a similar device was successfully employed in a centrifugal pump (Sloteman *et* al. 1984).

Some measurements of the axial and swirl velocities just upstream of an axial inducer are presented in figure 24. This data is taken from del Valle *et* al (1992), though very similar velocity profiles have been reported by Badowski (1969, 1970) (see also Janigro and Ferrini 1973), and the overall features of the flow are similar whether the pump is shrouded or unshrouded, axial or centrifugal (see, for example, Stepanoff 1948, Okamura and Miyashiro 1978, Breugelmans and Sen 1982, Sloteman *et* al. 1984). Measurements are shown in figure 24 for two distances upstream of the inlet plane (half a radius and

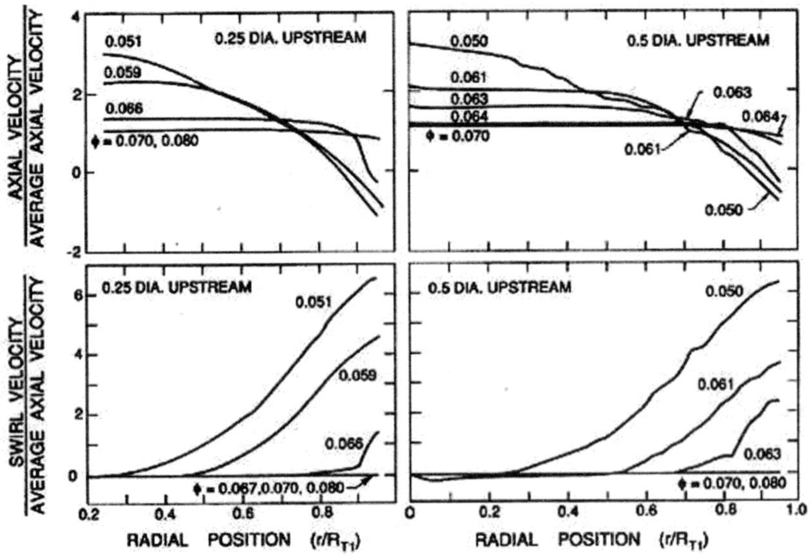

Figure 24. Axial and swirl velocity profiles in the inlet duct 0.25 diameters (left) and 0.5 diameters (right) upstream of the inlet plane of an inducer (Impeller VI) for various flow coefficients as shown (from del Valle, Braisted and Brennen 1992).

one radius upstream), and for a number of flow coefficients, ϕ. Note from the axial flow velocity profiles that, as the flow coefficient is decreased, the backflow reaches a half radius upstream at about $\phi \approx 0.066$, and one radius upstream at about $\phi \approx 0.063$. The size of the backflow region grows as ϕ is decreased. It is particularly remarkable that at $\phi \approx 0.05$, nearly 30% of the inlet area is experiencing reverse flow! We can further observe from the swirl velocity data that, in the absence of backflow, the inlet flow has zero swirl. Kelvin's theorem tells us this must be the case because the flow far upstream has no swirl.

Obviously the backflow has a high swirl velocity imparted to it by the impeller blades. But what is also remarkable is that this vorticity is rapidly spread to the core of the main inlet flow, so that at $\phi = 0.05$, for example, almost the entire inlet flow has a nonzero swirl velocity. The properties of swirling flows discussed above are not violated, since the origin of the vorticity is the pump itself and the vorticity is transmitted to the inflow via the backflow. The rapidity with which the swirl vorticity is diffused to the core of the incoming flow remains something of a mystery, for it is much too rapid to be caused by normal viscous diffusion (Braisted 1979). It seems likely that the inherent unsteadiness of the backflow (with a strong blade passing frequency component) creates extensive mixing which effects this rapid diffusion. However it arises, it is clear that this "backflow-induced swirl", or "pre-rotation", will clearly affect the incidence angles and, therefore, the performance of the pump.

Before leaving the subject of backflow, it is important to emphasize that this phe-

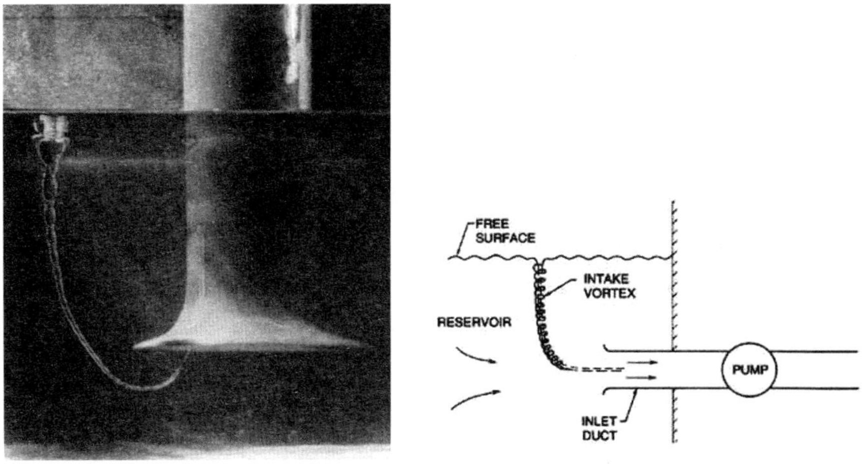

Figure 25. Right: sketch of a typical inlet vortex associated with prerotation. Left: Photograph of an air-filled inlet vortex from Wijdieks (1965) reproduced with permission of the Delft Hydraulics Laboratory.

nomenon also occurs at flow rates below design in centrifugal as well as axial flow pumps, and with shrouded as well as unshrouded impellers (see, for example, Okamura and Miyashiro 1978, Makay 1980). The detailed explanation may differ from one device to another, but the fundamental tendency for an impeller to exhibit this kind of secondary flow at larger angles of incidence seems to be universal.

But there is another, quite separate origin for prerotation, and this is usually manifest in practice when the fluid is being drawn into the pump from an "inlet bay" or reservoir with a free surface (figure 25). Under such circumstances, it is almost inevitable that the large scale flow in the reservoir has some nonuniformity that constitutes axial vorticity or circulation in the frame of reference of the pump inlet. Even though the fluid velocities associated with this nonuniformity may be very small, when the vortex lines are stretched as the flow enters the inlet duct, the vorticity is greatly amplified, and the inlet flow assumes a significant preswirl or "inlet prerotation". The effect is very similar to the bathtub vortex. Once the flow has entered an inlet duct of constant cross-sectional area, the magnitude of the swirl usually remains fairly constant over the short lengths of inlet ducting commonly used.

Often, the existence of "inlet prerotation" can have unforeseen consequences for the suction performance of the pump. Frequently, as in the case of the bathtub vortex, the core of the vortex runs from the inlet duct to the free surface of the reservoir, as shown in figure 25. Due to the low pressure in the center of the vortex, air is drawn into the core and may even penetrate to the depth of the duct inlet, as illustrated by the photograph in figure 25 taken from the work of Wijdieks (1965). When this occurs, the pump inlet suddenly experiences a two-phase air/water flow rather than the single-phase liquid inlet flow expected. This can lead, not only to a significant reduction in the performance of the pump, but also to the vibration and unsteadiness that often accompany two-phase flow. Even without air entrainment, the pump performance is almost always deteriorated by these suction vortices. Indeed this is one of the prime suspects when the expected performance is not realized in a particular installation. These intake vortices are very similar to those which can occur in aircraft engines (De Siervi et al. 1982).

6 Cavitation parameters and inception

6.1 Introduction

This section will deal with the parameters that are used to describe cavitation, and the circumstances that govern its inception. In subsequent sections, we address the deleterious effects of cavitation, namely cavitation damage, noise, the effect of cavitation on hydraulic performance, and cavitation-induced instabilities.

6.2 Cavitation Parameters

Cavitation is the process of the formation of vapor bubbles in low pressure regions within a flow. One might imagine that vapor bubbles are formed when the pressure in the liquid reaches the vapor pressure, p_V, of the liquid at the operating temperature. While many complicating factors discussed later cause deviations from this hypothesis, nevertheless it is useful to adopt this as a criterion for the purpose of our initial discussion. In practice, it can also provide a crude initial guideline.

The static pressure, p, in any flow is normally nondimensionalized as a pressure coefficient, C_p, defined as

$$C_p = (p - p_1)/\frac{1}{2}\rho U^2 \tag{67}$$

where p_1 is some reference static pressure for which we shall use the pump inlet pressure and U is some reference velocity for which we shall use the inlet tip speed, ΩR_{T1}. It is important to note that, for the flow of an incompressible liquid within rigid boundaries, C_p is only a function of the geometry of the boundaries and of the Reynolds number, Re, which, for present purposes, can be defined as $2\Omega R_{T1}^2/\nu$ where ν is the kinematic viscosity of the fluid. It is equally important to note that, in the absence of cavitation, the fluid velocities and the pressure coefficient are *independent* of the level of the pressure. Thus, for example, a change in the inlet pressure, p_1, will simply result in an equal change in all the other pressures, so that C_p is unaffected. It follows that, in any flow with prescribed fluid velocities, geometry and Reynolds number, there will be a particular location at which the pressure, p, is a minimum and that the difference between this minimum pressure, p_{min}, and the inlet pressure, p_1 is given by

$$C_{pmin} = (p_{min} - p_1)/\frac{1}{2}\rho U^2 \tag{68}$$

where C_{pmin} is some negative number which is a function only of the geometry of the device (pump) and the Reynolds number. If the value of C_{pmin} could be obtained either experimentally or theoretically, then we could establish the value of the inlet pressure, p_1, at which cavitation would first appear (assuming that this occurs when $p_{min} = p_V$)

as p_1 is decreased, namely

$$(p_1)_{\substack{CAVITATION\\APPEARANCE}} = p_V + \frac{1}{2}\rho U^2\left(-C_{pmin}\right) \tag{69}$$

which for a given device, given fluid, and given fluid temperature, would be a function only of the velocity, U.

Traditionally, several special dimensionless parameters are utilized in evaluating the potential for cavitation. Perhaps the most fundamental of these is the cavitation number, σ, defined as

$$\sigma = (p_1 - p_V)/\frac{1}{2}\rho U^2 \tag{70}$$

Clearly every flow has a value of σ whether or not cavitation occurs. There is, however, a particular value of σ corresponding to the particular inlet pressure, p_1, at which cavitation first occurs as the pressure is decreased. This is called the cavitation inception number, and is denoted by σ_i:

$$\sigma_i = \left[(p_1)_{\substack{CAVITATION\\APPEARANCE}} - p_V\right]/\frac{1}{2}\rho U^2 \tag{71}$$

If cavitation inception occurs when $p_{min} = p_V$, then, combining equations (69) and (71), it is clear that this criterion corresponds to a cavitation inception number of $\sigma_i = -C_{pmin}$. On the other hand, a departure from this criterion results in values of σ_i different from $-C_{pmin}$.

Several variations in the definition of cavitation number occur in the literature. Often the inlet tip velocity, ΩR_{T1}, is employed as the reference velocity, U, and this version will be used in this monograph unless otherwise stated. Sometimes, however, the relative velocity at the inlet tip, w_{T1}, is used as the reference velocity, U. Usually the magnitudes of w_{T1} and ΩR_{T1} do not differ greatly, and so the differences in the two cavitation numbers are small.

In the context of pumps and turbines, a number of other, surrogate cavitation parameters are frequently used in addition to some special terminology. The $NPSP$ (for net positive suction pressure) is an acronym used for $(p_1^T - p_V)$, where p_1^T is the inlet total pressure given by

$$p_1^T = p_1 + \frac{1}{2}\rho v_1^2 \tag{72}$$

For future purposes, note from equations (72), (70), and (21) that

$$\left(p_1^T - p_V\right) = \frac{1}{2}\rho\Omega^2 R_{T1}^2\left(\sigma + \phi_1^2\right) \tag{73}$$

Also, the $NPSE$, or net positive suction energy, is defined as $(p_1^T - p_V)/\rho$, and the $NPSH$, or net positive suction head, is $(p_1^T - p_V)/\rho g$. Furthermore, a nondimensional version of these quantities is defined in a manner similar to the specific speed as

$$S = \Omega Q^{\frac{1}{2}}/(NPSE)^{\frac{3}{4}} \tag{74}$$

and is called the "suction specific speed". Like the specific speed, N, the suction specific speed, is a dimensionless number, and should be computed using a consistent set of units, such as Ω in rad/s, Q in ft^3/s and $NPSE$ in ft^2/s^2. Unfortunately, it is traditional U.S. practice to use Ω in rpm, Q in gpm, and to use the $NPSH$ in ft rather than the $NPSE$. As in the case of the specific speed, one may obtain the traditional U.S. evaluation by multiplying the rational suction specific speed used in this monograph by 2734.6.

The suction specific speed is similar in concept to the cavitation number in that it represents a nondimensional version of the inlet or suction pressure. Moreover, there will be a certain critical value of the suction specific speed at which cavitation first appears. This special value is termed the inception suction specific speed, S_i. The reader should note that frequently, when a value of the "suction specific speed" is quoted for a pump, the value being given is some critical value of S that may or may not correspond to S_i. More frequently, it corresponds to S_a, the value at which the degradation in the head rise reaches a certain percentage value (see section 6.4).

The suction specific speed, S, may be obtained from the cavitation number, σ, and vice versa, by noting that, from the relations (21), (70), (72), and (74), it follows that

$$S = \left[\pi\phi_1\left(1 - R_{H1}^2/R_{T1}^2\right)\right]^{\frac{1}{2}} / \left[\frac{1}{2}\left(\sigma + \phi_1^2\right)\right]^{\frac{3}{4}} \tag{75}$$

We should also make note of a third nondimensional parameter, called Thoma's cavitation factor, σ_{TH}, which is defined as

$$\sigma_{TH} = \left(p_1^T - p_V\right) / \left(p_2^T - p_1^T\right) \tag{76}$$

where $(p_2^T - p_1^T)$ is the total pressure rise across the pump. Clearly, this is connected to σ and to S by the relation

$$\sigma_{TH} = \frac{\sigma + \phi_1^2}{\psi} = \left(\frac{N}{S}\right)^{\frac{4}{3}} \tag{77}$$

Since cavitation usually occurs at the inlet to a pump, σ_{TH} is not a particularly useful parameter since $(p_2^T - p_1^T)$ is not especially relevant to the phenomenon.

6.3 Cavitation Inception

For illustrative purposes in the last section, we employed the criterion that cavitation occurs when the minimum pressure in the flow just reaches the vapor pressure, $\sigma_i = -C_{pmin}$. If this were the case, the prediction of cavitation inception would be a straightforward matter. Unfortunately, large departures from this criterion can occur in practice, and this has lead to an extensive body of literature. Since cavitation inception is treated elsewhere in this course of lectures we will not cover that subject here; for further study the reader is referred to the classic references by Knapp, Daily and Hammit (1970), Acosta and Parkin (1975), Arakeri (1979), Brennen (1994, 1995).

6.4 Pump Performance

The performance of a pump when presented nondimensionally will take the generic form sketched in figure 26. As discussed earlier, the noncavitating performance will consist of the head coefficient, ψ, as a function of the flow coefficient, ϕ, where the design

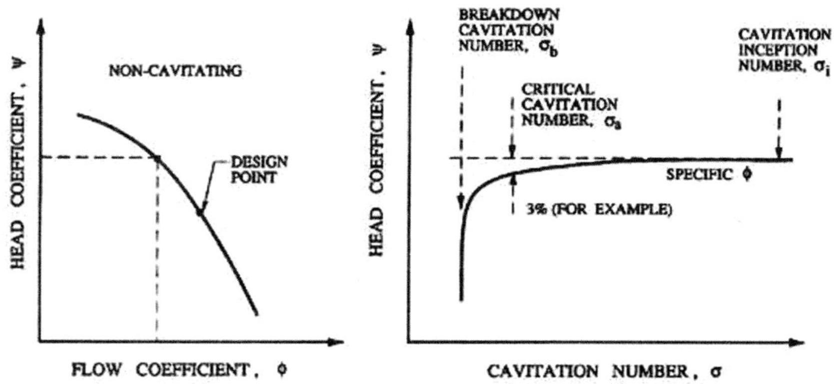

Figure 26. Schematic of noncavitating performance, $\psi(\phi)$, and cavitating performance, $\psi(\phi,\sigma)$, showing the three key cavitation numbers.

conditions can be identified as a particular point on the $\psi(\phi)$ curve. The noncavitating characteristic should be independent of the speed, Ω, though at lower speeds there may be some deviation due to viscous or Reynolds number effects. The *cavitating performance*, as illustrated on the right in figure 26, is presented as a family of curves, $\psi(\phi,\sigma)$, each for a specific flow coefficient, in a graph of the head coefficient against cavitation number, σ. Frequently, of course, both performance curves are presented dimensionally; then, for example, the NPSH is often used instead of the cavitation number as the abscissa for the cavitation performance graph.

It is valuable to identify three special cavitation numbers in the cavitation performance graph. Consider a pump operating at a particular flow rate or flow coefficient, while the inlet pressure, $NPSH$, or cavitation number is gradually reduced. As discussed in the previous chapter, the first critical cavitation number to be reached is that at which cavitation first appears; this is called the cavitation inception number, σ_i. Often the occurrence of cavitation is detected by the typical crackling sound that it makes (see section 7.2). As the pressure is further reduced, the extent (and noise) of cavitation will increase. However, it typically requires a further, substantial decrease in σ before any degradation in performance is encountered. When this occurs, the cavitation number at which it happens is often defined by a certain percentage loss in the head rise, H, or head coefficient, ψ, as shown in figure 26. Typically a critical cavitation number, σ_a, is defined at which the head loss is 2, 3 or 5%. Further reduction in the cavitation number will lead to major deterioration in the performance; the cavitation number at which this occurs is termed the breakdown cavitation number, and is denoted by σ_b.

It is important to emphasize that these three cavitation numbers may take quite different values, and to confuse them may lead to serious misunderstanding. For example, the cavitation inception number, σ_i, can be an order of magnitude larger than σ_a or σ_b. There exists, of course, a corresponding set of critical suction specific speeds that we denote by S_i, S_a, and S_b. Good examples of the differences between S_i and S_b are given

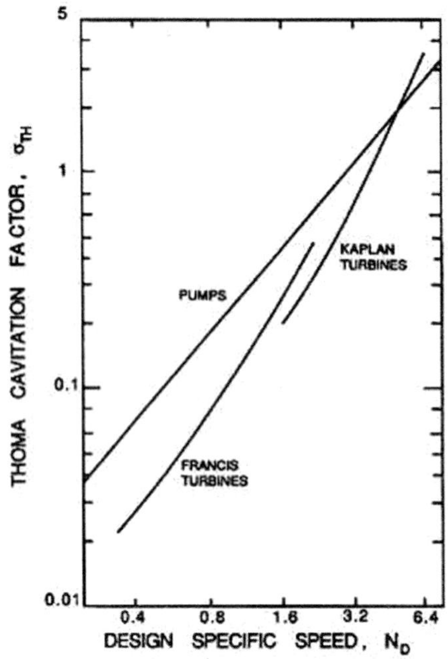

Figure 27. The Hydraulic Institute standards for the operation of pumps and turbines (Hydraulic Institute 1965).

by McNulty and Pearsall (1979) for many different types of pumps.

Perhaps the most common misunderstanding concerns the recommendation of the Hydraulic Institute that is reproduced in figure 27. This suggests that a pump should be operated with a Thoma cavitation factor, σ_{TH}, in excess of the value given in the figure for the particular specific speed of the application. The line, in fact, corresponds to a critical suction specific speed of 3.0. Frequently, this is erroneously interpreted as the value of S_i. In fact, it is more like S_a; operation above the line in figure 27 does *not* imply the absence of cavitation, or of cavitation damage.

The next section will include a qualitative description of the various forms of cavitation that can occur in a pump.

6.5 Types of Impeller Cavitation

Since cavitation in a pump impeller can take a variety of forms (see, for example, Wood 1963), it is appropriate at this stage to attempt some description and classification of these types of cavitation. It should be borne in mind that any such classification is necessarily somewhat arbitrary, and that types of cavitation may occur that do not readily fall within the classification system. Figure 28 includes sketches of some of the forms of cavitation that can be observed in an unshrouded axial flow impeller. As the inlet pressure is decreased, inception almost always occurs in the tip vortex generated by the corner where the leading edge meets the tip. Figure 29 includes a photograph of a typical cavitating tip vortex from tests of Impeller IV (the scale model of the SSME low pressure LOX turbopump shown in figure 11). Note that the backflow causes the flow in the vicinity of the vortex to have an upstream velocity component. Careful smoothing of the transition from the leading edge to the tip can reduce σ_i, but it will not eliminate the vortex, or vortex cavitation.

Usually the cavitation number has to be lowered quite a bit further before the next development occurs, and often this takes the form of traveling bubble cavitation on the suction surfaces of the blades. Nuclei in the inflow grow as they are convected into the regions of low pressure on the suction surfaces of the blades, and then collapse as they move into regions of higher pressure. For convenience, this will be termed "bubble cavitation." It is illustrated in figure 30 which shows bubble cavitation on a single hydrofoil.

With further reduction in the cavitation number, the bubbles may combine to form large attached cavities or vapor-filled wakes on the suction surfaces of the blades. In a more general context, this is known as "attached cavitation". In the context of pumps, it is often called "blade cavitation". Figure 31 presents an example of blade cavitation in a centrifugal pump.

When blade cavities (or bubble or vortex cavities) extend to the point on the suction surface opposite the leading edge of the next blade, the increase in pressure in the blade passage tends to collapse the cavity. Consequently, the surface opposite the leading edge of the next blade is a location where cavitation damage is often encountered.

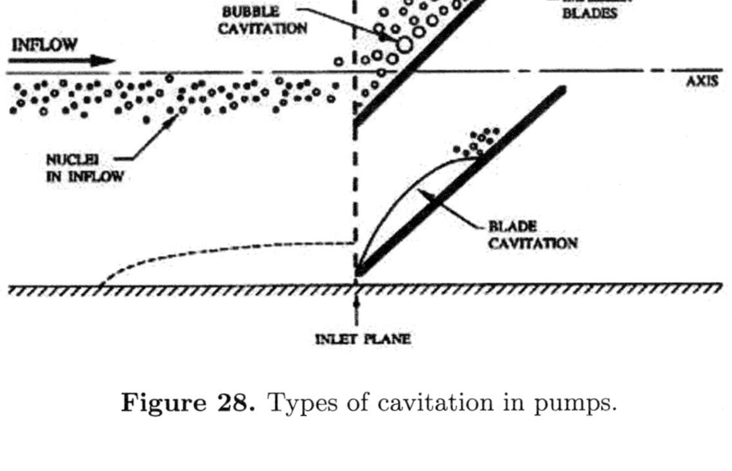

Figure 28. Types of cavitation in pumps.

Figure 29. Tip vortex cavitation on Impeller IV, the scale model of the SSME low pressure LOX turbopump (see figure 11) at an inlet flow coefficient, ϕ_1, of 0.07 and a cavitation number, σ, of 0.42 (from Braisted 1979).

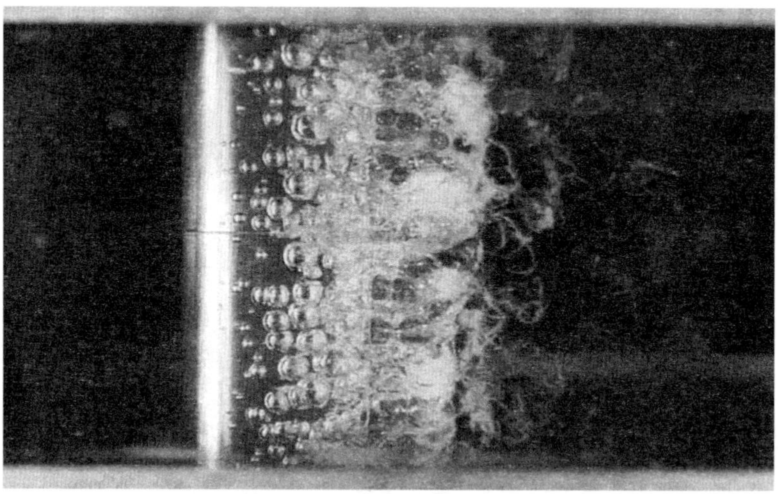

Figure 30. Bubble cavitation on the surface of a NACA 4412 hydrofoil at zero incidence angle, a speed of 13.7 m/s and a cavitation number of 0.3. The flow is from left to right and the leading edge of the foil is just to the left of the white glare patch on the surface (Kermeen 1956).

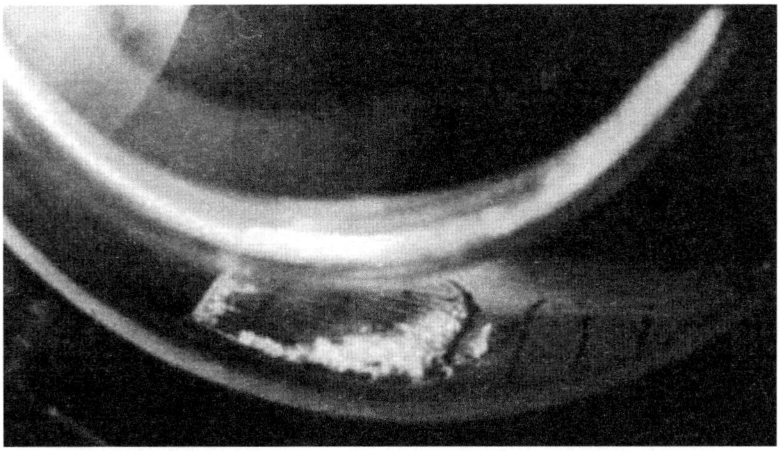

Figure 31. Blade cavitation on the suction surface of a blade in a centrifugal pump. The relative flow is from left to right and the cavity begins at the leading edge of the blade which is toward the left of the photograph. From Sloteman, Cooper, and Graf (1991), courtesy of Ingersoll-Dresser Pump Company.

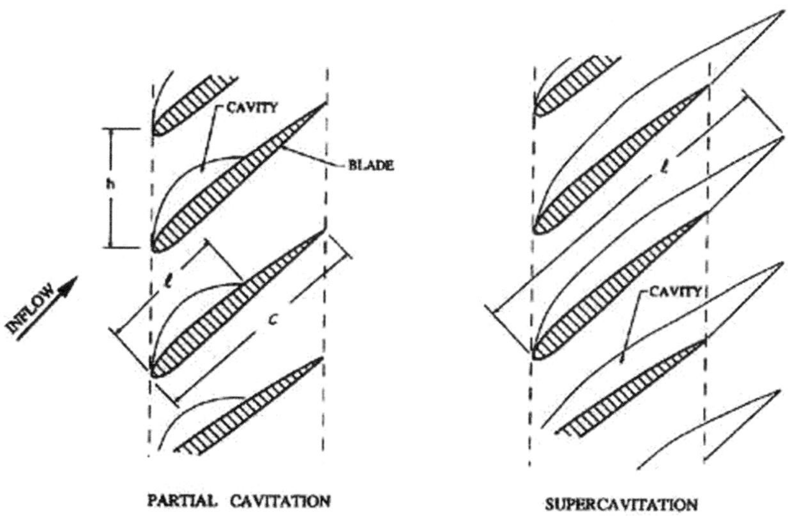

Figure 32. Partially cavitating cascade (left) and supercavitating cascade (right).

Figure 33. As figure 29, but here showing typical backflow cavitation.

Blade cavitation that collapses on the suction surface of the blade is also referred to as "partial cavitation", in order to distinguish it from the circumstances that occur at very low cavitation numbers, when the cavity may extend into the discharge flow downstream of the trailing edge of the blade. These long cavities, which are clearly more likely to occur in lower solidity machines, are termed "supercavities". Figure 32 illustrates the difference between partial cavitation and supercavitation. Some pumps have even been designed to operate under supercavitating conditions (Pearsall 1963). The potential advantage is that bubble collapse will then occur downstream of the blades, and cavitation damage might thus be minimized.

Finally, it is valuable to create the catch-all term "backflow cavitation" to refer to the cavitating bubbles and vortices that occur in the annular region of backflow upstream of the inlet plane when the pump is required to operate in a loaded condition below the design flow rate (see section 5.3). The increased pressure rise across the pump under these circumstances may cause the tip clearance flow to penetrate upstream and generate a backflow that can extend many diameters upstream of the inlet plane. When the pump also cavitates, bubbles and vortices are swept up in this backflow and, to the observer, can often represent the most visible form of cavitation. Figure 33 includes a photograph illustrating the typical appearance of backflow cavitation upstream of the inlet plane of an inducer.

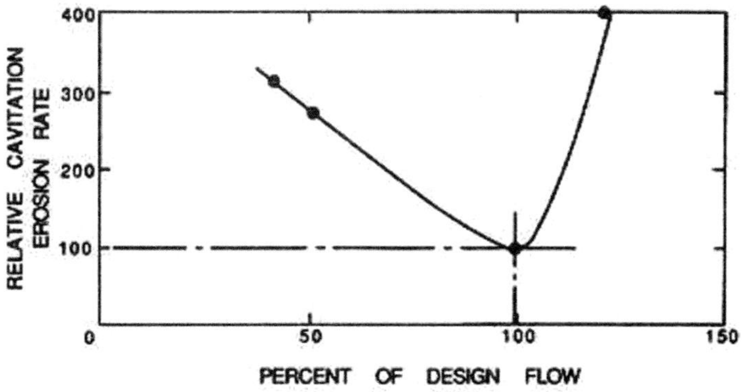

Figure 34. Cavitation erosion rates in a centrifugal pump as a function of the flow rate relative to the design flow rate (Pearsall 1978 from Grist 1974).

7 Cavitation damage and noise

7.1 Introduction

We now turn to the characteristics of cavitation for $\sigma < \sigma_i$ and to the deleterious effects that occur as soon as there is any cavitation, namely cavitation damage and cavitation noise. Cavitation damage is covered elsewhere in this collection of lectures and we bypass that subject here, only noting that the damage is the result of the violent collapse of cavitation bubbles close to a solid surface. Quantitative data on cavitation erosion rates in pumps is very limited because of the length of time necessary to make such measurements. The data that does exist (Mansell 1974) demonstrates that the rate of erosion is a strong function of the operating point as given by the cavitation number and the flow coefficient. The influence of the latter is illustrated in figure 34. At off-design conditions, the increased angle of incidence leads to increased cavitation and, therefore, increased damage rate.

One of the most frequently used surrogates that help to indicate the potential for cavitation damage in pumps is the measurement of cavitation noise and we pause here to consider that subject.

7.2 Cavitation Noise

The violence of cavitation bubble collapse also produces noise. In many practical circumstances, the noise is important not only because of the vibration that it may cause, but also because it advertizes the presence of cavitation and, therefore, the likelihood of cavitation damage. Indeed, the magnitude of cavitation noise is often used as a crude measure of the rate of cavitation erosion. For example, Lush and Angell (1984) have shown that, in a given flow at a given cavitation number, the rate of weight loss due to cavitation damage is correlates with the noise as the velocity of the flow is changed.

Prior to any discussion of cavitation noise, it is useful to identify the frequency with

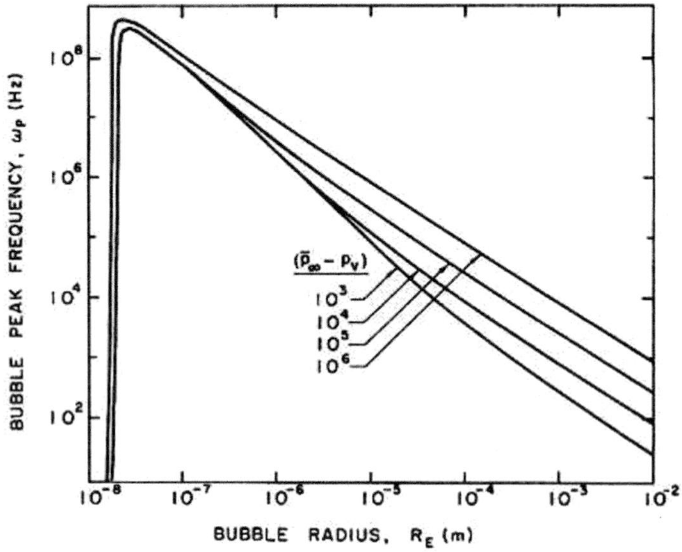

Figure 35. Bubble natural frequency, ω_P, in Hz as a function of the bubble radius and the difference between the equilibrium pressure and the vapor pressure (in $kg/m\ sec^2$) for water at $300°K$.

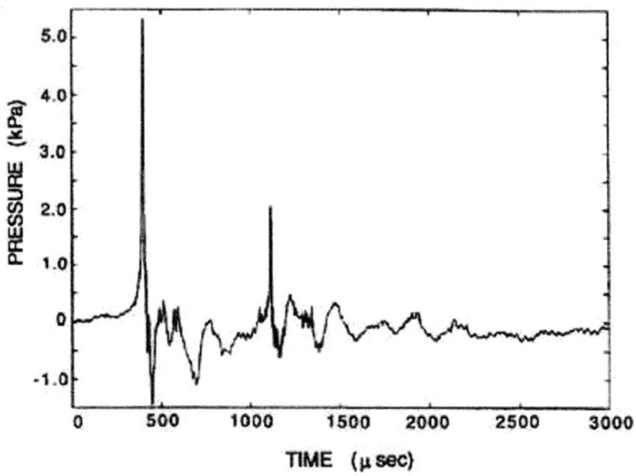

Figure 36. Typical acoustic signal from a single collapsing bubble (from Ceccio and Brennen 1991).

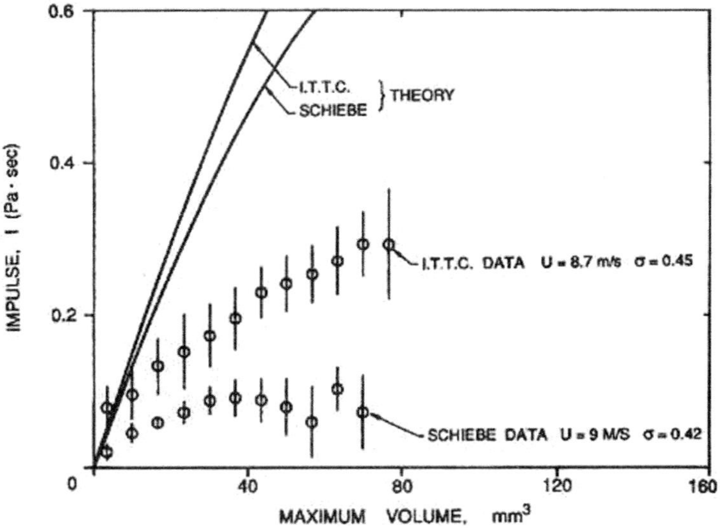

Figure 37. The acoustic impulse, I, produced by the collapse of a single cavitation bubble. Data is shown for two axisymmetric bodies as a function of the maximum volume prior to collapse. Also shown are the equivalent results from solutions of the Rayleigh-Plesset equation (from Ceccio and Brennen 1991).

which individual bubbles resonate in a quiescent liquid. This resonant frequency, ω_P, can be obtained from the Rayleigh-Plesset equation (see Brennen 1995) and is a function of the difference between the mean ambient pressure, $\bar{p}$, and the vapor pressure, p_V, the equilibrium bubble radius, R_E, and the liquid properties:

$$\omega_P = \left[\frac{3(\bar{p} - p_V)}{\rho_L R_E^2} + \frac{4\mathcal{S}}{\rho_L R_E^3} - \frac{8\nu^2}{R_E^4} \right]^{\frac{1}{2}} \tag{78}$$

The results of this calculation for bubbles in water at $300°K$ are presented in figure 35 for various mean pressure levels, $\bar{p}$. Note that the bubbles below about 0.02 μm are supercritically damped, and have no resonant frequency. Typical cavitation nuclei of size $10 \rightarrow 100$ μm have resonant frequencies in the range $10 \rightarrow 100$ kHz. Even though the nuclei are excited in a highly nonlinear way by the cavitation, one might expect that the spectrum of the noise that this process produces would have a broad maximum at the peak frequency corresponding to the size of the most numerous nuclei participating in the cavitation. Typically, this would correspond to the radius of the critical or most numerous nuclei. For example, if the critical nuclei size were of the order of $10 - 100$ μm, then, according to figure 35, one might expect to see cavitation noise frequencies of the order of $10 - 100$ kHz. This is, indeed, the typical range of frequencies produced by cavitation.

Fitzpatrick and Strasberg (1956) were the first to make extensive use of the Rayleigh-Plesset equation to predict the noise from individual collapsing bubbles and the spectra

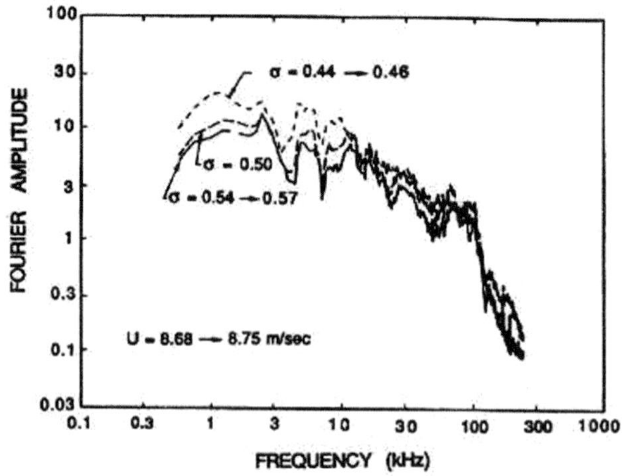

Figure 38. Typical spectra of noise from bubble cavitation for various cavitation numbers as indicated (Ceccio and Brennen 1991).

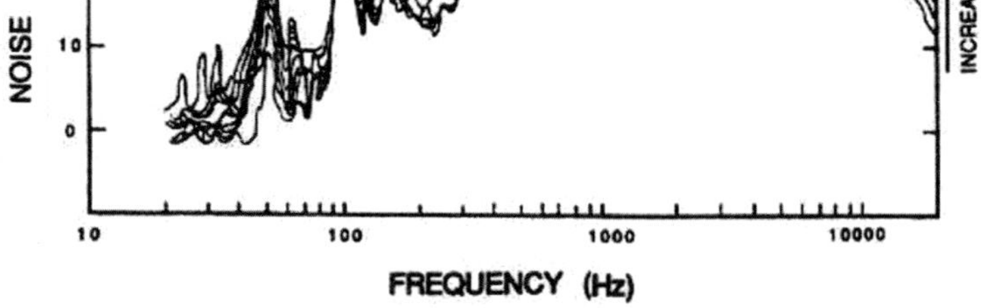

Figure 39. Typical spectra showing the increase in noise with increasing cavitation in an axial flow pump (Lee 1966).

that such a process would produce. More recently, Ceccio and Brennen (1991) have recorded the noise from individual cavitation bubbles in a flow. A typical acoustic signal is reproduced in figure 36. The large positive pulse at about 450 μs corresponds to the first collapse of the bubble. Since the radiated acoustic pressure, p_A, in this context is related to the second derivative of the volume of the bubble, $V(t)$, by

$$p_A = \frac{\rho_L}{4\pi\ell} \frac{d^2V}{dt^2} \tag{79}$$

(where ℓ is the distance of the measurement from the center of the bubble), the pulse corresponds to the very large and positive values of d^2V/dt^2 that occur when the bubble is close to its minimum size in the middle of the collapse. The first pulse is followed in figure 36 by some facility-dependent oscillations, and by a second pulse at about 1100 μs. This corresponds to the second collapse; no further collapses were observed in these particular experiments.

A good measure of the magnitude of the collapse pulse in figure 36 is the acoustic impulse, I, defined as the area under the curve or

$$I = \int_{t_1}^{t_2} p_A dt \tag{80}$$

where t_1, t_2 are the times before and after the pulse when $p_A = 0$. The acoustic impulses for cavitation on two axisymmetric headforms are compared in figure 37 with impulses predicted from integration of the Rayleigh-Plesset equation. Since these theoretical calculations assume that the bubble remains spherical, the discrepancy between the theory and the experiments is not too surprising. Indeed, the optimistic interpretation of figure 37 is that the theory can provide an order of magnitude estimate of the noise produced by a single bubble. This could then be combined with the nuclei number density distribution to obtain a measure of the amplitude of the noise (Brennen 1994).

The typical single bubble noise shown in figure 36 leads to the spectrum shown in figure 38. If the cavitation events are randomly distributed in time, this would also correspond to the overall cavitation noise spectrum. It displays a characteristic frequency content in the range of $1 \rightarrow 50\ kHz$ (the rapid decline at about 80 kHz represents the limit of the hydrophone used to make these measurements). Typical measurements of the noise produced by cavitation in an axial flow pump are illustrated in figure 39, and exhibit the same features demonstrated in figure 38. The signal in figure 39 also clearly contains some shaft or blade passage frequencies that occur in the absence of cavitation, but may be amplified or attenuated by cavitation. Figure 40 contains data obtained for cavitation noise in a centrifugal pump. Note that the noise at a frequency of 40 kHz shows a sharp increase with the onset of cavitation; on the other hand, the noise at the shaft and blade passage frequencies show only minor changes with cavitation number. The decrease in the 40 kHz cavitation noise as breakdown is approached is also a common feature in cavitation noise measurements.

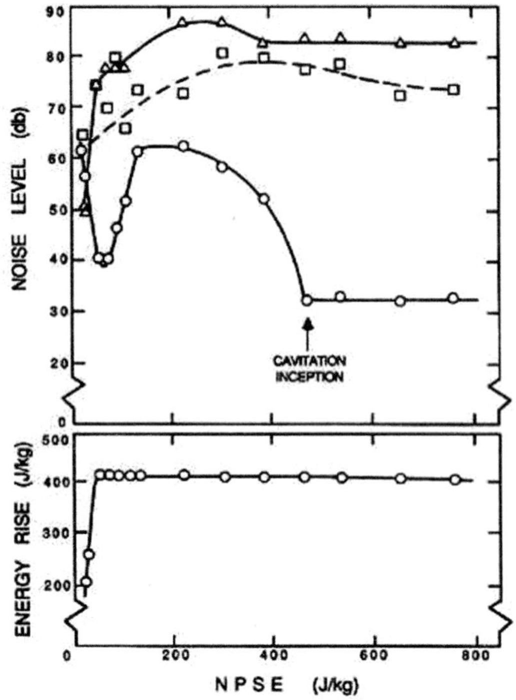

Figure 40. The relation between the cavitation performance, the noise and vibration produced at three frequency levels in a centrifugal pump, namely the shaft frequency (□), the blade passage frequency (△) and 40 kHz (○) (Pearsall 1966-67).

8 Cavitation and pump performance

8.1 Introduction

In this section we turn our attention to another of the deleterious consequences of cavitation, namely its effect upon the steady state hydraulic performance of a pump. In the next subsection we present several examples of the effect of cavitation on conventional pumps. This is followed by a discussion of the performance and design of cavitating inducers which are devices added to conventional pumps for the purpose of improving the cavitation performance. Subsequent subsections deal with the analytical methods available for the evaluation of cavitation performance and with the thermodynamic effects of the phase change process on that performance.

8.2 Typical Pump Performance Data

A typical non-cavitating performance characteristic for a centrifugal pump is shown in figure 41 for the Impeller X/Volute A combination (Chamieh 1983) described in subsection 3.8. The design flow coefficient for this pump is $\phi_2 = 0.092$ but we note that

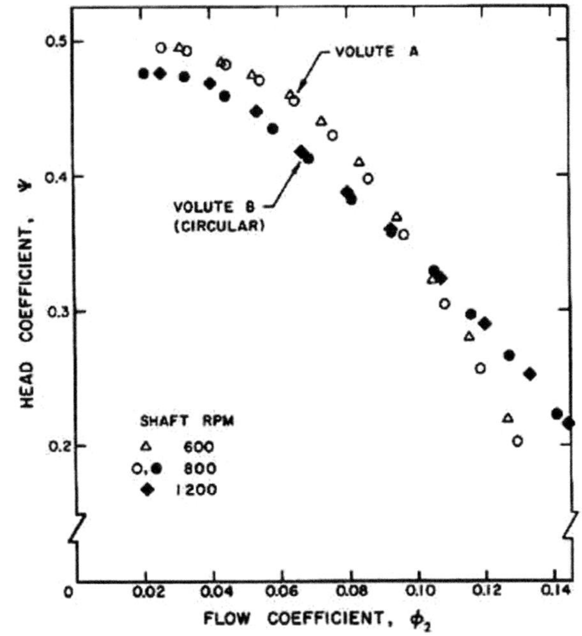

Figure 41. Typical non-cavitating performance for a centrifugal pump, namely Impeller X (see section 3.8) with Volute A and a circular volute of uniform cross-section (from Chamieh 1983).

it performs reasonably well down to about 30% of this design flow. This flexibility is characteristic of centrifugal pumps. Data is presented for three different shaft speeds, namely 600, 800 and 1200*rpm*; since these agree closely we can conclude that there is no perceptible effect of Reynolds number for this range of speeds. The effect of a different volute is also illustrated by the data for Volute B which is a circular volute of circumferentially uniform area. In theory this circular volute is not well matched to the impeller discharge flow and the result is that, over most of the range of flow coefficient, the hydraulic performance is inferior to that with Volute A. However, Volute B is superior at high flow coefficients. This suggests that the flow in Volute A may be more pathological than one would like at these high flow coefficients. It further serves to emphasize the importance of a volute (or diffuser) and the need for an understanding of the flow in a volute at both design and off-design conditions.

Typical cavitation performance characteristics for a centrifugal pump are presented in figure 42 for the Impeller X/Volute A combination. The breakdown cavitation numbers in the range $\sigma = 0.1 \rightarrow 0.4$ are typical of this type of pump. Note that the cavitation head loss occurs more gradually at high flow coefficients than at low values. This is a common feature of the cavitation performance of many pumps, both centrifugal and axial.

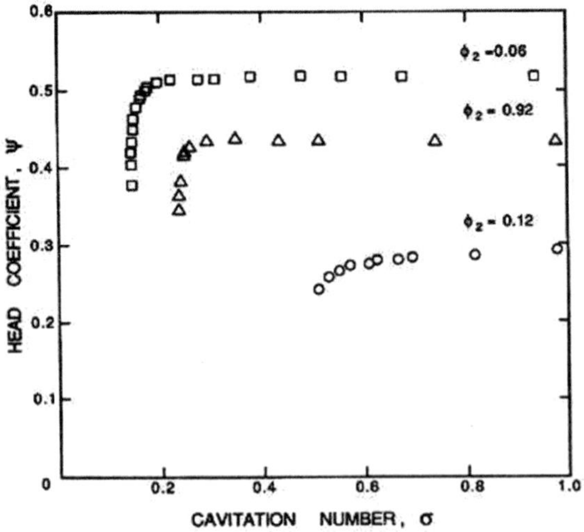

Figure 42. Cavitation performance for the Impeller X/Volute A combination (from Franz *et* al. 1989, 1990).

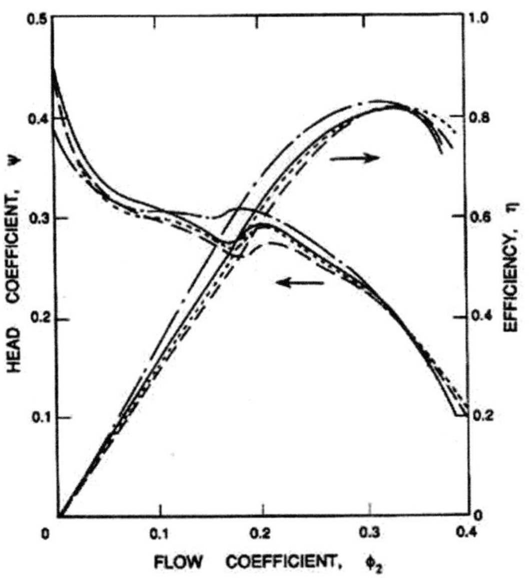

Figure 43. Typical non-cavitating performance characteristics for a four-bladed axial flow pump with tip blade angle, β_{bT}, of about 18°, a hub-tip ratio, R_H/R_T, of 0.483, a solidity of 0.68 and four different blade profiles (yielding the set of four performance curves). Adapted from Oshima and Kawaguchi (1963).

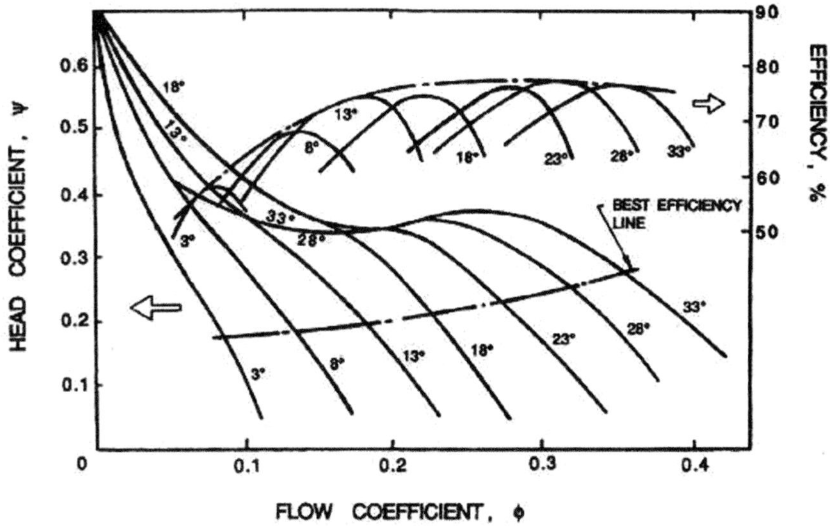

Figure 44. Head and efficiency characteristics for an axial flow pump with different tip blade angles, β_{bT} (from Peck 1966).

Now consider some examples of axial and mixed flow pumps. Figure 43 presents the non-cavitating characteristics for four similar axial flow pumps with slightly different blade profiles. The kinks in the curves are caused by flow separation which differs significantly from one profile to another. Note also that there are small regions of positive slope in the head characteristics. This often leads to instability and to fluctuating pressures and flow rates through the excitation of the surge and stall mechanisms discussed in the sections which follow. Sometimes the region of positive slope in the head characteristic can be even more marked. As a final example of non-cavitating performance we include in figure 44, the effect of the blade angle in an axial flow pump; note that angles of the order of 20° to 30° seem to be optimal for many purposes.

The cavitation characteristics of axial flow pumps are typified by the data in figure 45. The data of Guinard *et* al. (1953) provides a particularly well-documented example of the effect of cavitation on an axial flow pump. Note first from figure 45 that the cavitation inception number is smallest at the design flow and increases as ϕ is decreased; the decrease at very low ϕ does not, however, have an obvious explanation.

The cavitation data of figure 45 also help to illustrate several other characteristic phenomena. Note the significant increase in the head just prior to the decrease associated with breakdown. The cavitation data of figure 45 also illustrates the fact that breakdown at low flow coefficients occurs at higher cavitation numbers and is usually more abrupt than at higher flow coefficients.

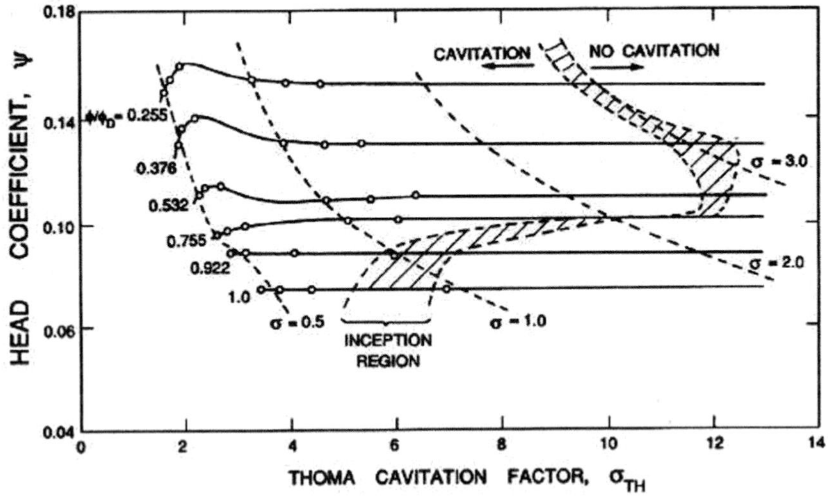

Figure 45. Cavitation performance characteristics of an axial flow pump. Adapted from Guinard *et* al. (1953).

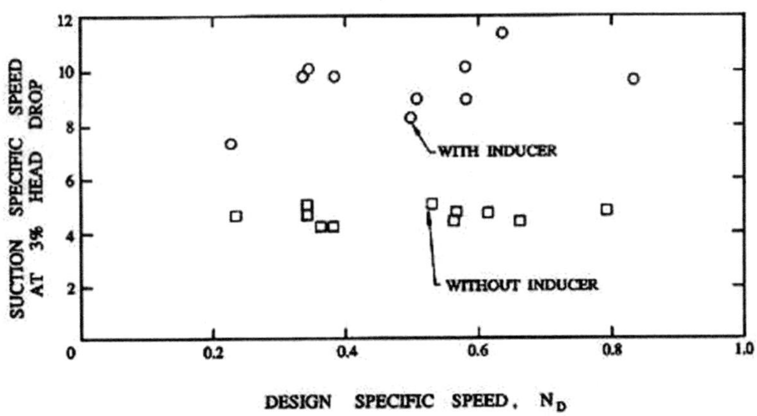

Figure 46. Comparison of the suction specific speed at 3% head drop for process pumps with and without inducer (from Janigro and Ferrini 1973).

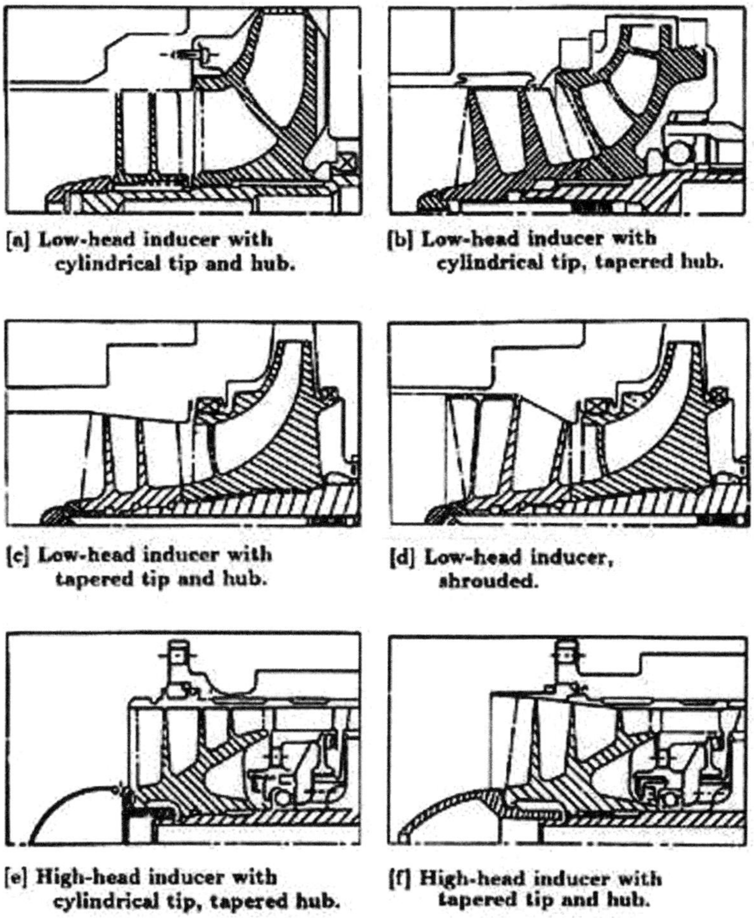

Figure 47. Various geometries of cavitating inducers (from Jakobsen 1971).

8.3 Inducer Designs

Axial flow inducers are intended to improve the cavitation performance of centrifugal or mixed flow pumps by increasing the inlet pressure to the pump to a level at which it can operate without excessive loss of performance due to cavitation. Typically they consist of an axial flow stage placed just upstream of the inlet to the main impeller. They are designed to operate at small incidence angles and to have thin blades so that the perturbation to the flow is small in order to minimize the production of cavitation and its deleterious effect upon the flow. The objective is to raise the pressure very gradually to the desired level. The typical advantage gained by the addition of an inducer is illustrated in figure 46 taken from Janigro and Ferrini (1973). This compares the cavitation performance of a class of process pumps with and without an inducer.

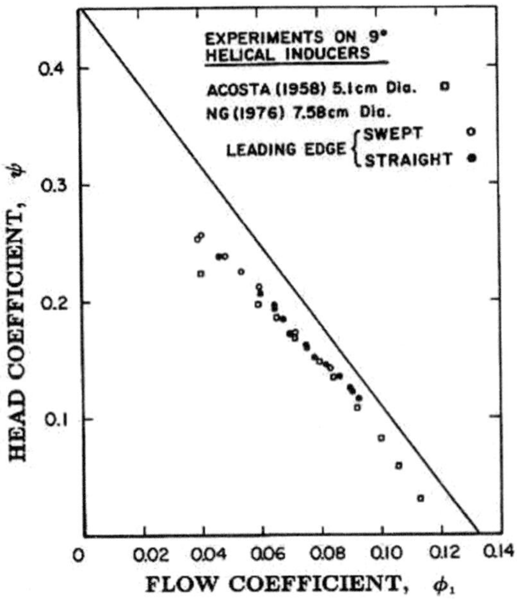

Figure 48. Non-cavitating performance of 9° helical inducers of two different sizes and with and without swept leading edges. Also shown is the theoretical performance prediction in the absence of losses (from Ng and Brennen 1978).

Various types of inducer design are documented in figure 47 taken from Jakobsen (1971). Most inducers of recent design seem to be of types (a) or (b). They are unshrouded, with a swept leading edge and often with a forward cant to the blades as in the case of the low pressure LOX pump in the SSME (figure 11). This blade cant has the effect of causing the leading edge to be located at a single axial plane counteracting the effect of the sweep given to the leading edge. They are also designed to function at an incidence angle of a few degrees. The reason that the design incidence angle is not zero is that under these conditions cavitation could form on either the pressure or suction surfaces or it could oscillate between the two. It is preferable to use a few degrees of incidence to eliminate this uncertainity and ensure suction surface cavitation.

8.4 Inducer Performance

Typical inducer performance characteristics are presented in figures 48 to 49. The non-cavitating performance of simple 9° helical inducers (see figure 11) is presented in figure 48. The data for the 5.1 *cm* and 7.6 *cm* diameter models appear to coincide indicating very little Reynolds number effect. Furthermore, the non-cavitating performance is the same whether the leading edge is swept or straight. Also included in the figure are the results of a lossless performance prediction (Brennen 1994). The agreement with the experiments is about as good as one could expect. It is most satisfactory close to the

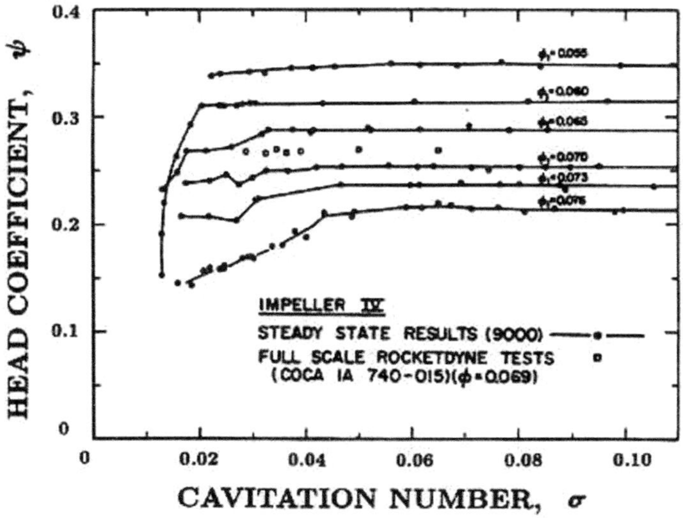

Figure 49. Cavitation performance of an SSME low pressure LOX pump model at 9000 *rpm* and various flow coefficients. Also shown are full scale test data from Rocketdyne (from Ng and Brennen 1978).

zero incidence flow coefficient of about 0.09 → 0.10 where one would expect the viscous losses to be a minimum. The comparison also suggests that the losses increase as one either increases or decreases the flow from that zero incidence value.

An example of cavitating inducer performance is presented in figure 49, in this case for an SSME low pressure LOX pump model, along with some data from full scale tests. Note that the head tends to be somewhat erratic at the lower cavitation numbers. Such behavior is typical of most axial flow inducer data and is probably due to hydraulic losses caused by unsteadiness in the flow.

8.5 Effects of Inducer Geometry

In this section we comment on several geometric factors for which the data suggests optimum values. Clearly, the solidity, s, needs to be as small as possible and yet large enough to achieve the desired discharge flow angle. Data on the effect of the solidity on the performance of a 3-bladed, 9° helical inducer has been obtained by Acosta (1958) and on a 4-bladed, $8\frac{1}{2}^\circ$ helical inducer by Henderson and Tucker (1962). The latter data is shown in figure 50. This data suggests an optimum value of s of about 1.5.

The same two studies also investigated the effect of the tip clearance and the data of Henderson and Tucker (1962) is reproduced in figure 51. As was the case with the solidity, the non-cavitating performance is less sensitive to changes in the tip clearance than is the cavitation performance. Note from figure 51 that the non-cavitating performance is relatively insensitive to the clearance unless the latter is increased above 2% of the chord when the performance begins to decline more rapidly. The cavitating performance shows

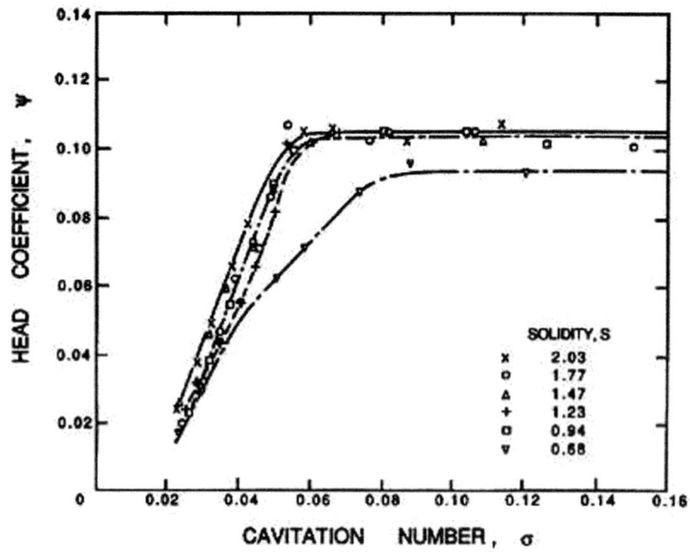

Figure 50. The effect of solidity on the cavitation performance of a cavitating inducer (Janigro and Ferrini 1973 from Henderson and Tucker 1962).

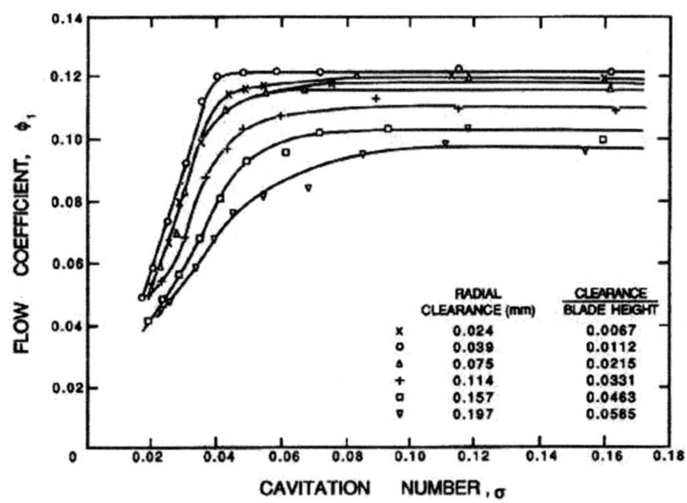

Figure 51. The effect of tip clearance on the cavitation performance of a cavitating inducer (from Henderson and Tucker 1962 as given by Janigro and Ferrini 1973).

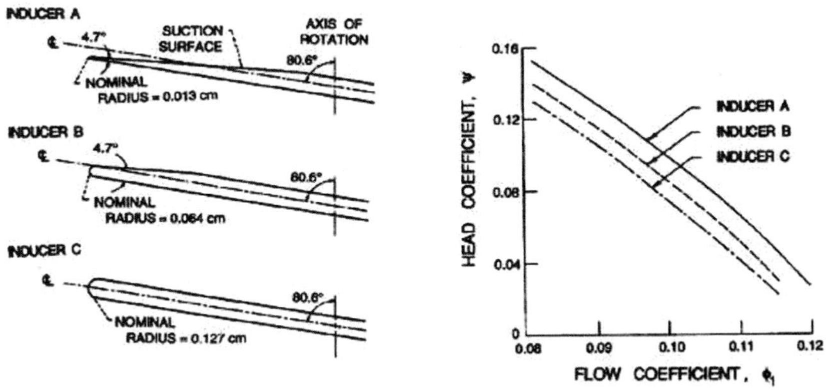

Figure 52. Non-cavitating performance of three 9.4° helical inducers with different leading edges as shown. Tests performed with liquid hydrogen (from Moore and Meng 1970b).

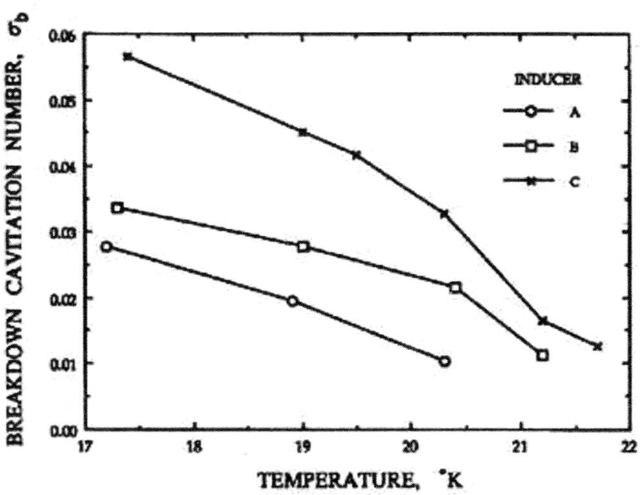

Figure 53. The breakdown cavitation numbers, σ_b (defined in this case by a 30% head drop) as a function of temperature for three shapes of leading edge (see figure 52) on 9.4° helical inducers operating in liquid hydrogen (from Moore and Meng 1970a,b).

a similar dependence though the fractional changes in the performance are larger. Note that the performance near the knee of the curve indicates an optimum clearance of about 1% of the chord.

Moore and Meng (1970a,b) have made a study of the effect of the leading edge geometry on inducer performance and their results are depicted graphically in figures 52 and 53.

Note that the leading edge geometry has a significant effect on the non-cavitating performance and on the breakdown cavitation number. Simply stated, the sharper the leading edge the better the hydraulic performance under both cavitating and non-cavitating conditions. There is, however, a trade-off to be made here for very thin leading edges may flutter (see Brennen 1994). Incidentally, figure 53 also demonstrates the thermal effect on cavitation performance which is discussed in section 8.7.

8.6 Analyses of Cavitation in Pumps

In this and the subsections which follow we shall try to give a brief overview of the various kinds of models which have been developed for the analysis of developed cavitation in a pump. Clearly different types of cavitation require different analytical models. We begin in this subsection with the various attempts which have been made to model traveling bubble cavitation in a pump and to extract from such a model information regarding the damage potential, noise or performance decrement caused by that cavitation. In a later subsection we shall outline the methods developed for attached blade cavitation. As for the other types of cavitation which are typically associated with the secondary flows (*e.g.*, tip vortex cavitation, backflow cavitation) there is little that can be added to what has already been described in the last section. Much remains to be understood concerning secondary flow cavitation, perhaps because some of the more important effects involve highly unsteady and transient cavitation.

To return to a general discussion of traveling bubble cavitation, it is clear that, given the pressure and velocity distribution along a particular streamline in a reference frame fixed in the impeller, one can input that information into the Rayleigh-Plesset equation. The equation can then be integrated to find the size of the bubble at each point along its trajectory. Since the first applications of the Rayleigh-Plesset equation to traveling bubble cavitation by Plesset (1949) and Parkin (1952) there have been many such investigations, most of which are reviewed by Holl (1969). A notable example is the work of Johnson and Hsieh (1966). While most of the literature discusses single bubble solutions of this type for flows around simple headforms the same programs can readily be used for the flow around a pump blade provided the pressure distributions on streamlines are known either from an analytic or numerical solution or from experimental measurements of the flow in the absence of cavitation. Such investigations would allow one to examine both the location and intensity of bubble collapse in order to learn more about the potential for cavitation damage.

These methods, however, have some serious limitations. First, the Rayleigh-Plesset equation is only valid for spherical bubbles and collapsing bubbles lose their spherical symmetry. Consequently any investigation of damage requires considerations beyond those of the Rayleigh-Plesset equation. Secondly, the analysis described above assumes that the concentration of bubbles is sufficiently small so that bubbles do not interact and are not sufficiently numerous to change the flow field from that for non-cavitating flow. This means that they are of little value in predicting the effect of cavitation on pump performance since such an effect implies interactions between the bubbles and the flow field.

It follows that to model the performance loss due to traveling bubble cavitation one must use a two-phase or two-component flow model which implicitly includes interaction between the bubbles and the liquid flow field. One of the first models of this kind was investigated by Cooper (1967) and there have been a number of similar investigations for two-component flows in pumps, for example, that by Rohatgi (1978). While these investigations are useful, they are subject to serious limitations. In particular, they assume that the two-phase mixture is in thermodynamic equilibrium. Such is certainly not the case in cavitation flows where an expression like the Rayleigh-Plesset equation is needed to describe the dynamics of disequilibrium. Nevertheless the models of Cooper and others have value as the first coherent attempts to evaluate the effects of traveling bubble cavitation on pump performance.

Rather than the assumption of thermodynamic equilibrium, two-phase bubble flow models need to be developed in which the bubble dynamics are included through appropriate use of the Rayleigh-Plesset equation. In recent years a number of investigators have employed such models to investigate the dynamics and acoustics of clouds of cavitation bubbles in which the bubbles and the flow interact (see, for example, Chahine 1982, d'Agostino and Brennen 1983, 1989, d'Agostino *et* al. 1988, Biesheuval and van Wijngaarden 1984, Omta 1987). Among other things these investigations demonstrate that a cloud of bubbles has a set of natural frequencies of its own, separate from (but related to) the bubble natural frequency and that the bubble and flow interaction effects become important when the order of magnitude of the parameter $\alpha A^2/R^2$ exceeds unity where α is the void fraction and A and R are the dimensions of the cloud and bubbles respectively. These more appropriate models for travelling bubble cavitation have not, as yet, been used to investigate cavitation effects in pumps.

Several other concepts should be mentioned before we leave the subject of bubbly cavitation in pumps. One such concept which has not received the attention it deserves was put forward by Jakobsen (1964). He attempted to merge the free streamline models (which are discussed later in section 8.8) with his observations that attached cavities on the suction surfaces of impeller blades tend to break up into bubbly mixtures near the closure or reattachment point of the attached cavity. Jakobsen suggested that condensation shocks occur in this bubbly mixture and constitute a mechanism for head breakdown.

One area in which the Rayleigh-Plesset analysis has been useful is in evaluating the differences in the cavitation occurring in different liquids and in the same liquid at different temperatures. These issues will be addressed in the next subsection. In the subsequent subsection we turn our attention to the free streamline methods which have been developed to model the flows which occur when large attached cavities or gas-filled voids occur on the blades of a turbomachine.

8.7 Thermal Effect on Pump Performance

Changes in the temperature of the liquid being pumped will clearly affect the vapor pressure, p_V, and therefore the NPSH or cavitation number. This effect has, of course, already been incorporated in the analysis or presentation of the performance by using the difference between the inlet pressure and the vapor pressure rather than the absolute value of the inlet pressure as a flow parameter. But there is another effect of the liquid temperature which is not so obvious and requires some discussion and analysis. It is illustrated by figure 54 which includes cavitation performance data for a centrifugal pump (Arndt 1981 from Chivers 1969) operating with water at different inlet temperatures. Note that the cavitation breakdown decreases substantially with increasing temperature. Somewhat counter-intuitively the performance actually improves as the temperature gets greater! The variation of the breakdown cavitation number, σ_b, with the inlet temperature in Chivers' (1969) experiments is shown in figure 55 which includes data for two different speeds and shows a consistent decrease in σ_b with increasing temperature. The data for the two speeds deviate somewhat at the lowest temperatures. To illustrate that the thermal effect occurs in other liquids and in other kinds of pumps, we include in figure 56 data reported by Gross (1973) from tests of the Saturn J-2 liquid oxygen inducer pump. This shows the same pattern manifest in figure 55. Other data of this kind has been obtained by Stepanoff (1961), Spraker (1965) and Salemann (1959) for a variety of other liquids.

The explanation for this effect is most readily given by making reference to traveling bubble cavitation though it can be extended to other forms of cavitation. However, for simplicity, consider a single bubble (or nucleus) which begins to grow when it enters a region of low pressure. Liquid on the surface of the bubble will vaporize to provide the increase in volume of vapor filling the bubble. Consider, now, what happens at two different temperatures, one "high" and one "low." At "low" temperatures the density of the saturated vapor is low and, therefore, the mass rate of evaporation of liquid needed is small. Consequently, the rate at which heat is needed as latent heat to effect this vaporization is low. Since the heat will be conducted from the bulk of the liquid and since the rate of heat transfer is small, this means that the amount by which the temperature

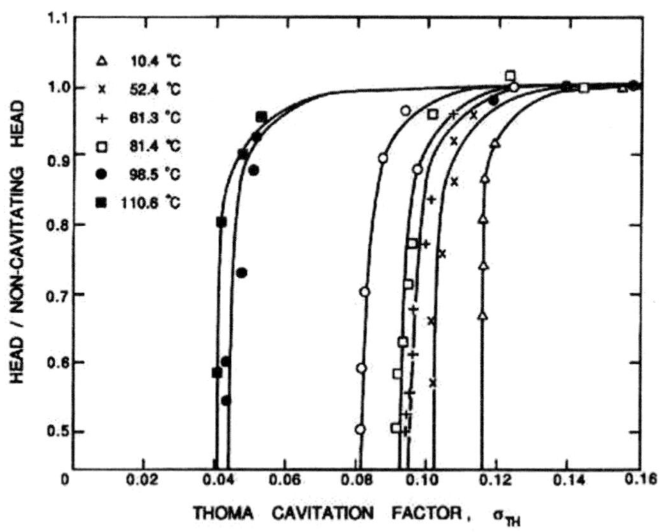

Figure 54. Typical cavitation performance characteristics for a centrifugal pump pumping water at various temperatures as indicated (Arndt 1981 from Chivers 1969).

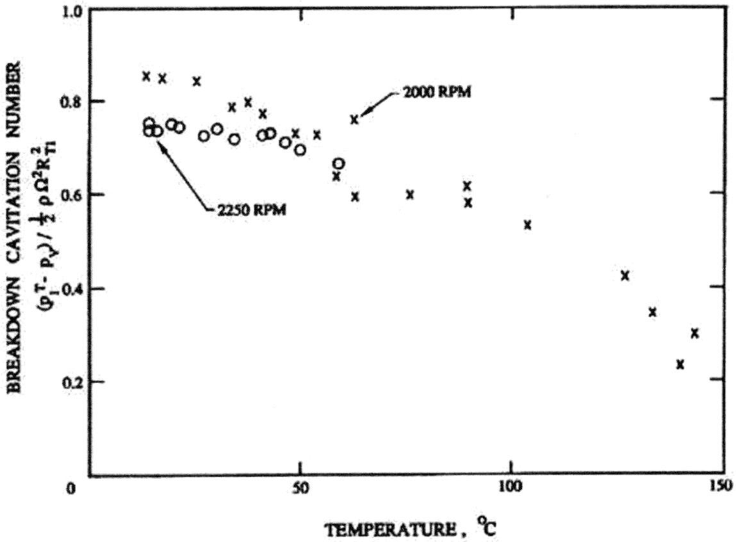

Figure 55. Thermodynamic effect on cavitation breakdown for a commercial centrifugal pump (data from Chivers 1969).

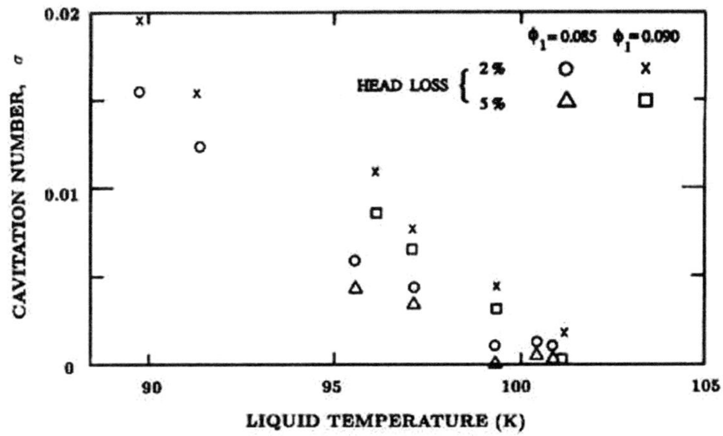

Figure 56. Effect of temperature on the cavitation performance of the J-2 liquid oxygen inducer pump (adapted from Gross 1973).

of the interface falls below the bulk liquid temperature is also small. Consequently the vapor pressure in the cavity only falls slightly below the value of the vapor pressure at the bulk liquid temperature. Therefore, the driving force behind the bubble growth, namely the difference between the internal pressure (vapor pressure) and the pressure far from the bubble, is not much influenced by thermal effects.

Now, consider the same phenomenon occuring at the "high" temperature. Since the vapor density can be many orders of magnitude larger than at the "low" temperature, the mass rate of evaporation for the same volume growth rate is much larger. Thus the heat which must be conducted to the interface is much larger which means that a substantial thermal boundary layer builds up in the liquid at the interface. This causes the temperature in the bubble to fall well below that of the bulk liquid and this, in turn, means that the vapor pressure within the bubble is much lower than otherwise might be expected. Consequently, the driving force behind the bubble growth is reduced. This reduction in the rate of bubble growth due to thermal effects is the origin of the thermal effect on the cavitation performance in pumps. Since the cavitation head loss is primarily due to disruption of the flow by volumes of vapor growing and collapsing within the pump, any reduction in the rate of bubble growth will lessen the disruption and result in improved performance.

This thermal effect can be extended to attached or blade cavities with only minor changes in the details. At the downstream end of a blade cavity, vapor is entrained by the flow at a certain volume rate which will depend on the flow velocity and other geometric parameters. At higher temperatures this implies a larger rate of entrainment of mass of vapor due to the larger vapor density. Since vaporization to balance this entrainment is occurring over the surface of the cavity, this implies a larger temperature difference at the higher temperature. And this implies a lower vapor pressure in the cavity than might otherwise be expected and hence a larger "effective" cavitation number. Consequently the cavitation performance is improved at the higher temperature.

Both empirical and theoretical arguments have been put forward in attempts to quantify these thermal effects. We shall begin with the theoretical arguments put forward by Ruggeri and Moore (1969) and by Brennen (1973). These explicitly apply to bubble cavitation and proceed as follows.

At the beginning of bubble growth, the rate of growth rapidly approaches the constant asymptotic value of dR/dt given by the Rayleigh-Plesset equation in the absence of thermal effects (Brennen 1994, 1995). On the other hand, the thermal term in that equation, which is initially zero, will grow like $t^{\frac{1}{2}}$. Consequently there will be a critical time, t_C, at which the thermal term will become important and will begin to reduce the rate of bubble growth. Thereafter the growth rate of the bubble radius will be reduced and, indeed, will decline with time like $t^{-\frac{1}{2}}$. This dependence on time is typical of that observed in boiling.

Now consider a nucleus or bubble passing through the pump which it will do in a time of order $1/\Omega\phi$. It follows that if $\Omega\phi \gg 1/t_C$ then the bubble growth will not be inhibited by thermal effects and explosive cavitating bubble growth will occur with the potential of causing substantial disruption of the flow and degradation of pump performance. On the other hand, if $\Omega\phi \ll 1/t_C$, most of the bubble growth will be thermally inhibited and the cavitation performance will be much improved.

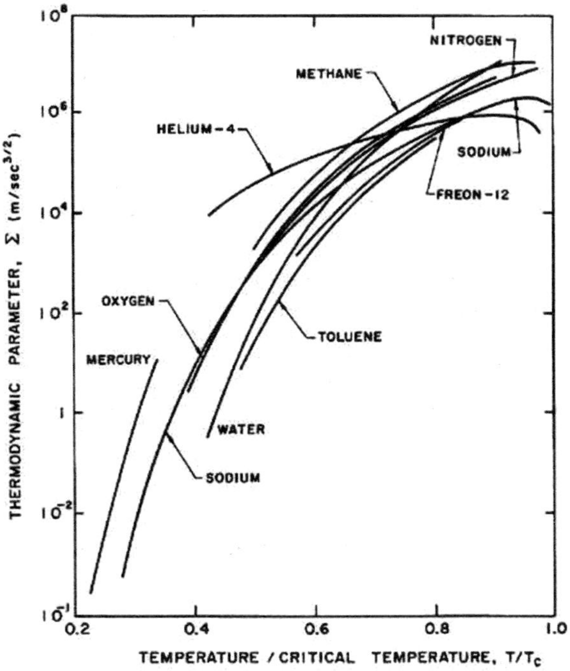

Figure 57. Thermodynamic parameter, Σ, as a function of temperature for various saturated fluids.

To calculate t_C we need values of a quantity, Σ, which is a function only of the liquid temperature; t_C is roughly inversely proportional to Σ^2. Typical values of Σ for a variety of liquids are presented in figure 57 as a function of temperature (the ratio of temperature to critical temperature is used in order to show all the fluids on the same graph). The large changes in the value of Σ are caused primarily by the change in the vapor density with temperature.

As an example, consider a cavitating flow of water in which the typical pressure difference causing bubble growth is of the order of $10^4 kg/m\ s^2$ or $0.1bar$. Water at $20°C$ has a value of Σ of about $1m/s^{\frac{3}{2}}$, and the value of t_C is then of the order of $10s$. Thus, in virtually all pumps, $\Omega\phi$ will be much greater than $1/t_C$ and no thermal effect will occur. On the other hand at $100°C$, the value of Σ for water is about $10^3 m/s^{\frac{3}{2}}$ and it follows that $t_C = 10\mu s$. Thus in virtually all cases $\Omega\phi \ll 1/t_C$ and a strong thermal effect can be expected. In fact, in a given application there will exist a "critical" temperature above which one should expect a thermal effect on cavitation. For a water pump rotating at $3000rpm$ this "critical" temperature is about $70°C$, a value which is consistent with the experimental measurements of pump performance.

The principal difficulty with the above approach is in finding some way to accurately evaluate t_C. Alternatively, the experimental data could be examined for guidance in establishing a criterion based on the above model. Brennen (1994) proposes that the

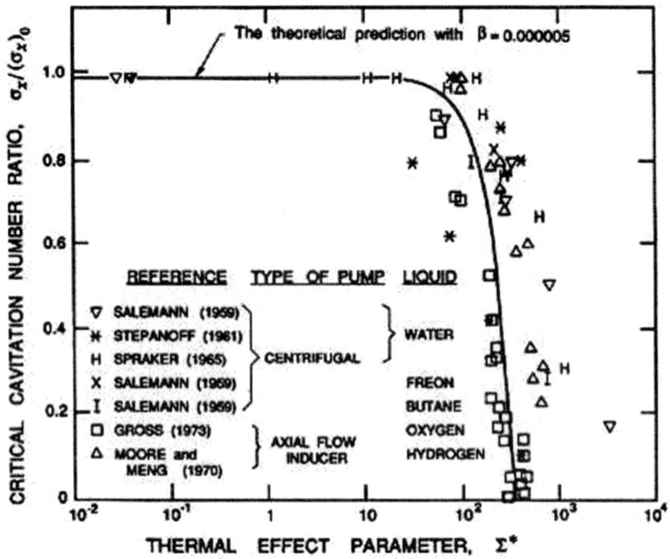

Figure 58. The ratio of the critical cavitation number σ_x (σ_a or σ_b) to $(\sigma_x)_0$ (the value of σ_x in the absence of any thermal effect) as a function of the thermal effect parameter, Σ^*. Data is shown for a variety of pumps and liquids.

ratio of critical cavitation numbers, $\sigma_x/(\sigma_x)_0$, should be a simple function of the modified thermal effect parameter, Σ^*, defined by

$$\Sigma^* = \Sigma / \left\{ R_T^2 \Omega^3 \phi (\sigma_x)_0 \right\}^{\frac{1}{2}} \tag{81}$$

To test this hypothesis, data from a range of different experiments is presented in figure 58. It can be seen that the data correspond very roughly to some kind of common curve for all these different pumps and liquids. Consequently this attempt to model the thermal effect has succeeded to a limited degree.

A number of purely empirical approaches to the same problem have been suggested in the past. All these empirical methods seek to predict the change in $NPSH$, say $\Delta NPSH$, due to the thermal effect. This quantity $\Delta NPSH$ is the increment by which the cavitation performance characteristic would be shifted to the left as a result of the thermal effect. The method suggested by Stahl and Stepanoff (1956) and Stepanoff (1961, 1964) is widely used; it is based on the premise that the cavitation characteristic of a particular pump operating at a particular speed with two different liquids (or with two different temperatures in the same liquid) would be horizontally shifted by

$$\Delta NPSH = H_{T1} - H_{T2} \tag{82}$$

where the quantities H_{T1} and H_{T2} only depend on the thermodynamic properties of the two individual fluids considered separately. This generic property is denoted by H_T.

For convenience, Stepanoff also uses the symbol B' to denote the group $\rho_L^2 c_{PL} T_\infty / \rho_V^2 \mathcal{L}^2$ which also occurs in Σ and almost all analyses of the thermal effect. Then, by examining data from a number of single-stage 3500 *rpm* pumps, Stahl and Stepanoff arrived at an empirical relation between H_T and the thermodynamic properties of the following form:

$$H_T \quad (\text{in } m) = 28.9\rho_L g/p_V (B')^{\frac{4}{3}} \tag{83}$$

where $p_V/g\rho_L$ is the vapor pressure head (in m) and B' is in m^{-1}. Clearly equation (83) can be used to find H_T for the desired liquid and operating temperature and for the reference liquid at the reference operating temperature. Then the cavitation performance under the desired conditions can be obtained by application of the shift given by equation (82) to the known cavitation performance under the reference conditions.

8.8 Free Streamline Methods

The diversity of types of cavitation in a pump and the complexity of the two-phase flow which it generates mean that reliable analytical methods for predicting the cavitating performance characteristics are virtually non-existent. However if the cavity flow can be approximated by single, fully developed or attached cavities on each blade, then this allows recourse to the methods of free streamline theory for which the reader may wish to consult the reviews by Tulin (1964) and Wu (1972) or the books by Birkhoff and Zarantonello (1957) and Brennen (1995). The analytical approaches can be subdivided into linear theories which are applicable to slender, streamlined flows (Tulin 1964) and non-linear theories which are more accurate but can be mathematically much more complex (Wu 1972). Both approaches to free streamline flows have been used in a wide range of cavity flow problems and it is necessary to restrict the present discussion to some of the solutions of relevance to attached cavitation in pumps.

It is instructive to begin by quoting some of the results obtained for single hydrofoils for which the review by Acosta (1973) provides an excellent background. In particular we will focus on the results of approximate linear theories for a partially cavitating or supercavitating flat plate hydrofoil. The partially cavitating solution (Acosta 1955) yields a lift coefficient

$$C_L = \pi\alpha \left[1 + (1-\ell)^{-\frac{1}{2}}\right] \tag{84}$$

where ℓ is the ratio of the cavity length to the chord of the foil and is related to the cavitation number, σ, by

$$\frac{\sigma}{2\alpha} = \frac{2 - \ell + 2(1-\ell)^{\frac{1}{2}}}{\ell^{\frac{1}{2}}(1-\ell)^{\frac{1}{2}}} \tag{85}$$

Thus, for a given cavity length, ℓ, and a given angle of incidence, α, the cavitation number follows from equation (85) and the lift coefficient from equation (84). Note that as $\ell \to 0$ the value of C_L tends to the theoretical value for a non-cavitating flat plate, namely $2\pi\alpha$. The corresponding solution for a supercavitating flat plate was given by Tulin (1953) in his pioneering paper on linearized cavity flows. In this case

$$C_L = \pi\alpha\ell \left[\ell^{\frac{1}{2}}(\ell-1)^{-\frac{1}{2}} - 1\right] \tag{86}$$

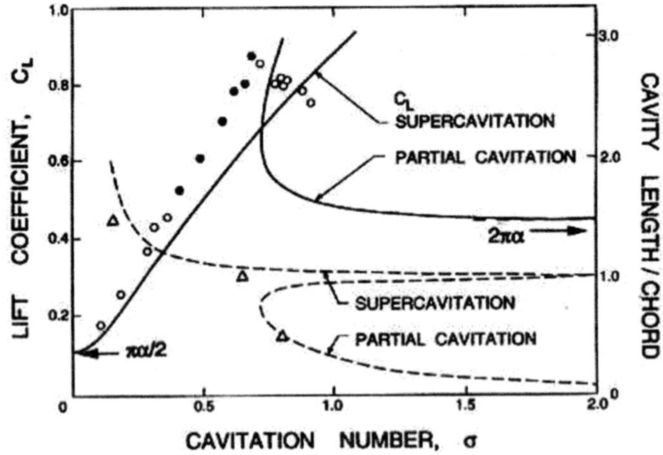

Figure 59. Typical results from the linearized theories for a cavitating flat plate at an angle of incidence of 4°. The lift coefficients, C_L (solid lines), and the ratios of cavity length to chord, ℓ (dashed lines), are from the supercavitation theory of Tulin (1953) and the partial cavitation theory of Acosta (1955). Also shown are the experimental results of Wade and Acosta (1966) for ℓ (△) and for C_L (○ and •) where the open symbols represent points of stable operation and the solid symbols denote points of unstable cavity operation.

$$\alpha\left(\frac{2}{\sigma}+1\right) = (\ell-1)^{\frac{1}{2}} \tag{87}$$

where now, of course, $\ell > 1$.

The lift coefficient and the cavity length from equations (84) to (87) are plotted against cavitation number in figure 59 for a typical angle of incidence of $\alpha = 4°$. Note that as $\sigma \to \infty$ the fully wetted lift coefficient, namely $2\pi\alpha$, is recovered from the partial cavitation solution and that as $\sigma \to 0$ the lift coefficient tends to $\pi\alpha/2$. Notice also that both the solutions become pathological when the length of the cavity approaches the chord length ($\ell \to 1$). However, if some small portion of each curve close to $\ell = 1$ were eliminated, then the characteristic decline in the performance of the hydrofoil as the cavitation number is decreased is readily observed. It also compares well with the experimental observations as illustrated by the favorable comparison with the data of Wade and Acosta (1966) included in figure 59. Consequently, as the cavitation number is decreased, a single foil exhibits only a small change in the performance or lift coefficient until some critical value of σ (about 0.7 in the case of figure 59) is reached. Below this critical value the performance begins to "breakdown" quite rapidly. Thus, even a single foil mirrors the typical cavitation performance experienced in a pump. On a more detailed level, note that the small increase in the supercavitating lift coefficient which occurs as the cavitation number is decreased toward the critical value of σ is, in fact, observed experimentally with many single hydrofoils (for example, Wade and Acosta 1966) as well

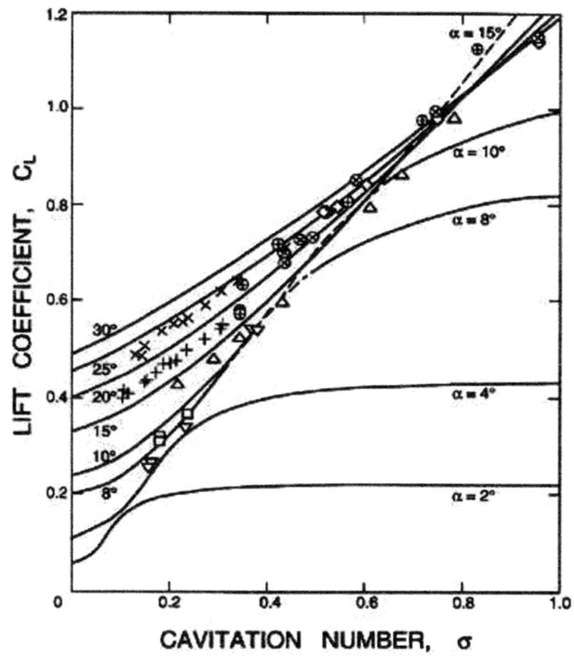

Figure 60. Lift coefficients for a flat plate from the non-linear theory of Wu (1962). The experimental data (Parkin 1958) is for angles of incidence as follows: 8° (▽), 10° (□), 15° (△), 20° (⊕), 25° (⊗), and 30° (◇). Also shown is some data of Silberman (1959) in a free jet tunnel: 20° (+) and 25° (×).

as in some pumps. The peculiar behaviour of the analytical solutions close to the critical cavitation number is related to an instability which is observed when the cavity length is of the same order as the chord of the foil.

Before leaving the subject of the single cavitating foil we should note that more exact, non-linear solutions for a flat plate or an arbitrarily shaped profile have been generated by Wu (1956, 1962), Mimura (1958) and others. As an example of these non-linear results, the lift and drag coefficients at various cavitation numbers and angles of incidence are presented in figures 60 and 61 where they are compared with the experimental data of Parkin (1958) and Silberman (1959). Data both for supercavitating and partially cavitating conditions are shown in these figures, the latter occurring at the higher cavitation numbers and lower incidence angles (the dashed parts of the curves represent a somewhat arbitrary smoothing through the critical region in which the cavity lengths are close to the chord length). This comparison demonstrates that the non-linear theory yields values which are in good agreement with the experimental measurements. In the case of circular-arc hydrofoils, Wu and Wang (1964) have shown similar agreement with the data of Parkin (1958) for this type of profile. For a recent treatment of supercavitating single foils the reader is referred to the work of Furuya and Acosta (1973).

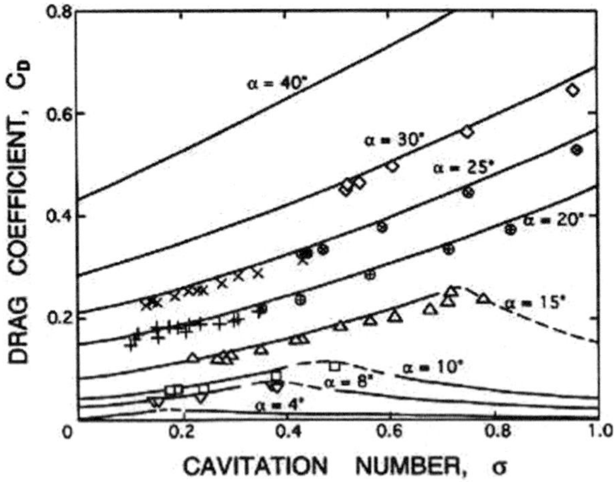

Figure 61. Drag coefficients corresponding to the lift coefficients of figure 60.

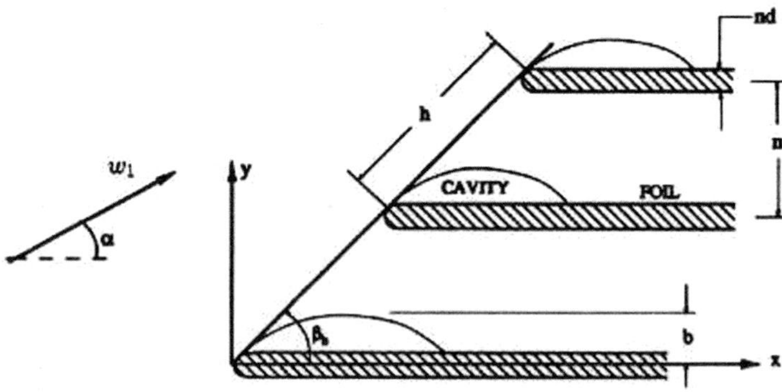

Figure 62. Schematic of partially cavitating cascade of flat blades of thickness nd (Brennen and Acosta 1973).

8.9 Partially Cavitating Cascades

In the context of pumps, the solutions by Acosta and Hollander (1959) and Stripling and Acosta (1962) of partial cavitation in a semi-infinite cascade of infinitely thin blades and the solution by Wade (1967) of a finite cascade of partially cavitating foils provide a particularly valuable means of analyzing the performance of two-dimensional cascades with blade cavities. More recently the three dimensional aspects of these solutions have been explored by Furuya (1974). As a complement to purely analytical methods, more heuristic approaches are possible in which the conventional cascade analyses (see sections 4.2, 4.5) are supplemented by lift and drag data for blades operating under cavitating conditions.

Partly for the purposes of example and partly because the results are useful, we shall recount here the results of the free-streamline solution of Brennen and Acosta (1973). This is a slightly modified version of the Acosta and Hollander solution for partial cavitation in a cascade of infinitely thin, flat blades. The modification was to add finite thickness to the blades. As we shall see, this can be important in terms of the relevance of the theory.

A sketch of the cascade geometry is shown in figure 62. A single parameter is introduced to the solution in order to yield finite blade thickness. This parameter implies a ratio, d, of the blade thickness far downstream to the normal spacing between the blades. It also implies a radius of curvature of the parabolic leading edge of the blade, κ, given by

$$\frac{1}{\kappa} \approx \frac{d^2 \beta_b^3}{\pi h(1+\sigma_c)} \tag{88}$$

where σ_c is the choked cavitation number (see below). Equation (88) and the fact that the ultimate thickness is not reached until about half a blade spacing downstream, both imply very sharp leading edges.

One of the common features of all of these free streamline solutions is that there exists a certain minimum cavitation number at which the cavity becomes infinitely long and below which there are no solutions. This minimum cavitation number is called the choked cavitation number, σ_c. Were such a flow to occur in practice, it would permit large deviation angles at discharge and a major degradation of performance. Consequently the choked cavitation number, σ_c, is often considered an approximation to the breakdown cavitation number, σ_b, for the pump flow which the cascade solution represents. The Brennen and Acosta solution yields a choked cavitation number given by

$$\sigma_c = \left[1 + 2\sin\frac{\alpha}{2}\sec\frac{\beta_b}{2}\sin\frac{(\beta_b-\alpha)}{2} + 2d\sin^2\frac{\beta_b}{2}\right]^2 - 1 \tag{89}$$

which, since the solution is only valid for small incidence angles, α, and since β_b is normally small, yields

$$\sigma_c \approx \alpha(\beta_b - \alpha) + \beta_b^2 d \tag{90}$$

Furthermore, at a general cavitation number, σ, the maximum thickness, b, of the cavity is given by

$$\frac{b}{n} = 2\pi\left[d - (1+\sigma)^{\frac{1}{2}} + \sin(\beta_b - \alpha)/\sin\beta_b\right] \tag{91}$$

or

$$\frac{b}{n} \approx 2\pi\left[d - \frac{\alpha}{\beta_b} - \frac{\sigma}{2}\right] \tag{92}$$

As a rough example, consider a 10° helical inducer ($\beta_b = 10°$) with a fractional blade thickness of $d = 0.15$ operating at a flow coefficient, $\phi = 0.08$, so that the incidence angle, $\alpha = 4°$. Then, according to the relation (90), the choked cavitation number is $\sigma_c = 0.0119$ which is close to the experimentally observed breakdown cavitation number. It is important to note the role played by the blade thickness in this typical calculation because with $d = 0$ the result is $\sigma_c = 0.0073$. Note also that with infinitely thin blades,

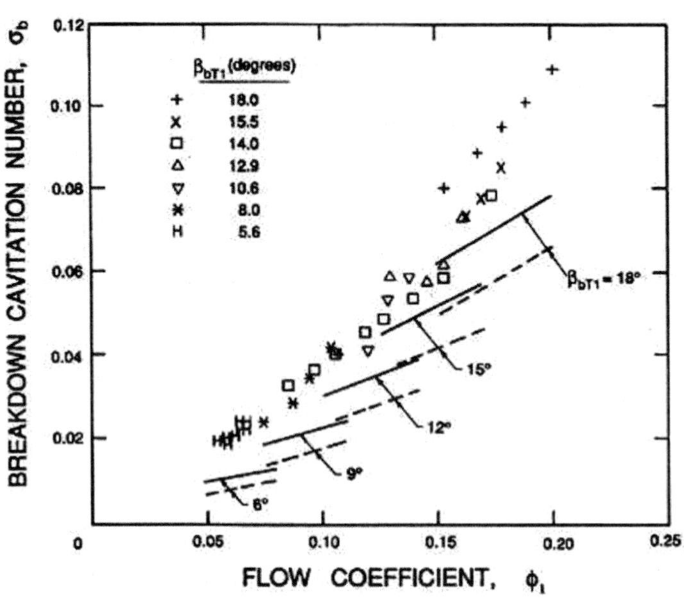

Figure 63. The breakdown cavitation numbers for a series of inducers by Stripling (1962) plotted against inlet flow coefficient. The inducers have inlet blade angles at the tip, β_{bT1} (in degrees), as indicated. Also shown are the results of the cascade analysis (equation (90)) applied at the rms radius, R_{RMS}, with blade thickness (solid line) and without blade thickness (dashed line).

the expression (90) predicts $\sigma_c = 0$ at zero incidence. Thus, the blade thickness is important in estimating the choked cavitation number in any pump.

Most pumps or inducer designs incorporate significant variations in α, β_b and d over the inlet plane and hence the above analysis has to be performed as a function of the inlet radial position as described by Brennen (1994).

Perhaps the most exhaustive experimental investigation of breakdown cavitation numbers for inducers is the series of experiments reported by Stripling (1962) in which inducers with blade angles at the tip, β_{bT1}, varying from 5.6° to 18°, various leading edge geometries, blade numbers of 3 and 4 and two hub-to-tip ratios were investigated. Some of Stripling's experimental data is presented in figure 63 where the σ_b values are plotted against the flow coefficient, ϕ_1. In his paper Stripling argues that the data correlate with the parameter $\phi_1 \sin \beta_{bT1}/(1 + \cos \beta_{bT1})$ but, in fact, the experimental data are much better correlated with ϕ_1 alone as demonstrated in figure 63. There is no satisfactory explanation for the fact that σ_b correlates better with ϕ_1.

Stripling correlates his data with the theoretical values of the choked cavitation number which one would obtain from the above theory in the case of infinitely thin blades. (In this limit the expression for σ_c is more easily obtained by simultaneous solution of the Bernoulli equation and an equation for the momentum parallel with the blades as Stripling demonstrates.) More specifically, Stripling uses the blade angles, β_{b1}, and incidence angles at the rms radius, R_{RMS}, where

$$R_{RMS} = \left[\frac{1}{2}\left(R_{T1}^2 + R_{H1}^2\right)\right]^{\frac{1}{2}} \tag{93}$$

His theoretical results then correspond to the dashed lines in figure 63. When the blade thickness term is added as in equation (90) the choked cavitation numbers are given by the solid lines in figure 63 which are considerably closer to the experimental values of σ_b than the dashed lines. The remaining discrepancy could well be due to the fact that the σ_c values are larger at some radius other than R_{RMS} and hence breakdown occurs first at that other radius.

Up to this point we have only discussed the calculation of the choked or breakdown cavitation number from the analysis of a partially cavitating cascade. There remains the issue of how to predict the degradation in the head or the cavitation head losses prior to breakdown. The problem here is that calculation of the lift from these analyses produces little for, as one could anticipate, a small partial cavity will not significantly alter the performance of a cascade of higher solidity since the discharge, with or without the cavity, is essentially constrained to follow the direction of the blades. The hydraulic losses which one seeks are additional (or possibly negative) frictional losses generated by the disruption to the flow caused by the cavitation. A number of authors, including Stripling and Acosta (1962), have employed modifications to cascade analyses in order to evaluate the loss of head, ΔH, due to cavitation. One way to view this loss is to recognize that the presence of a cavity in the blade passage causes a reduction in the cross-sectional area available to the liquid flow. When the cavity collapses this area increases creating a "diffuser" which is not otherwise present. Hydraulic losses in this "diffuser" flow could be considered responsible for the cavitation head loss and could be derived from knowledge of the cavity blockage, b/n.

8.10 Cavitation Performance Correlations

Finally we provide brief mention of several of the purely empirical methods which are used in practice to generate estimates of the cavitation head loss in pumps. These often consist of an empirical correlation between the cavitation head loss, ΔH, the net positive suction head, $NPSH$, and the suction specific speed, S. Commonly this correlation is written as

$$\Delta H = P(S) \times NPSH \tag{94}$$

where the dimensionless parameter, $P(S)$, is established by prior experience. A typical function, $P(S)$, is presented in figure 64. Such methods can only be considered approximate; there is no fundamental reason to believe that $\Delta H/NPSH$ is a function only of the suction specific speed, S, for all pumps though it will certainly correlate with that parameter for a given pump and a given liquid at a given Reynolds number and a given temperature. A more informed approach is to select a value of the cavitation number, σ_W, which is most fundamental to the interaction of the flow and the pump blade namely

$$\sigma_W = (p_1 - p_V)/\frac{1}{2}\rho_L w_1^2 \tag{95}$$

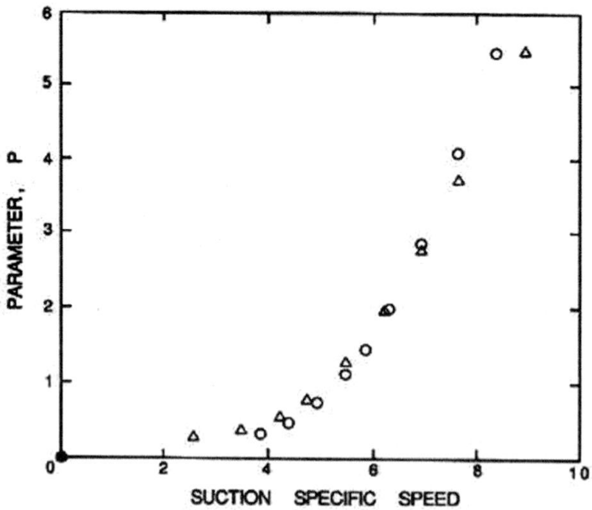

Figure 64. Some data on the cavitation head loss parameter, $P = \Delta H/NPSH$, for axial inducer pumps. The two symbols are for two different pumps.

Then, using the definition of $NPSH$ (section 6.2) and the velocity triangle,

$$NPSH = \left((1+\sigma_W)v_{m1}^2 + \sigma_W \Omega^2 R_{T1}^2\right)/2g \tag{96}$$

It is interesting to observe that the estimate of the cavitation-free $NPSH$ for mixed flow pumps obtained empirically by Gongwer (1941) namely

$$\left(1.8v_{m1}^2 + 0.23\Omega^2 R_{T1}^2\right)/2g \tag{97}$$

and his estimate of the breakdown $NPSH$ namely

$$\left(1.49v_{m1}^2 + 0.085\Omega^2 R_{T1}^2\right)/2g \tag{98}$$

correspond quite closely to specific values of σ_W, namely $\sigma_W \approx 0.3$ and $\sigma_W \approx 0.1$ respectively.

9 Pump vibration

9.1 Introduction

The trend toward higher speed, high power density liquid turbomachinery has inevitably increased the potential for fluid/structure interaction problems, and the severity of those problems. Even in the absence of cavitation and its complications, these fluid structure interaction phenomena can lead to increased wear and, under the worst conditions, to structural failure. Exemplifying this trend, the Electrical Power Research

Institute (Makay and Szamody 1978) has recognized that the occurrence of these problems in boiler feed pumps has contributed significantly to downtime in conventional power plants.

Unlike the cavitation issues, unsteady flow problems in liquid turbomachines do not have a long history of research. In some ways this is ironic since, as pointed out by Ek (1957) and Dean (1959), the flow within a turbomachine must necessarily be unsteady if work is to be done on or by the fluid. Yet many of the classical texts on pumps or turbines barely make mention of unsteady flow phenomena or of design considerations that might avoid such problems. In contrast to liquid turbomachinery, the literature on unsteady flow problems in gas turbomachinery is considerably more extensive, and there are a number of review papers that provide a good survey of the subject (for example, McCroskey 1977, Cumpsty 1977, Mikolajczak *et* al. 1975, Platzer 1978, Greitzer 1981). We will not attempt a review of this literature but we will try, where appropriate, to indicate areas of useful cross-reference. It is also clear that this subject incorporates a variety of problems ranging, for example, from blade flutter to fluid-induced rotordynamic instability. Because of this variety and the recent vintage of the fundamental research, no clear classification system for these problems has yet evolved and there may indeed be some phenomena that have yet to be properly identified. It follows that the classification system that we will attempt here will be tentative, and not necessarily comprehensive. Nevertheless, it seems that three different categories of flow oscillation can occur, and that there are a number phenomena within each of the three categories. We briefly list them here and return to some in the sections that follow.

[A] **Global Flow Oscillations.** A number of the identified vibration problems involve large scale oscillations of the flow. Specific examples are:

- [**A1**] Rotating stall or rotating cavitation occurs when a turbomachine is required to operate at a high incidence angle close to the value at which the blades may stall. It is often the case that stall will first be manifest on a small number of adjacent blades and that this "stall cell" will propagate circumferentially at some fraction of the main impeller rotation speed. This phenomenon is called rotating stall and is usually associated with turbomachines having a substantial number of blades (such as compressors). It has, however, also been reported in centrifugal pumps. When the turbomachine cavitates the same phenomenon may still occur, perhaps in some slightly altered form. Such circumstances will be referred to as "rotating stall with cavitation." But there is also a different phenomenon which can occur in which one or two blades manifest a greater degree of cavitation and this "cell" propagates around the rotor in a manner superficially similar to the propagation of rotating stall. This phenomenon is known as "rotating cavitation."
- [**A2**] Surge is manifest in a turbomachine that is required to operate under highly loaded circumstances where the slope of the head rise/flow rate curve is positive. It is a system instability to which the dynamics of all the components of the system (reservoirs, valves, inlet and suction lines and turbomachine) contribute. It results in pressure and flow rate oscillations throughout the system. When cavitation is present the phenomenon is termed "cavitation surge" and can occur even when the slope of the head rise/flow rate curve is

negative.

[A3] Partial cavitation or supercavitation can become unstable when the length of the cavity approaches the length of the blade so that the cavity collapses in the region of the trailing edge. Such a circumstance can lead to violent oscillations in which the cavity length oscillates dramatically.

[A4] Line resonance occurs when one of the blade passing frequencies in a turbomachine happens to coincide with one of the acoustic modes of the inlet or discharge line. The pressure oscillation magnitudes associated with these resonances can often cause substantial damage.

[A5] It has been speculated that an axial balance resonance could occur if the turbomachine is fitted with a balance piston (designed to equalize the axial forces acting on the impeller) and if the resonant frequency of the balance piston system corresponds with the rotating speed or some blade passing frequency. Though there exist several apocryphal accounts of such resonances, the phenomenon has yet to be documented experimentally.

[A6] Cavitation noise can sometimes reach a sufficient amplitude to cause resonance with structural frequencies of vibration.

[A7] The above items all assume that the turbomachine is fixed in a non-accelerating reference frame. When this is not the case the dynamics of the turbomachine may play a crucial role in generating an instability that involves the vibration of that machine as a whole. Such phenomena, of which the Pogo instabilities are, perhaps, the best documented examples, are described further in section 9.11.

[B] Local Flow Oscillations. Several other vibration problems involve more localized flow oscillations and vibration of the blades:

[B1] Blade flutter. As in the case of airfoils, there are circumstances under which an individual blade may begin to flutter (or diverge) as a consequence of the particular flow condition (incidence angle, velocity), the stiffness of the blade, and its method of support.

[B2] Blade excitation due to rotor-stator interaction. While [B1] would occur in the absence of excitation it is also true that there are a number of possible mechanisms of excitation in a turbomachine that can cause significant blade vibration. This is particularly true for a row of stator blades operating just downstream of a row of impeller blades or vice versa. The wakes from the upstream blades can cause a serious vibration problem for the downstream blades at blade passing frequency or some multiple thereof. Non-axisymmetry in the inlet, the volute, or housing can also cause excitation of impeller blades at the impeller rotation frequency.

[B3] Blade excitation due to vortex shedding or cavitation oscillations. In addition to the excitation of [B2], it is also possible that vortex shedding or the oscillations of cavitation could provide the excitation for blade vibrations.

[C] Radial and Rotordynamic Forces. Global forces perpendicular to the axis of rotation can generate several types of problem:

[C1] Radial forces are forces perpendicular to the axis of rotation caused by circumferential nonuniformities in the inlet flow, casing, or volute. While these may be stationary in the frame of the pump housing, the loads that act on the impeller and, therefore, the bearings can be sufficient to create wear, vibration, and even failure of the bearings.

[C2] Fluid-induced rotordynamic forces occur as the result of movement of the axis of rotation of the impeller-shaft system of the turbomachine. Contributions to these rotordynamic forces can arise from the seals, the flow through the impeller, leakage flows, or the flows in the bearings themselves. Sometimes these forces can cause a reduction in the critical speeds of the shaft system, and therefore an unforeseen limitation to its operating range. One of the common characteristics of a fluid-induced rotordynamic problem is that it often occurs at subsynchronous frequency.

Radial and rotordynamic forces are not addressed in this review; the reader is referred to Brennen (1994) for coverage of that subject. System dynamic analyses and instabilities are addressed in the next section. The remainder of this section will briefly describe some of the other unsteady problems encountered in liquid turbomachines.

Before leaving the issue of classification, it is important to emphasize that many of the phenomena that cause serious vibration problems in turbomachines involve an interaction between two or more of the above mentioned items. Perhaps the most widely recognized of these resonance problems is that involving an interaction between blade passage excitation frequencies and acoustic modes of the suction or discharge lines. But the literature contains other examples. For instance, Dussourd (1968) describes flow oscillations which involve the interaction of rotating stall and acoustic line frequencies. Another example is given by Marscher (1988) who investigated a resonance between the rotordynamic motions of the shaft and the subsynchronous unsteady flows associated with flow recirculation at the inlet to a centrifugal impeller.

9.2 Frequencies of Oscillation

One of the diagnostics which is often, but not always, useful in addressing a turbomachine vibration problem is to examine the dominant frequencies and to investigate how they change with rotating speed. Table 1 is intended as a rough guide to the kinds of frequencies at which the above problems occur. We have attempted to place the phenomena in rough order of increasing frequency partly in order to illustrate the fact that the frequencies can range all the way from a few Hz up to tens of kHz. Some of the phenomena scale with the impeller rotating speed, Ω. Others, such as surge, may vary somewhat with Ω but not linearly; still others, like cavitation noise, will be largely independent of Ω.

Before proceeding to a discussion of the specific vibrational problems outlined above, it may be valuable to illustrate the spectral content of the shaft vibration of a typical centrifugal pump in normal, nominally steady operation. Figure 65 presents examples of the spectra (for two frequency ranges) taken from the shaft of the five-bladed centrifugal Impeller X operating in the vaneless Volute A (no stator blades) at 300 rpm (5 Hz). Clearly the synchronous vibration at the shaft fundamental of 5 Hz dominates

Table 1. Typical frequency ranges of pump vibration problems.

VIBRATION CATEGORY		FREQUENCY RANGE
A2	Surge	System dependent, $3 - 10\ Hz$ in compressors
A2	Cavitation Surge	System dependent, $0.1 - 0.4\Omega$
A1	Rotor rotating stall	$0.5\Omega - 0.7\Omega$
A1	Vaneless diffuser stall	$0.05\Omega - 0.25\Omega$
A1	Rotating cavitation	$1.1\Omega - 1.2\Omega$
A3	Partial cavitation oscillation	$< \Omega$
C1	Excessive radial force	Some fraction of Ω
C2	Rotordynamic vibration	Fraction of Ω when critical speed is approached.
A4	Blade passing excitation (or B2)	Multiples of blade passing frequencies
B1	Blade flutter	Natural frequencies of blade in liquid
B3	Vortex shedding	Frequency of vortex shedding
A6	Cavitation noise	$1\ kHz - 20\ kHz$

the low frequencies; this excitation may be caused by mechanical imperfections in the shaft such as an imbalance or by circumferential nonuniformities in the flow such as might be generated by the volute. It is also clear that the most dominant harmonic of shaft frequency occurs at 5Ω because there are 5 impeller blades. Note, however, that there are noticeable peaks at 2Ω and 3Ω arising from significantly nonsinusoidal excitation at the shaft frequency, Ω. The other dominant peaks labelled $1 \rightarrow 4$ represent structural resonant frequencies unaffected by shaft rotational speed. At higher rotational speeds, more coincidence with structural frequencies occurs and the spectra contain more noise.

9.3 Unsteady Flows

Many of the phenomena listed in subsection 9.1 require some knowledge of the unsteady flows corresponding to the steady cascade flows discussed in subsections 4.2 and 4.5. In the case of non-cavitating axial cascades, a large volume of literature has been generated in the context of gas turbine engines, and there exist a number of extensive reviews including those by Woods (1961), McCroskey (1977), Mikolajczak *et* al. (1975) and Platzer (1978). Much of the analytical work utilizes linear cascade theory, for example, Kemp and Sears (1955), Woods (1955), Schorr and Reddy (1971), and Kemp and Ohashi (1975). Some of this has been applied to the analysis of unsteady flows in pumps and extended to cover the case of radial or mixed flow machines. For example, Tsukamoto and Ohashi (1982) utilized these methods to model the start-up transients in centifugal pumps and Tsujimoto *et* al. (1986) extended the analysis to evaluate the unsteady torque in mixed flow machines.

However, most of the available methods are restricted to lightly loaded cascades and

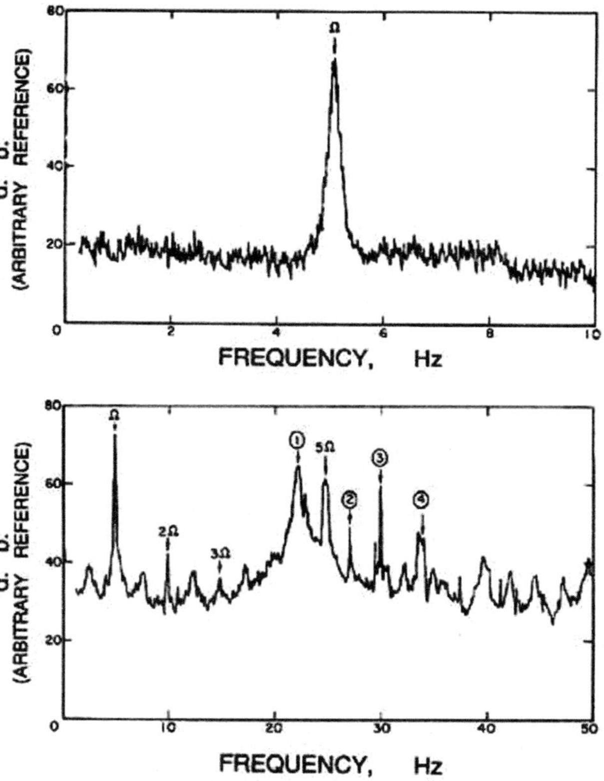

Figure 65. Typical spectra of vibration for a centrifugal pump (Impeller X/Volute A) operating at 300 *rpm* (Chamieh et al. 1985).

impellers at low angles of incidence. Other, more complex, theories (for example, Adamczyk 1975) are needed at larger angles of incidence and for highly cambered cascades when there is a strong coupling between the steady and unsteady flow (Platzer 1978). Moreover, most of the early theories were only applicable to globally uniform unsteady flows in which the blades all move in unison. Samoylovich (1962) appears to have been the first to consider oscillations with arbitrary interblade phase differences, the kind of analysis needed for flutter investigations (see below).

When the incidence angles are large so that the blades stall, one must resort to unsteady free streamline methods in order to model the flows (Woods 1961). Apart from the work of Sisto (1967), very little analytical work has been done on this problem which is of considerable importance in the context of turbomachinery. One of the fluid mechanical complexities is the unsteady or dynamic response of a separated flow that may lead to significant departures from the sucession of events one might construct based on a quasistatic approach. Some progress has been made in understanding the "dynamic stall" for a single foil (see, for example, Ham 1968). However, it would appear that more

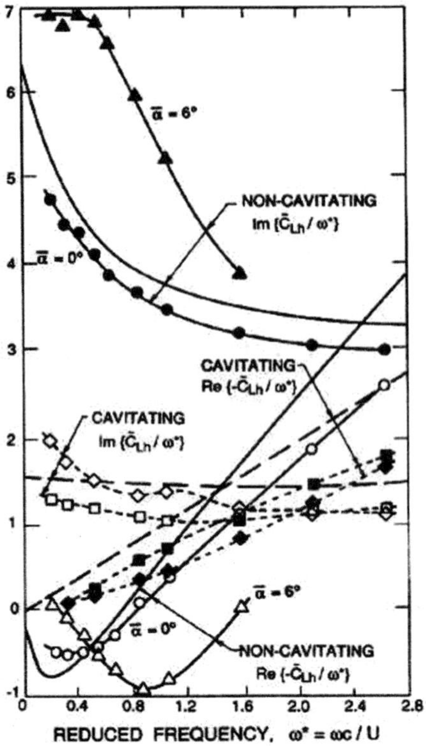

Figure 66. Fluctuating lift coefficient, $\tilde{C}_{Lh}$, for foils undergoing heave oscillations at a reduced frequency, $\omega^* = \omega c/U$. Real and imaginary parts of $\tilde{C}_{Lh}/\omega^*$ are presented for (a) non-cavitating flow at mean incidence angles of 0° and 6° (b) cavitating data for a mean incidence of 8°, for very long choked cavities (□) and for cavities 3 chords in length (◇). Adapted from Acosta and DeLong (1971).

work is needed to understand the complex dynamic stall phenomena in turbomachines.

Unsteady free streamline analyses can be more confidentally applied to the analysis of cavitating cascades because the cavity or free streamline pressure is usually known and constant whereas the corresponding pressure for the wake flows may be varying with time in a way that is difficult to predict. Thus, for example, the unsteady response for a single supercavitating foil (Woods 1957, Martin 1962, Parkin 1962) has been compared with experimental measurements by Acosta and DeLong (1971). As an example, we present (figure 66) some data from Acosta and DeLong on the unsteady forces on a single foil undergoing heave oscillations at various reduced frequencies, $\omega^* = \omega c/2U$. The oscillating heave motion, d, is represented by

$$d = Re\left\{\tilde{d}e^{j\omega t}\right\} \tag{99}$$

where the complex quantity, $\tilde{d}$, contains the amplitude and phase of the displacement. The resulting lift coefficient, C_L, is decomposed (using the notation of the next chapter) into

$$C_L = \bar{C}_L + Re\left\{\tilde{C}_{Lh} e^{j\omega t}\right\} \tag{100}$$

and the real and imaginary parts of $\tilde{C}_{Lh}/\omega*$ which are plotted in figure 66 represent the unsteady lift characteristics of the foil. It is particularly important to note that substantial departures from quasistatic behaviour occur for reduced frequencies as low as 0.2, though these departures are more significant in the noncavitating flow than in the cavitating flow. The lines without points in figure 66 present results for the corresponding linear theories and we observe that the agreement between the theory and the experiments is fairly good. Notice also that the $Re\{-\tilde{C}_{Lh}\}$ for noncavitating foils is negative at low frequencies but becomes positive at larger ω whereas the values in the cavitating case are all positive. Similar data for cavitating cascades would be necessary in order to analyse the potential for instability in cavitating, axial flow pumps. The author is not aware of any such data or analysis.

The information is similarly meagre for all of the corresponding dynamic characteristics of radial rather than axial cascades and, consequently, our ability to model dynamic instabilities in centrifugal pumps is very limited indeed.

9.4 Rotating Stall

Rotating stall is a phenomenon which may occur in a cascade of blades when these are required to operate at a high angle of incidence close to that at which the blades will stall. In a pump this usually implies that the flow rate has been reduced to a point close to or below the maximum in the head characteristic (see, for example, figure 44). Emmons *et* al. (1955) first provided a coherent explanation of propagating stall. The cascade in figure 67 will represent a set of blades (a rotor or a stator) operating at a high angle of incidence. Then, if blade B were stalled, this generates a separated wake and therefore increased blockage to the flow in the passage between blades B and A. This, in turn, would tend to deflect the flow away from this blockage in the manner indicated in the figure. The result would be an increase in the angle of incidence on blade A and a decrease in the angle of incidence on blade C. Thus, blade A would tend to stall while any stall on blade C would tend to diminish. Consequently, the stall "cell" would tend to move upwards in the figure or in a direction away from the oncoming flow. Of course, the stall cell could consist of a larger number of blades with more than one exhibiting increased separation or stall. The stall cell will rotate around the axis and hence the name "rotating stall." Moreover, the speed of propagation will clearly be some fraction of the circumferential component of the relative velocity, either $v_{\theta 1}$ in the case of a stator or $w_{\theta 1}$ in the case of a rotor. Consequently, in the case of a rotor, the stall rotates in the same direction as the rotor but with 50-70% of the rotor angular velocity.

In distinguishing between rotating stall and surge, it is important to note that the disturbance depicted in figure 67 does not necessarily imply any oscillation in the total mass flow rate through the turbomachine. Rather it implies a redistribution of that flow. On the other hand, it is always possible that the perturbation caused by rotating stall

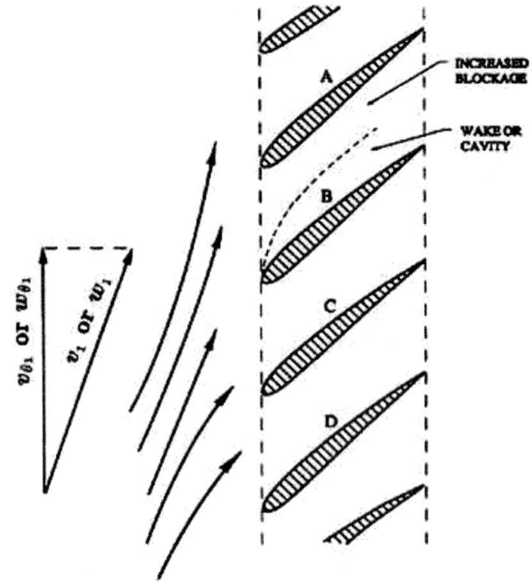

Figure 67. Schematic of a stall cell in rotating stall or rotating cavitation.

could resonate with, say, one of the acoustic modes in the inlet or discharge lines, in which case significant oscillation of the mass flow rate could occur.

While rotating stall can occur in any turbomachine, it is most frequently observed and most widely studied in compressors with large numbers of blades. Excellent reviews of this literature have been published by Emmons et al. (1959) and more recently by Greitzer (1981). Both point to a body of work designed to predict both the onset and consequences of rotating stall. A useful approximate criterion is that rotating stall in the rotor occurs when one approaches a maximum in the total head rise as the flow coefficient decreases. This is, however, no more than a crude approximation and Greitzer (1981) quotes a number of cases in which rotating stall occurs while the slope of the performance curve is still negative. A more sophisticated criterion that is widely used is due to Leiblein (1965), and involves the diffusion factor, Df, defined previously in equation (58). Experience indicates that rotating stall may begin when Df is increased to a value of about 0.6.

Though most of the observations of rotating stall have been made for axial compressors, Murai (1968) observed and investigated the phenomenon in a typical axial flow pump with 18 blades, a hub/tip radius ratio of 0.7, a tip solidity of 1.15, and a tip blade angle of 20°. His data on the rotating speed of the stall cell are reproduced in figure 68. Note that the onset of the rotating stall phenomenon occurs when the flow rate is reduced to a point below the maximum in the head characteristic. Notice also that the stall cell propagation velocities have typical values between 0.45 and 0.6 of the rotating speed. Rotating stall has not, however, been reported in pumps with a small number

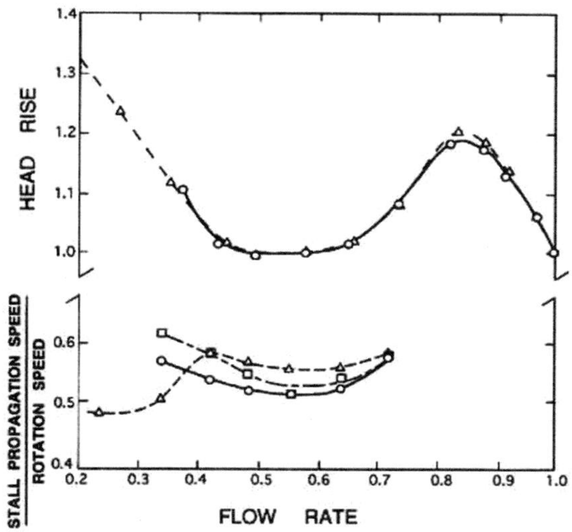

Figure 68. The head characteristic for an 18-bladed axial flow pump along with the measurements of the propagation velocity of the rotating stall cell relative to the shaft speed. Adapted from Murai (1968). Data is shown for three different inlet pressures. Flow and head scales are dimensionless.

of blades perhaps because Df will not approach 0.6 for typical axial pumps or inducers with a small number of blades. Most of the stability theories (for example, Emmons *et al.* 1959) are based on actuator disc models of the rotor in which it is assumed that the stall cell is much longer than the distance between the blades. Such an assumption would not be appropriate in an axial flow pump with three or four blades.

Murai (1968) also examined the effect of limited cavitation on the rotating stall phenomenon and observed that the cavitation did cause some alteration in the propagation speed as illustrated by the changes with inlet pressure seen in figure 68. It is, however, important to emphasize the difference between the phenomenon observed by Murai in which cavitation is secondary to the rotating stall and the phenomenon to be discussed below, namely rotating cavitation, which occurs at a point on the head-flow characteristic at which the slope is negative and stable, and at which rotating stall would not occur.

Turning now to centrifugal pumps, there have been a number of studies in which rotating stall has been observed either in the impeller or in the diffuser/volute. Hergt and Benner (1968) observed rotating stall in a vaned diffuser and conclude that it only occurs with some particular diffuser geometries. Lenneman and Howard (1970) examined the blade passage flow patterns associated with rotating stall and present data on the ratio, Ω_{RS}/Ω, of the propagation velocity of the stall cell to the impeller speed, Ω. They observed ratios ranging from 0.54 to 0.68 with, typically, lower values of the ratio at lower impeller speeds and at higher flow coefficients.

Perhaps the most detailed study is the recent research of Yoshida *et al.* (1991)

who made the following observations on a 7-bladed centrifugal impeller operating with a variety of diffusers, with and without vanes. Rotating stall with a single cell was observed to occur in the impeller below a certain critical flow coefficient which depended on the diffuser geometry. In the absence of a diffuser, the cell speed was about $80 - 90\%$ of the impeller rotating speed; with diffuser vanes, this cell speed was reduced to the range $50 - 80\%$. When impeller rotating stall was present, they also detected the presence of some propagating disturbances with 2, 3 and 4 cells rather than one. These are probably due to nonlinearities and an interaction with blade passage excitation. Rotating stall was also observed to occur in the vaned diffuser with a speed less than 10% of the impeller speed. It was most evident when the clearance between the impeller and diffuser vanes was large. As this clearance was decreased, the diffuser rotating stall tended to disappear.

Even in the absence of blades, it is possible for a diffuser or volute to exhibit a propagating rotating "stall". Jansen (1964) and van der Braembussche (1982) first described this flow instability and indicate that the flow pattern propagates with a speed in the range of $5-25\%$ of the impeller speed. Yoshida *et* al. (1991) observed a four-cell rotating stall in their vaneless diffusers over a large range of flow coefficients and measured its velocity as about 20% of the impeller speed.

Finally, we note that rotating stall may resonate with an acoustic mode of the inlet or discharging piping to produce a serious pulsation problem. Dussourd (1968) identified such a problem in a boiler feed system in which the rotating stall frequency was in the range $0.15\Omega \rightarrow 0.25\Omega$, much lower than usual. He also made use of frequency domain methods in modelling the dynamics of this multistage centrifugal pump system. This represents a good example of one of the many hybrid problems that can arise in systems with turbomachines.

9.5 Rotating Cavitation

Inducers or impellers in pumps that do not show any sign of rotating stall while operating under noncavitating conditions may exhibit a superficially similar phenemenon known as "rotating cavitation" when they are required to operate at low cavitation numbers. However, it is important to emphasize the fundamental difference in the two phenomena. Rotating stall occurs at locations along the head-flow characteristic at which the blades may stall, usually at flow rates for which the slope of the head/flow characteristic is positive and therefore unstable in the sense discussed in the next section. On the other hand, rotating cavitation is observed to occur at locations where the slope is negative. These would normally be considered stable operating points in the absence of cavitation. Consequently, the dynamics of the cavitation are essential to rotating cavitation. Another difference between the phenomena is the difference in the propagating speeds.

Rotating cavitation was first explicitly identified by Kamijo, Shimura and Watanabe (1977) (see also 1980), though some evidence of it can be seen in the shaft vibration measurements of Rosemann (1965). When it has been observed, rotating cavitation generally occurs when the cavitation number, σ, is reduced to a value at which the head is beginning to be affected by the cavitation as seen in figure 69 taken from Kamijo *et* al. (1977). Rosenmann (1965) reported that the vibrations (that we now recognize as rotat-

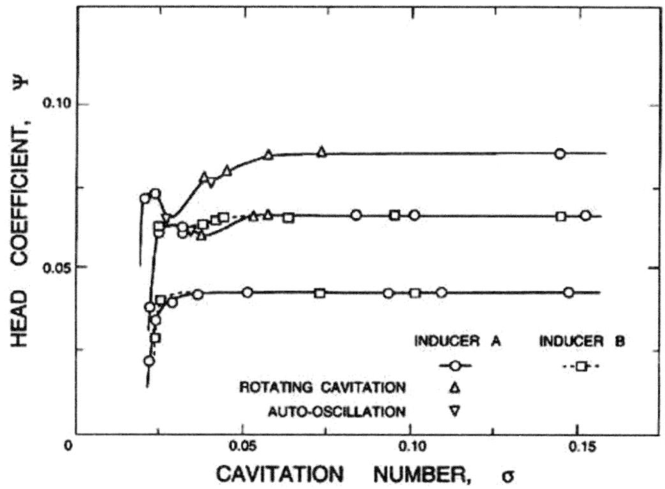

Figure 69. Occurrence of rotating cavitation and cavitation surge in the performance of the cavitating inducer tested by Kamijo et al. (1977).

ing cavitation) occurred for cavitation numbers between 2 and 3 times the breakdown value and were particularly evident at the lower flow coefficients at which the inducer was more heavily loaded.

Usually, further reduction of σ below the value at which rotating cavitation occurs will lead to cavitation surge or surge (see below and figure 69). It is not at all clear why some inducers and impellers do not exhibit rotating cavitation at all but proceed directly to cavitation surge if that instability is going to occur.

Unlike rotating stall whose rotational velocities are less than that of the rotor, rotating cavitation is characterized by a propagating velocity that is slightly larger than the impeller speed. Kamijo et al. (1977) (see also Kamijo et al. 1992) observed propagating velocities $\Omega_{RC}/\Omega \approx 1.15$, and this is very similar to one of the somewhat ambigous propagating disturbance velocities of 1.1Ω reported by Rosemann (1965).

Recently, Tsujimoto et al. (1992) have modelled the dynamics of rotating cavitation. They have shown that the cavitation compliance and mass flow gain factor (see section 10.6) play a crucial role in determining the instability of rotating cavitation in much the same way as these parameters influence the stability of an entire system which includes a cavitating pump (see section 9.7). Also note that the analysis of Tsujimoto et al. (1992) predicts supersynchronous propagating speeds in the range $\Omega_{RC}/\Omega = 1.1$ to 1.4, consistent with the experimental observations.

9.6 Surge

Surge and cavitation surge (see next subsection) are system instabilities that involve not just the characteristics of the pump but those of the rest of the pumping system. They

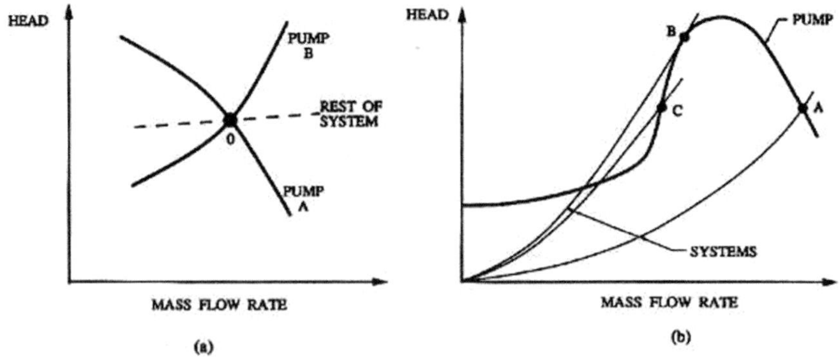

Figure 70. Quasistatically stable and unstable operation of pumping systems.

result in pressure and flow rate oscillations that can not only generate excessive vibration and reduce performance but also threaten the structural integrity of the turbomachine or other components of the system. In chapter 10 we provide more detail on general analytical approaches to this class of system instabilities. But for present purposes, it is useful to provide a brief outline of some of the characteristics of these system instabilities. To do so, consider first figure 70(a) in which the steady-state characteristic of the pump (head *r*ise versus mass flow rate) is plotted together with the steady-state characteristic of the rest of the system to which the pump is connected (head *d*rop versus mass flow rate). In steady-state operation the head rise across the pump must equal the head drop for the rest of the system, and the flow rates must be the same so that the combination will operate at the intersection, O. Consider, now, the response to a small decrease in the flow to a value just below this equilibrium point, O. Pump A will then produce more head than the head drop in the rest of the system, and this discrepancy will cause the flow rate to increase, causing a return to the equilibrium point. Therefore, because the slope of the characteristic of Pump A is less than the slope of the characteristic of the rest of the system, the point O represents a quasistatically stable operating point. On the other hand, the system with Pump B is quasistatically unstable. Perhaps the best known example of this kind of instability occurs in multistage compressors in which the characteristics generally take the shape shown in figure 70(b). It follows that the operating point A is stable, point B is neutrally stable, and point C is unstable. The result of the instability at points such as C is the oscillation in the pressure and flow rate known as "compressor surge."

While the above description of quasistatic stability may help in visualizing the phenomenon, it constitutes a rather artificial separation of the total system into a "pump" and "the rest of the system." A more general analytical perspective is obtained by defining a resistance, R_i^*, for each of the series components of the system (one of which would be the pump) distinguished by the subscript, i:

$$R_i^* = \frac{d(\Delta H)}{dm} \tag{101}$$

where ΔH is the quasistatic head drop across that component (inlet head minus discharge head) and is a function of the mass flow rate, m. By this definition, the slope of the pump characteristic in figure 70(a) is $-R^*_{PUMP}$, and the slope of the characteristic of the rest of the system is R^*_{SYSTEM}. It follows that the earlier established criterion for stability is equivalent to

$$\sum_i R_i^* > 0 \tag{102}$$

In other words, the system is quasistatically stable if the total system resistance is positive.

Perhaps the most satisfactory interpretation of the above formulation is in terms of the energy balance of the total system. The net flux of energy out of each of the elements of the system is $m(\Delta H)_i$. Consequently, the net energy flux out of the system is

$$m \sum_i (\Delta H)_i = 0 \tag{103}$$

which is zero at a steady state operating point.

Suppose the stability of the system is now explored by inserting somewhere in the system a hypothetical perturbing device which causes an increase in the flow rate by dm. Then the new net energy flux out of the system, E^*, is given by

$$E^* = dm \left[\sum_i (\Delta H_i) + m \frac{d \sum_i (\Delta H_i)}{dm} \right] \tag{104}$$

$$= m \, dm \sum_i R_i^* \tag{105}$$

where the relations (101) and (103) have been used. The quantity E^* could be interpreted as the energy flux that would have to be supplied to the system through the hypothetical device in order to reestablish equilibrium. Clearly, then, if the required energy flux, E^*, is positive, the original system is stable. Therefore the criterion (102) is the correct condition for stability.

All of the above is predicated on the changes to the system being sufficiently slow for the pump and the system to follow the steady state operating curves. Thus the analysis is only applicable to those instabilities whose frequencies are low enough to lie within some quasistatic range. At higher frequency, it is necessary to include the inertia and compressibility of the various components of the flow. Greitzer (1976) (see also 1981) has developed such models for the prediction of both surge and rotating stall in axial flow compressors.

It is important to observe that, while quasisteady instabilities will certainly occur when $\sum_i R_i^* < 0$, there may be other dynamic instabilities that occur even when the system is quasistatically stable. One way to view this possibility is to recognize that the resistance of any flow is frequently a complex function of frequency once a certain quasisteady frequency has been exceeded. Consequently, the resistances, R_i^*, may be different at frequencies above the quasistatic limit. It follows that there may be operating points at which the total dynamic resistance over some range of frequencies is negative. Then the system would be dynamically unstable even though it may still be quasistatically stable. Such a description of dynamic instability is instructive but overly simplistic and a more systematic approach to this issue must await the methodologies of chapter 10.

9.7 Cavitation Surge

In many installations involving a pump that cavitates, violent oscillations in the pressure and flow rate in the entire system occur when the cavitation number is decreased to values at which the head rise across the pump begins to be affected (Braisted and Brennen 1980, Kamijo *et* al. 1977, Sack and Nottage 1965, Natanzon *et* al. 1974, Miller and Gross 1967, Hobson and Marshall 1979). These oscillations can also cause substantial radial forces on the shaft of the order of 20% of the axial thrust (Rosenmann 1965). This surge phenomenon which was also called "auto-oscillation" is known as cavitation surge and can lead to very large flow rate and pressure fluctuations in the system. In boiler feed systems, discharge pressure oscillations with amplitudes as high as 14 *bar* have been reported informally. It is a genuinely dynamic instability in the sense described in the last section, for it occurs when the slope of the pump head rise/flow rate curve is still strongly negative. Another characteristic of cavitation surge is that it appears to occur more readily when the inducer is more heavily loaded; in other words at lower flow coefficients. These are also the circumstances under which backflow will occur. Indeed, Badowski (1969) puts forward the hypothesis that the dynamics of the backflow are responsible

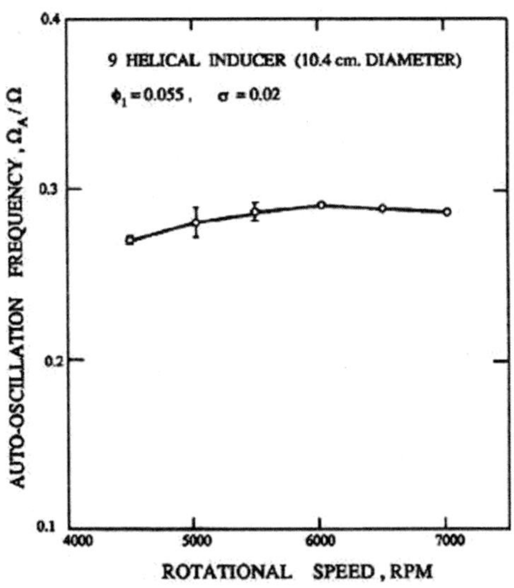

Figure 71. Data from Braisted and Brennen (1980) on the ratio of the cavitation surge frequency to the shaft frequency as a function of the latter for a 9° helical inducer operating at a cavitation number of 0.02 and a flow coefficient of 0.055.

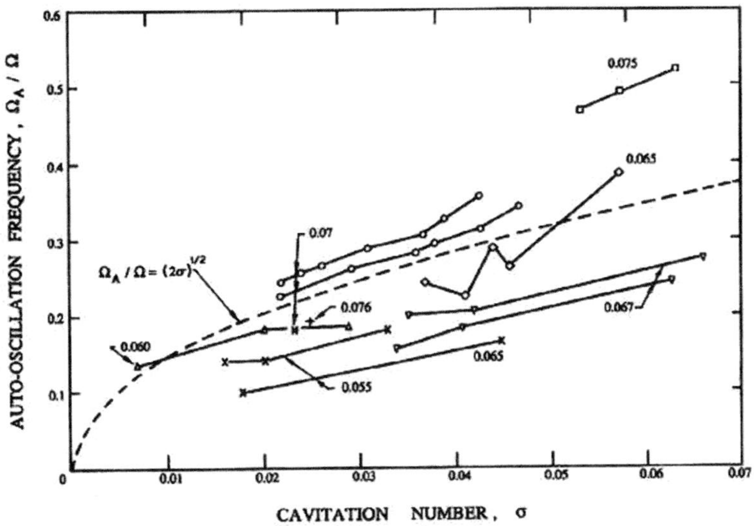

Figure 72. Data from Braisted and Brennen (1980) on the ratio of the frequency of cavitation surge to the frequency of shaft rotation for several inducers: SSME Low Pressure LOX Pump models: 7.62 *cm* diameter: $\times$ (9000 *rpm*) and + (12000 *rpm*), 10.2 *cm* diameter: $\circ$ (4000 *rpm*) and $\Box$ (6000 *rpm*); 9° helical inducers: 7.58 *cm* diameter: $*$ (9000 *rpm*): 10.4 *cm* diameter: $\bigtriangledown$ (with suction line flow straightener) and $\triangle$ (without suction line flow straightener). The flow coefficients, ϕ_1, are as labelled.

for cavitating inducer instability. Further evidence of this connection is provided by Hartmann and Soltis (1960) but with an atypical inducer that has 19 blades. It is certainly the case that the limit cycle associated with a strong cavitation surge appears to involve large periodic oscillations in the backflow. Consequently, it would seem that any nonlinear model purporting to predict the magnitude of cavitation surge should incorporate the dynamics of the backflow. While most of the detailed investigations have focussed on axial pumps and inducers, Yamamoto (1991) has observed and investigated cavitation surge occurring in cavitating centrifugal pumps. He also noted the important role played by the backflow in the dynamics of the cavitation surge.

Unlike compressor surge, the frequency of cavitation surge, Ω_A, usually scales with the shaft speed of the pump. Figure 71 demonstrates this by plotting Ω_A/Ω against the shaft *rpm* ($60\Omega/2\pi$) for a particular helical inducer. Figure 72 (also from Braisted and Brennen 1980) shows how this reduced cavitation surge frequency, Ω_A/Ω, varies with flow coefficient, ϕ, cavitation number, σ, and impeller geometry. While still noting that the frequency, Ω_A, will, in general, be system dependent, nevertheless the expression

$$\Omega_A/\Omega = (2\sigma)^{\frac{1}{2}} \tag{106}$$

appears to provide a crude estimate of the cavitation surge frequency.

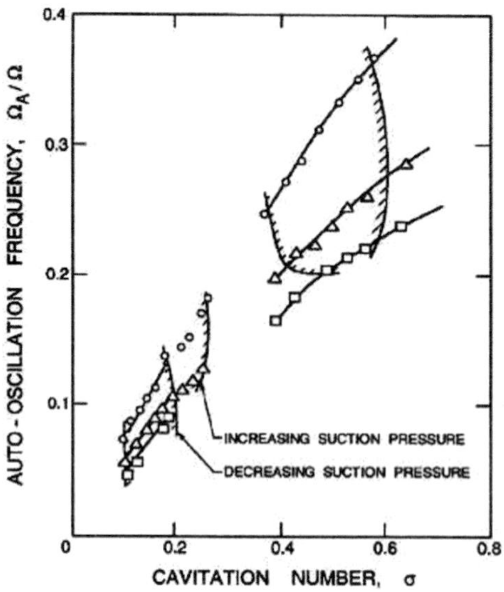

Figure 73. Data from Yamamoto (1991) on the ratio of the frequency of cavitation surge to the frequency of shaft rotation for a centrifugal pump. Data is shown for three different lengths of suction pipe: ○ (2.7 m), △ (4.9 m) and □ (7.1 m). Regions of instability are indicated by the hatched lines.

Some data from Yamamoto (1991) on the frequencies of cavitation surge of a cavitating centrifugal pump are presented in figure 73. This data exhibits a dependence on the length of the suction pipe that reinforces the understanding of cavitation surge as a system instability. The figure also shows the limits of instability observed by Yamamoto; these are unusual in that there appear to be two separate regions or zones of instability. Finally, it is clear that the data of figures 72 and 73 show a similar dependence of the cavitation surge frequency on the cavitation number, σ, though the magnitudes of σ differ considerably. However, it is likely that the relative sizes of the cavities are similiar in the two cases, and therefore that the correlation between the cavitation surge frequency and the relative cavity size might be closer than the correlation with cavitation number.

As previously stated, cavitation surge occurs when the region of cavitation head loss is approached as the cavitation number is decreased. Figure 74 provides an example of the limits of cavitation surge taken from the work of Braisted and Brennen (1980). However, since the onset is even more dependent than the cavitation surge frequency on the detailed dynamic characteristics of the system, it would not even be wise to quote any approximate guideline for onset. Our current understanding is that the methodologies of chapter 10 are essential for any prediction of cavitation surge.

It should be noted that section 10 describes linear perturbation models that can predict the limits of oscillation but not the amplitude of the oscillation once it occurs. There do not appear to be any accepted analytical models that can make this important

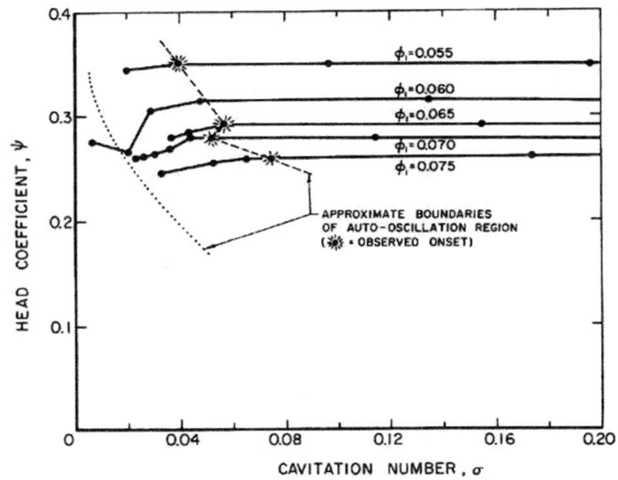

Figure 74. Cavitation performance of the SSME low pressure LOX pump model, Impeller IV, showing the onset and approximate desinence of the cavitation surge at 6000 *rpm* (from Braisted and Brennen 1980).

prediction. Furthermore, the energy dissipated in the large amplitude oscillations within the pump can lead to a major change in the mean (time averaged) performance of the pump (Brennen 1994).

9.8 Developed Cavity Oscillation

There are several circumstances in which developed cavities can exhibit self-sustained oscillations in the absence of any external excitation. One of these is the instability associated with a partial cavity whose length is approximately equal to the chord of the foil. Experimentally, it is observed that when the cavitation number is decreased to the level at which the attached partial cavity on a single hydrofoil approaches about 0.7 of the chord, c, of the foil, the cavity will begin to oscillate violently (Wade and Acosta 1966). It will grow to a length of about $1.5c$, at which point the cavity will be pinched off at about $0.5c$, and the separated cloud will collapse as it is convected downstream. This collapsing cloud of bubbles carries with it shed vorticity, so that the lift on the foil oscillates at the same time. This phenomenon is called "partial cavitation oscillation." It persists with further decrease in cavitation number until a point is reached at which the cavity collapses at some critical distance downstream of the trailing edge that is usually about $0.3c$. For cavitation numbers lower than this, the flow again becomes quite stable. The frequency of partial cavitation oscillation on a single foil is usually less than $0.1U/c$, where U is the velocity of the oncoming stream, and c is the chord length of the foil. In cascades or pumps, supercavitation is usually only approached in machines of low solidity, but, under such circumstances, partial cavitation oscillation can occur, and can be quite violent. Wade and Acosta (1966) were the first to observe partial cavitation oscillation in a cascade. During another set of experiments on cavitating cascades, Young, Murphy, and Reddcliff (1972) observed only "random unsteadiness of the cavities."

A second circumstance in which a fully developed cavity may exhibit natural oscillations occurs when the cavity is formed by introducing air to the wake of a foil in order to form a "ventilated cavity." When the flow rate of air exceeds a certain critical level, the cavity may begin to oscillate, large pockets of air being shed at the rear of the main cavity during each cycle of oscillation. This problem was studied by Silberman and Song (1961) and by Song (1962). The typical radian frequency for these oscillations is about $6U/\ell$, based on the length of the cavity, ℓ. Clearly, this second phenomenon is less relevant to pump applications.

9.9 Acoustic Resonances

In the absence of cavitation or flow-induced vibration, flow noise generated within the turbomachine itself is almost never an issue when the fluid is a liquid. One reason for this is that the large wavelength of the sound in the liquid leads to internal acoustic resonances that are too high in frequency and, therefore, too highly damped to be important. This contrasts with the important role played by internal resonances in the production of noise in gas turbines and compressors (Tyler and Sofrin 1962, Cumpsty 1977). In noncavitating liquid turbomachinery, the higher acoustic velocity and the smaller acoustic damping mean that pipeline resonances play the same kind of role that the internal resonances play in the production of noise in gas turbomachinery.

In liquid turbomachines, resonances exterior to the machine or resonances associated with cavitation do create a number of serious vibration problems. As mentioned in the introduction, pipeline resonances with the acoustic modes of the inlet or discharge piping can occur when one of the excitation frequencies produced by the pump or hydraulic turbine happens to coincide with one of the acoustic modes of those pipelines. Jaeger (1963) and Strub (1963) document a number of cases of resonance in hydropower systems. Many of these do not involve excitation from the turbine but some do involve excitation at blade passing frequencies (Strub 1963). One of the striking features of these phenomena is that very high harmonics of the pipelines can be involved (20th harmonics have been noted) so that damage occurs at a whole series of nodes equally spaced along the pipeline. The cases described by Jaeger involve very large pressure oscillations, some of which led to major failures of the installation. Sparks and Wachel (1976) have similarly documented a number of cases of pipeline resonance in pumping systems. They correctly identify some of these as system instabilities of the kind discussed in section 9.6 and in chapter 10.

Cavitation-induced resonances and vibration problems are dealt with in other sections of this chapter. But it is appropriate in the context of resonances to mention one other possible cavitation mechanism even though it has not, as yet, been demonstrated experimentally. One might judge that the natural frequency, ω_P, of bubbles given by equation (78) (subsection 7.2), being of the order of kHz, would be too high to cause vibration problems. However, it transpires that a finite cloud of bubbles may have much smaller natural frequencies that could resonant, for example, with a blade passage frequency to produce a problem. d'Agostino and Brennen (1983) showed that the lowest natural frequency, ω_C, of a spherical cloud of bubbles of radius, A, consisting of bubbles of radius, R, and with a void fraction of α would be given by

$$\omega_C = \omega_P \left[1 + \frac{4}{3\pi^2} \frac{A^2}{R^2} \frac{\alpha}{1-\alpha} \right]^{-\frac{1}{2}} \tag{107}$$

It follows that, if $\alpha A^2/R^2 \gg 1$, then the cloud frequency will be significantly smaller than the bubble frequency. This requires that the void fraction be sufficiently large so that $\alpha \gg R^2/A^2$. However, this could be relatively easily achieved in large clouds of small bubbles. Though the importance of cloud cavitation in pumps has been clearly demonstrated, the role played by the basic dynamic characteristics of clouds has not, as yet, been elucidated.

9.10 Blade Flutter

Up to this point, all of the instabilities have been essentially hydrodynamic and would occur with a completely rigid structure. However, it needs to be observed that structural flexibility could modify any of the phenomena described. Furthermore, if a hydrodynamic instability frequency happens to coincide with the frequency of a major mode of vibration of the structure, the result will be a much more dangerous vibration problem. Though the hydroelastic behavior of single hydrofoils has been fairly well established (see the review by Abramson 1969), it would be virtually impossible to classify all of the possible fluid-structure interactions in a turbomachine given the number of possible hydrodynamic instabilities and the complexity of the typical pump structure. Rather, we shall confine attention to one of the simpler interactions and briefly discuss blade flutter. Though the rotor-stator interaction effects outlined above are more likely to cause serious blade vibration problems in turbomachines, it is also true that a blade may flutter and fail even in the absence of such excitation.

It is well known (see, for example, Fung 1955) that the incompressible, unstalled flow around a single airfoil will not exhibit flutter when permitted only one degree of freedom of flutter motion. Thus, classic aircraft wing flutter requires the coupling of two degrees of flutter motion, normally the bending and torsional modes of the cantilevered wing. Turbomachinery flutter is quite different from classic aircraft wing flutter and usually involves the excitation of a single structural mode. Several different phenomena can lead to single degree of freedom flutter when it would not otherwise occur in incompressible, unseparated (unstalled) flow. First, there are the effects of compressibility that can lead to phenomena such as supersonic flutter and choke flutter. These have been the subject of much research (see, for example, the reviews of Mikolajczak *et* al (1975), Platzer (1978), Sisto (1977), McCroskey (1977)), but are not of direct concern in the context of liquid turbomachinery, though the compressibility introduced by cavitation might provide some useful analogies. Of greater importance in the context of liquid turbomachinery is the phenomenon of stall flutter (see, for example, Sisto 1953, Fung 1955). A blade which is stalled during all or part of a cycle of oscillation can exhibit single degree of freedom flutter, and this type of flutter has been recognized as a problem in turbomachinery for many years (Platzer 1978, Sisto 1977). Unfortunately, there has been relatively little analytical work on stall flutter and any modern theory must at least consider the characteristics of dynamic stall (see McCroskey 1977). Like all single degree of freedom flutter problems, including those in turbomachines, the critical incident speed for the onset of stall flutter, U_C, is normally given by a particular value of a reduced speed, $U_{CR} = 2U_C/c\omega_F$, where c is the chord length and ω_F is the frequency of flutter or the natural frequency of the participating structural vibration mode. The inverse of U_{CR} is the reduced frequency, k_{CR}, or Strouhal number. Fung (1955) points out that the reduced frequency for stall flutter with a single foil is a function of the difference, θ, between the mean angle of incidence of the flow and the static angle of stall. A crude guide would be $k_{CR} = 0.3 + 4.5\theta$, $0.1 < k_{CR} < 0.8$. The second term in the expression for k_{CR} reflects the decrease in the critical speed with increasing incidence.

Of course, in a turbomachine or cascade, the vibration of one blade will generate forces on the neighboring blades (see, for example, Whitehead 1960), and these interactions can cause significant differences in the flutter analyses and critical speeds; often they have a large unfavorable effect on the flutter characteristics (McCroskey 1977). One must allow for various phase angles between neighboring blades, and examine waves which travel both forward and backward relative to the rotation of the rotor. A complete analysis of the vibrational modes of the rotor (or stator) must be combined with an unsteady fluid flow analysis (see, for example, Verdon 1985) in order to accurately predict the flutter boundaries in a turbomachine. Of course, most of the literature deals with structures that are typical of compressors and turbines. The lowest modes of vibration in a pump, on the other hand, can be very different in character from those in a compressor or turbine. Usually the blades have a much smaller aspect ratio so that the lowest modes involve localized vibration of the leading or trailing edges of the blades. Consequently, any potential flutter is likely to cause failure of portions of these leading or trailing edges.

The other major factor is the effect of cavitation. The changes which developed cavitation cause in the lift and drag characteristics of a single foil, also cause a fundamental change in the flutter characteristics with the result that a single cavitating foil can flutter (Abramson 1969). Thus a cavitating foil is unlike a noncavitating, nonseparated foil but qualitatively similar to a stalled foil whose flow it more closely resembles. Abramson (1969) provides a useful review of both the experiments and the analyses of flutter of rigid cavitating foils. However, as we previously remarked, the most likely form of flutter in a pump will not involve global blade motion but flexure of the leading or trail-

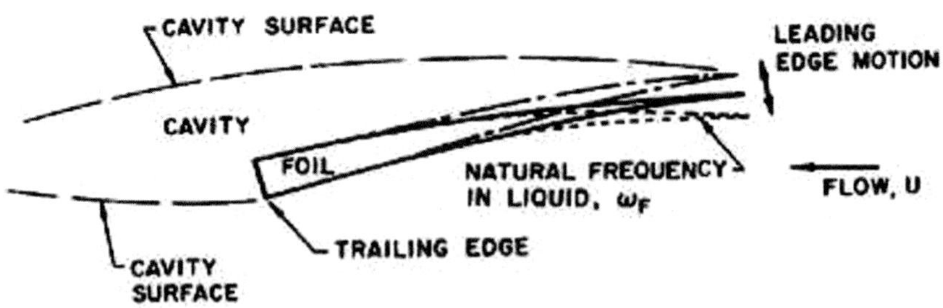

Figure 75. Sketch of the leading edge flutter of a cavitating hydrofoil or pump blade.

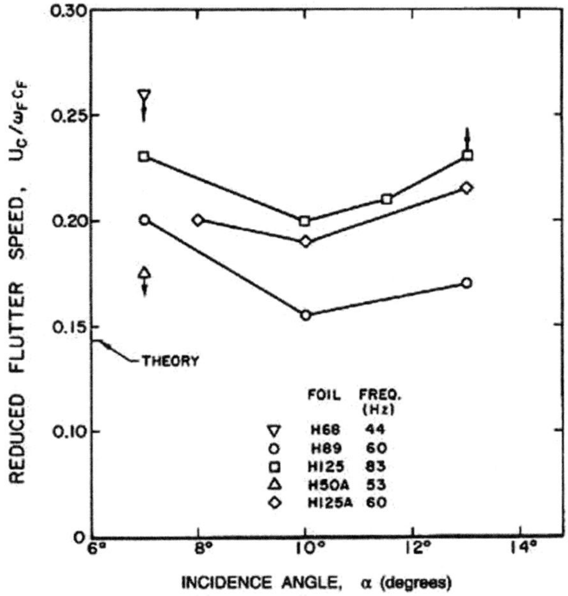

Figure 76. Dimensionless critical flutter speeds for single supercavitating hydrofoils at various angles of incidence, α, and very long cavities ($>$ 5 chord lengths) (from Brennen, Oey, and Babcock 1980).

ing edges. Since cavitation occurs at the leading edges, and since these are often made thin in order to optimize the hydraulic performance, leading edge flutter seems the most likely concern (figure 75). Data on this phenomenon was obtained by Brennen, Oey, and Babcock (1980), and is presented in figure 76. The critical incident fluid velocity, U_C, is nondimensionalized using ω_F, the lowest natural frequency of oscillation of the leading edge immersed in water, and a dimension, c_F, that corresponds to the typical chordwise length of the foil from the leading edge to the first node of the first mode of vibration. The data shows that $U_C/c_F\omega_F$ is almost independent of the incidence angle, and is consistent for a wide range of natural frequencies. Brennen *et* al. also utilize the unsteady lift and moment coefficients calculated by Parkin (1962) to generate a theoretical estimate of $U_C/c_F\omega_F$ of 0.14. From figure 76 this seems to constitute an upper design limit on the reduced critical speed. Also note that the value of 0.14 is much smaller than the values of $1-3$ quoted earlier for the stall flutter of a noncavitating foil. This difference emphasizes the enhanced flutter possibilities caused by cavitation. Brennen *et* al. also tested their foils under noncavitating conditions but found no sign of flutter even when the tunnel velocity was much larger than the cavitating flutter speed.

One footnote on the connection between the flutter characteristics of figure 76 and the partial cavitation oscillation of section 9.8 is worth adding. The data of figure 76 was obtained with long attached cavities, covering the entire suction surface of the foil as indicated in figure 75. At larger cavitation numbers, when the cavity length was decreased to about two chord lengths, the critical speed decreased markedly, and the leading edge flutter phenomenon began to metamorphose into the partial cavitation oscillation described in subsection 9.8.

9.11 Pogo Instabilities

All of the other discussion in this chapter has assumed that the turbomachine as a whole remains fixed in a nonaccelerating reference frame or, at least, that a vibrational degree of freedom of the machine is not necessary for the instability to occur. However, when a mechanism exists by which the internal flow and pressure oscillations can lead to vibration of the turbomachine as a whole, then a new set of possibilities are created. We refer to circumstances in which flow or pressure oscillations lead to vibration of the turbomachine (or its inlet or discharge pipelines) which in turn generate pressure oscillations that feed back to create instability. An example is the class of liquid-propelled rocket vehicle instabilities known as Pogo instabilities (NASA 1970). Here the longitudinal vibration of the rocket causes flow and pressure oscillation in the fuel tanks and, therefore, in the inlet lines. This, in turn, implies that the engines experience fluctuating inlet conditions, and as a result they produce a fluctuating thrust that promotes the longitudinal vibration of the vehicle. Rubin (1966) and Vaage *et* al. (1972) provide many of the details of these phenomena that are beyond the scope of this text. It is, however, important to note that the dynamics of the cavitating inducer pumps are crucial in determining the limits of these Pogo instabilities, and provide one of the main motivations for the measurements of the dynamic transfer functions of cavitating inducers described in section 10.

In closing, it is important to note that feedback systems involving vibrational motion of the turbomachine are certainly not confined to liquid propelled rockets. However, detailed examinations of the instabilities are mostly confined to this context. Brennen (1994) includes a brief introduction to the frequency domain methods which can be used to address problems involving oscillatory, translational or rotational motions of the whole hydraulic sytem.

10 Unsteady flow in hydraulic systems

10.1 Introduction

This section is devoted to a description of the methods available for the analysis of unsteady flows in pumps and their associated hydraulic systems. There are two basic approaches to the solution of unsteady internal flows: solution in the time domain or in the frequency domain. The traditional time domain methods for hydraulic systems are treated in depth elsewhere (for example, Streeter and Wylie 1967, 1974), and will only be touched upon here. They have the great advantage that they can incorporate the nonlinear convective inertial terms in the equations of fluid flow, and are best suited to evaluating the transient response of flows in long pipes in which the equations of the flow and the structure are fairly well established. However, they encounter great difficulties when either the geometry is complex (for example inside a pump), or the fluid is complex (for example in the presence of cavitation). Under these circumstances, frequency domain methods have distinct advantages, both analytically and experimentally. On the other hand, the nonlinear convective inertial terms cannot readily be included in the frequency-domain methodology and, consequently, these methods are only accurate for small perturbations from the mean flow. This does permit evaluation of stability limits, but not the evaluation of the amplitude of large unstable motions.

It should be stressed that many unsteady hydraulic system problems can and should be treated by the traditional time domain or "water-hammer" methods. However, since the focus of this review is on pumps and cavitation, we place an emphasis here on frequency domain methods. Sections 10.2 through 10.4 constitute an introduction to these frequency domain methods. This is followed by a summary of the transfer functions for simple components and for pumps, both noncavitating and cavitating.

10.2 Frequency Domain Methods

When the quasistatic assumption for a device like a pump or turbine becomes questionable, or when the complexity of the fluid or the geometry makes the construction of a set of differential equations impractical or uncertain, then it is clear that experimental information on the dynamic behavior of the device is necessary. In practice, such experimental information is most readily obtained by subjecting the device to fluctuations in the flow rate or head for a range of frequencies, and measuring the fluctuating quantities at inlet and discharge. Such experimental results will be presented later. For present purposes it is sufficient to recognize that one practical advantage of frequency domain methods is the capability of incorporation of experimentally obtained dynamic information and the greater simplicity of the experiments required to obtain the necessary dynamic data. Another advantage, of course, is the core of fundamental knowledge that exists regarding such methodology (see for example, Pipes 1940, Hennyey 1962, Paynter 1961, Brown 1967). As stated earlier, the disadvantage is that the methods are limited to small linear perturbations in the flow rate. When the perturbations are linear, Fourier analysis and synthesis can be used to convert from transient data to individual frequency components and vice versa. All the dependent variables such as the mean velocity, u, mass flow rate, m, pressure, p, or total pressure, p^T, are expressed as the sum of a mean component (denoted by an overbar) and a complex fluctuating component (denoted by a tilde) at a frequency, ω, which incorporates the amplitude and phase of the fluctuation:

$$p(s,t) = \bar{p}(s) + Re\left\{\tilde{p}(s,\omega)e^{j\omega t}\right\} \tag{108}$$

$$p^T(s,t) = \bar{p}^T(s) + Re\left\{\tilde{p}^T(s,\omega)e^{j\omega t}\right\} \tag{109}$$

$$m(s,t) = \bar{m}(s) + Re\left\{\tilde{m}(s,\omega)e^{j\omega t}\right\} \tag{110}$$

where j is $(-1)^{\frac{1}{2}}$ and Re denotes the real part. Since the perturbations are assumed linear ($|\tilde{u}| \ll \bar{u}, |\tilde{m}| \ll \bar{m}$, etc.), they can be readily superimposed, so a summation over many frequencies is implied in the above expressions. In general, the perturbation quantities will be functions of the mean flow characteristics as well as position, s, and frequency, ω.

We should note that there do exist a number of codes designed to examine the frequency response of hydraulic systems using frequency domain methods (see, for example, Amies and Greene 1977).

10.3 TRANSFER MATRICES

The transfer matrix for any component or device is the matrix which relates the fluctuating quantities at the discharge node to the fluctuating quantities at the inlet node. The earliest exploration of such a concept in electrical networks appears to be due to Strecker and Feldtkeller (1929) while the utilization of the idea in the context of fluid systems owes much to the pioneering work of Pipes (1940). The concept is the following. If the quantities at inlet and discharge are denoted by subscripts $i = 1$ and $i = 2$, respectively, and, if $\{\tilde{q}_i^n\}, n = 1, 2 \rightarrow N$ denotes the vector of independent fluctuating quantities at inlet and discharge for a system of order N, then the transfer matrix, $[T]$, is defined as

$$\{\tilde{q}_2^n\} = [T]\{\tilde{q}_1^n\} \tag{111}$$

It is a square matrix of order N. For example, for an order two system in which the independent fluctuating variables are chosen to be the total pressure, $\tilde{p}^T$, and the mass flow rate, $\tilde{m}$, then a convenient transfer matrix is

$$\left\{ \begin{matrix} \tilde{p}_2^T \\ \tilde{m}_2 \end{matrix} \right\} = \begin{bmatrix} T_{11} & T_{12} \\ T_{21} & T_{22} \end{bmatrix} \left\{ \begin{matrix} \tilde{p}_1^T \\ \tilde{m}_1 \end{matrix} \right\} \tag{112}$$

The words transfer function and transfer matrix are used interchangeably here to refer to the matrix $[T]$. In general it will be a function of the frequency, ω, of the perturbations and the mean flow conditions in the device.

The most convenient independent fluctuating quantities for a hydraulic system of order two are usually

1. Either the pressure, $\tilde{p}$, or the instantaneous total pressure, $\tilde{p}^T$. Note that these are related by

$$\tilde{p}^T = \tilde{p} + \frac{\bar{u}^2}{2}\tilde{\rho} + \bar{\rho}\bar{u}\tilde{u} + gz\tilde{\rho} \tag{113}$$

where $\bar{\rho}$ is the mean density, $\tilde{\rho}$ is the fluctuating density which is barotropically connected to $\tilde{p}$, and z is the vertical elevation of the system node. Neglecting the $\tilde{\rho}$ terms as is acceptable for incompressible flows

$$\tilde{p}^T = \tilde{p} + \bar{\rho}\bar{u}\tilde{u} \tag{114}$$

2. Either the velocity, $\tilde{u}$, the volume flow rate, $\bar{A}\tilde{u} + \bar{u}\tilde{A}$, or the mass flow rate, $\tilde{m} = \bar{\rho}\bar{A}\tilde{u} + \bar{\rho}\bar{u}\tilde{A} + \bar{u}\bar{A}\tilde{\rho}$. Incompressible flow at a system node in a rigid pipe implies

$$\tilde{m} = \bar{\rho}\bar{A}\tilde{u} \tag{115}$$

The most convenient choices are $\{\tilde{p}, \tilde{m}\}$ or $\{\tilde{p}^T, \tilde{m}\}$, and, for these two vectors, we will respectively use transfer matrices denoted by $[T^*]$ and $[T]$, defined as

$$\left\{ \begin{matrix} \tilde{p}_2 \\ \tilde{m}_2 \end{matrix} \right\} = [T^*] \left\{ \begin{matrix} \tilde{p}_1 \\ \tilde{m}_1 \end{matrix} \right\} \quad ; \quad \left\{ \begin{matrix} \tilde{p}_2^T \\ \tilde{m}_2 \end{matrix} \right\} = [T] \left\{ \begin{matrix} \tilde{p}_1^T \\ \tilde{m}_1 \end{matrix} \right\} \tag{116}$$

If the flow is incompressible and the cross-section at the nodes is rigid, then the $[T^*]$ and $[T]$ matrices are clearly connected by

$$T_{11} = T^*_{11} + \frac{\bar{u}_2}{A_2} T^*_{21} \quad ; \quad T_{12} = T^*_{12} - \frac{\bar{u}_1}{A_1} T^*_{11} + \frac{\bar{u}_2}{A_2} T^*_{22} - \frac{\bar{u}_1}{A_1}\frac{\bar{u}_2}{A_2} T^*_{21}$$
$$T_{21} = T^*_{21} \quad ; \quad T_{22} = T^*_{22} - \frac{\bar{u}_1}{A_1} T^*_{21} \tag{117}$$

and hence one is readily constructed from the other. Note that the determinants of the two matrices, $[T]$ and $[T^*]$, are identical.

10.4 Some Simple Transfer Matrices

The flow of an incompressible fluid in a straight, rigid pipe will be governed by the following equations:

$$\frac{\partial u}{\partial s} = 0 \tag{118}$$

$$\frac{\partial p^T}{\partial s} = -\frac{\rho f u|u|}{4a} - \rho \frac{\partial u}{\partial t} \tag{119}$$

If the velocity fluctuations are small compared with the mean velocity denoted by U (positive in direction from inlet to discharge), and the term $u|u|$ is linearized, then the above equations lead to the transfer function

$$[T] = \begin{bmatrix} 1 & -(R + j\omega L) \\ 0 & 1 \end{bmatrix} \tag{120}$$

where $(R + j\omega L)$ is the "impedance" made up of a "resistance", R, and an "inertance", L, given by

$$R = \frac{fU\ell}{2aA} \quad , \quad L = \frac{\ell}{A} \tag{121}$$

where A, a, and ℓ are the cross-sectional area, radius, and length of the pipe. A number of different pipes in series would then have

$$R = Q \sum_i \frac{f_i \ell_i}{2a_i A_i^2} \quad ; \quad L = \sum_i \frac{\ell_i}{A_i} \tag{122}$$

where Q is the mean flow rate. For a duct of non-uniform area

$$R = Q \int_0^\ell \frac{f(s)ds}{2a(s)(A(s))^2} \quad ; \quad L = \int_0^\ell \frac{ds}{A(s)} \tag{123}$$

A second, common hydraulic element is a simple "compliance", exemplified by an accumulator or a surge tank. It consists of a device installed in a pipeline and storing a volume of fluid, V_L, which varies with the local pressure, p, in the pipe. The compliance, C, is defined by

$$C = \rho \frac{dV_L}{dp} \tag{124}$$

In the case of a gas accumulator with a mean volume of gas, $\bar{V}_G$, which behaves according to the polytropic index, k, it follows that

$$C = \rho \bar{V}_G / k\bar{p} \tag{125}$$

where $\bar{p}$ is the mean pressure level. In the case of a surge tank in which the free surface area is A_S, it follows that

$$C = A_S/g \tag{126}$$

The relations across such compliances are

$$\tilde{m}_2 = \tilde{m}_1 - j\omega C\tilde{p}^T \quad ; \quad \tilde{p}_1^T = \tilde{p}_2^T = \tilde{p}^T \tag{127}$$

Therefore, using the definition (116), the transfer function [T] becomes

$$[T] = \begin{bmatrix} 1 & 0 \\ -j\omega C & 1 \end{bmatrix} \tag{128}$$

Clearly this component is equivalent to a capacitor to ground in an electrical circuit.

10.5 Non-cavitating Pumps

Consider now the questions associated with transfer functions for pumps or other turbomachines. In the simple fluid flows of section 10.4 we were able to utilize the known equations governing the flow in order to construct the transfer functions for those simple components. In the case of more complex fluids or geometries, one cannot necessarily construct appropriate one-dimensional flow equations, and therefore must resort to results derived from more global application of conservation laws or to experimental measurements of transfer matrices. Consider first the transfer matrix, $[TP]$, for incompressible flow through a pump (all pump transfer functions will be of the $[T]$ form defined in equation (116)) which will clearly be a function not only of the frequency, ω, but also of the mean operating point as represented by the flow coefficient, ϕ, and the cavitation number, σ. At very low frequencies one can argue that the pump will simply track up and down the performance characteristic, so that, for small amplitude perturbations and in the absence of cavitation, the transfer function becomes

$$[TP] = \begin{bmatrix} 1 & \frac{d(\Delta p^T)}{dm} \\ 0 & 1 \end{bmatrix} \tag{129}$$

where $d(\Delta p^T)/dm$ is the slope of the steady state operating characteristic of total pressure rise versus mass flow rate. Thus we define the pump resistance, $R_P = -d(\Delta p^T)/dm$, where R_P is usually positive under design flow conditions, but may be negative at low flow rates as discussed earlier (section 9.6). At finite frequencies, the elements TP_{21} and TP_{22} will continue to be zero and unity respectively, since the instantaneous flow rate into and out of the pump must be identical when the fluid and structure are incompressible and no cavitation occurs. Furthermore, TP_{11} must continue to be unity since, in an incompressible flow, the total pressure differences must be independent of the level of the pressure. It follows that the transfer function at higher frequencies will become

$$[TP] = \begin{bmatrix} 1 & -I_P \\ 0 & 1 \end{bmatrix} \tag{130}$$

where the pump impedence, I_P, will, in general, consist of a resistive part, R_P, and a reactive part, $j\omega L_P$. The resistance, R_P, and inertance, L_P, could be functions of both the frequency, ω, and the mean flow conditions. Such simple impedance models for pumps have been employed, together with transfer functions for the suction and discharge lines, to model the dynamics of pumping systems. For example, Dussourd (1968) used frequency domain methods to analyse pulsation problems in boiler feed pump systems. More recently, Sano (1983) used transfer functions to obtain natural frequencies for pumping systems that agree well with those observed experimentally.

The first fundamental investigation of the dynamic response of pumps seems to have been carried out by Ohashi (1968) who analyzed the oscillating flow through a cascade, and carried out some preliminary experimental investigations on a centrifugal pump. These studies enabled him to evaluate the frequency at which the response of the pump would cease to be quasistatic (see below). Fanelli (1972) appears to have been the first to explore the nature of the pump transfer function, while the first systematic measurements of the impedance of a noncavitating centrifugal pump are those of Anderson, Blade and Stevans (1971). Typical resistive and reactive component measurements from the work of Anderson, Blade and Stevans are reproduced in figure 77. Note that, though the resistance approaches the quasistatic value at low frequencies, it also departs significantly from this value at higher frequencies. Moreover, the reactive part is only roughly linear with frequency. The resistance and inertance are presented again in figure 78, where they are compared with the results of a dynamic model proposed by Anderson, Blade and Stevans. In this model, each pump impeller passage is represented by a resistance and an inertance, and the volute by a series of resistances and inertances. Since each impeller passage discharges into the volute at different locations relative to the volute discharge, each impeller passage flow experiences a different impedance on its way to the discharge. This results in an overall pump resistance and inertance that are frequency dependent as shown in figure 78. Note that the comparison with the experimental observations (which are also included in figure 78) is fair, but not completely satisfactory. Moreover, it should be noted, that the comparison shown is for a flow coefficient of 0.442 (above the design flow coefficient), and that, at higher flow coefficients, the model and experimental results exhibited poorer agreement.

Subsequent measurements of the impedance of non-cavitating axial and mixed flow pumps by Ng and Brennen (1978) exhibit a similar increase in the resistance with frequency (see next section). In both sets of dynamic data, it does appear that significant departure from the quasistatic values can be expected when the reduced frequency, (frequency/rotation frequency) exceeds about 0.02 (see figures 78 and 79). This is roughly consistent with the criterion suggested by Ohashi (1968) who concluded that non-quasistatic effects would occur above a reduced frequency of $0.05 Z_R \phi / \cos\beta$. For the inducers of Ng and Brennen, Ohashi's criterion yields values for the critical reduced frequency of about 0.015.

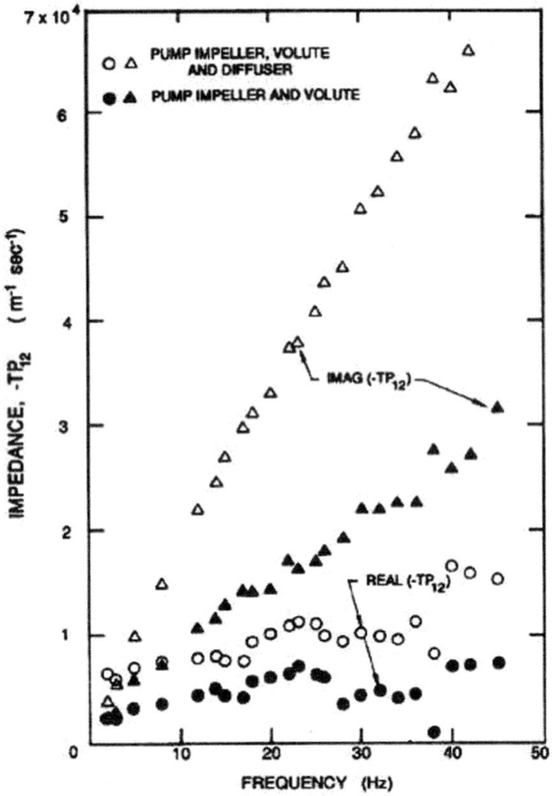

Figure 77. Impedance measurements made by Anderson, Blade and Stevans (1971) on a centrifugal pump (impeller diameter of 18.9 *cm*) operating at a flow coefficient of 0.442 and a speed of 3000 *rpm*. The real or resistive part of $(-T_{12})$ and the imaginary or reactive part of (T_{12}) are plotted against the frequency of the perturbation.

10.6 Cavitating Inducers

In the presence of cavitation, the transfer function for a pump or inducer will be considerably more complicated than that of equation (130). Even at low frequencies, the values of TP_{11} will become different from unity, because the head rise will change with the inlet total pressure, as manifest by the nonzero value of $d(\Delta p^T)/dp_1^T$ at a given mass flow rate, m_1. Furthermore, the volume of cavitation, $V_C(p_1^T, m_1)$, will vary with both the inlet total pressure, p_1^T (or $NPSH$ or cavitation number), and with the mass flow rate, m_1 (or with angle of incidence), so that

$$[TP] = \begin{bmatrix} 1 + \frac{d(\Delta p^T)}{dp_1^T}\Big|_{m_1} & \frac{d\Delta p^T}{dm_1}\Big|_{p_1^T} \\ j\omega\rho_L \frac{dV_C}{dp_1^T}\Big|_{m_1} & 1 + j\omega\rho_L \frac{dV_C}{dm_1}\Big|_{p_1^T} \end{bmatrix} \tag{131}$$

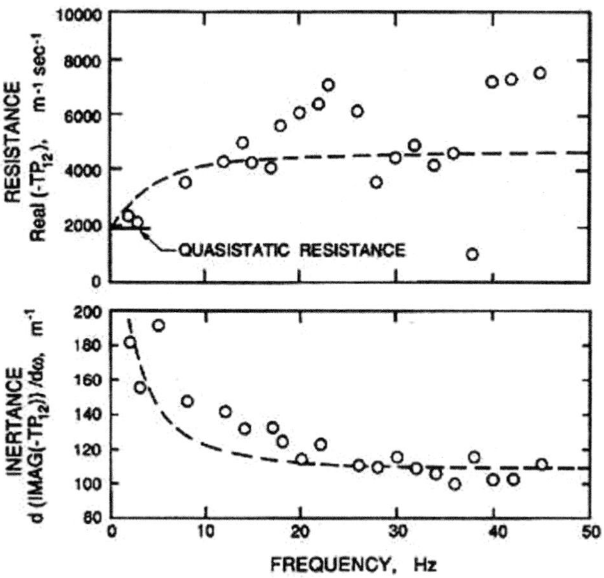

Figure 78. Typical inertance and resistance values from the centrifugal pump data of figure 77. Data do not include the diffuser contribution. The lines correspond to analytical values obtained as described in the text.

Brennen and Acosta (1973, 1975, 1976) identified this quasistatic or low frequency form for the transfer function of a cavitating pump, and calculated values of the cavitation compliance, $-\rho_L(dV_C/dp_1^T)_{m_1}$ and the cavitation mass flow gain factor, $-\rho_L(dV_C/dm_1)_{p_1^T}$, using the cavitating cascade solution discussed in section 8.9. Both the upper limit of frequency at which this quasistatic approach is valid and the form of the transfer function above this limit cannot readily be determined except by experiment. Though it was clear that experimental measurements of the dynamic transfer functions were required, these early investigations of Brennen and Acosta did highlight the importance of both the compliance and the mass flow gain factor in determining the stability of systems with cavitating pumps.

Ng and Brennen (1978) and Brennen et al. (1982) conducted the first experiments to measure the complete transfer function for cavitating inducers. Typical transfer functions (smoothed using polynomial curve fits - see below) are those for a 10.2 cm diameter model of the SSME low pressure LOX pump impeller, whose steady state cavitation performance was presented in figure 49. Transfer matrices for that inducer are presented in figure 79 as a function of frequency (up to 32 Hz), for a speed of 6000 rpm, a flow coefficient $\phi_1 = 0.07$ and for five different cavitation numbers ranging from data set A that was taken under noncavitating conditions, to data set C that showed a little cavitation, to data set H that was close to breakdown. The real and imaginary parts are represented by the solid and dashed lines, respectively. Note, first, that, in the absence of cavitation

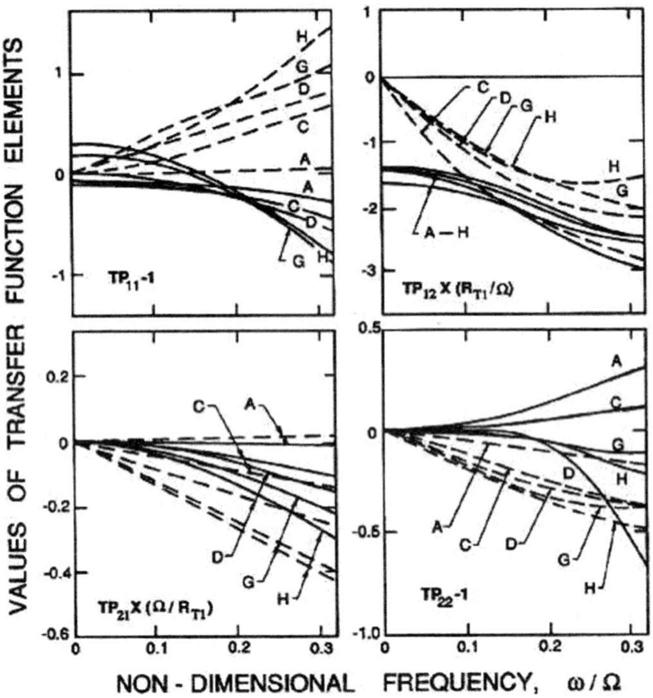

Figure 79. Polynomial curves fitted to typical transfer functions for a cavitating inducer obtained by Brennen *et* al. (1982) for a 10.2 *cm* diameter inducer (Impeller VI) operating at 6000 *rpm* and a flow coefficient of $\phi_1 = 0.07$. Data is shown for four different cavitation numbers, $\sigma =$ (A) 0.37, (C) 0.10, (D) 0.069, (G) 0.052 and (H) 0.044. Real and imaginary parts are denoted by the solid and dashed lines respectively. The quasistatic pump resistance is indicated by the arrow (adapted from Brennen *et* al. 1982).

(Case A), the transfer function is fairly close to the anticipated form of equation (130) in which $TP_{11} = TP_{22} = 1$, $TP_{21} = 0$. Also, the impedance (TP_{12}) is comprised of an expected inertance (the imaginary part of TP_{12} is linear in frequency) and a resistance (real part of $-TP_{12}$) which is consistent with the quasistatic resistance from the slope of the head rise characteristic namely $TP_{12}R_{T1}/\Omega = 1.07$). The resistance appears to increase with increasing frequency, a trend which is consistent with the centrifugal pump measurements of Anderson, Blade and Stevans (1971) which were presented in figure 78.

It is clear from figure 79 that, as the cavitation develops, the transfer function departs significantly from the form of equation (130). One observes that TP_{11} and TP_{22} depart from unity, and develop nonzero imaginary parts that are fairly linear with frequency. Also TP_{21} becomes nonzero, and, in particular, exhibits a compliance which clearly increases with decreasing cavitation number. The polynomials used to fit the data had

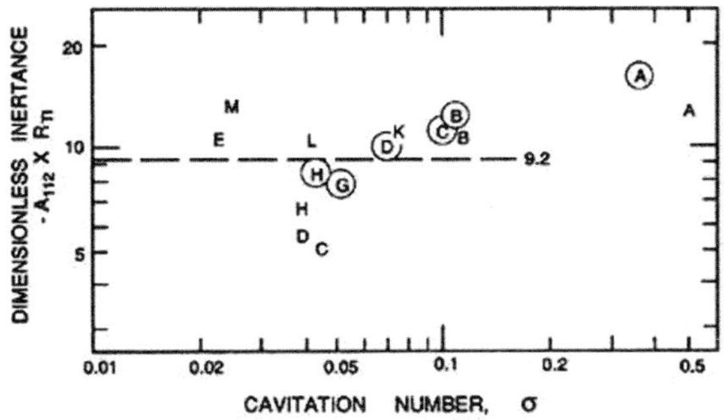

Figure 80. The inertance, $-A_{112}$, non-dimensionalized as $-A_{112}R_{T1}$, as a function of cavitation number for two axial inducer pumps (Impellers IV and VI) with the same geometry but different diameters. Data for the 10.2 *cm* diameter Impeller VI is circled and was obtained from the data of figure 79. The uncircled points are for the 7.58 *cm* diameter Impeller IV. Adapted from Brennen *et* al. (1982).

the form

$$TP_{ij} = \sum_{n=0}^{n^*} A_{nij}(j\omega)^n \tag{132}$$

and used values of n^* of 3 or 5.

We now proceed to examine several of the coefficients A_{nij} that are of particular interest (note that $A_{011} = A_{022} = 1$, $A_{021} = 0$ for reasons described earlier). We begin with the inertance, $-A_{112}$, which is presented nondimensionally in figure 80. Though there is significant scatter at the lower cavitation numbers, the two different sizes of inducer pump appear to yield similar inertances. Moreover, the data suggest some decrease in the inertance with decreasing σ. On the other hand, the corresponding data for the compliance, $-A_{121}$, which is presented in figure 81 seems roughly inversely proportional to the cavitation number. The same is true for both the mass flow gain factor, $-A_{122}$ presented in figure 82. All of these data appear to conform to the physical scaling implicit in the nondimensionalization of each of the dynamic characteristics.

It is also valuable to consider the results of figures 80 to 82 in the context of an analytical model for the dynamics of cavitating pumps (Brennen 1978). We present here a brief physical description of that model, the essence of which is depicted schematically in figure 83, which shows a developed, cylindrical surface within the inducer. The cavitation is modeled as a bubbly mixture which extends over a fraction, ϵ, of the length, c, of each blade passage before collapsing at a point where the pressure has risen to a value which causes collapse. The mean void fraction of the bubbly mixture is denoted by α_0. Thus far we have described a flow which is nominally steady. We must now consider perturbing both the pressure and the flow rate at inlet, since the relation between these perturbations, and those at discharge, determine the transfer function. Pressure perturbations at inlet will cause pressure waves to travel through the bubbly mixture and this part of the process is modeled using a mixture compressibility parameter, K, to

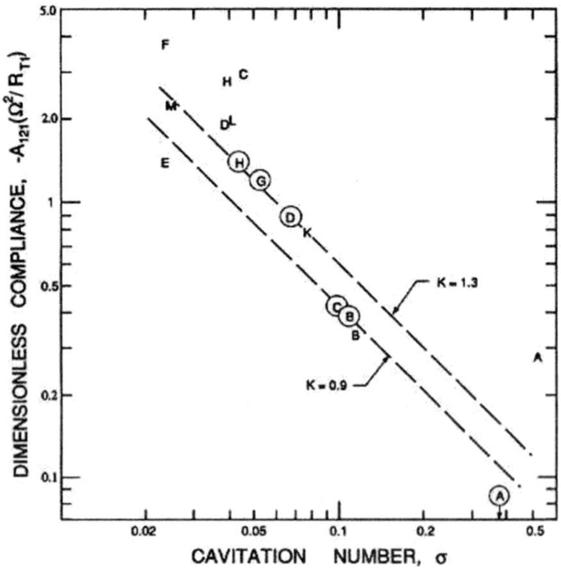

Figure 81. The compliance, $-A_{121}$, nondimensionalized as $-A_{121}\Omega^2/R_{T1}$ for the same circumstances as described in figure 80.

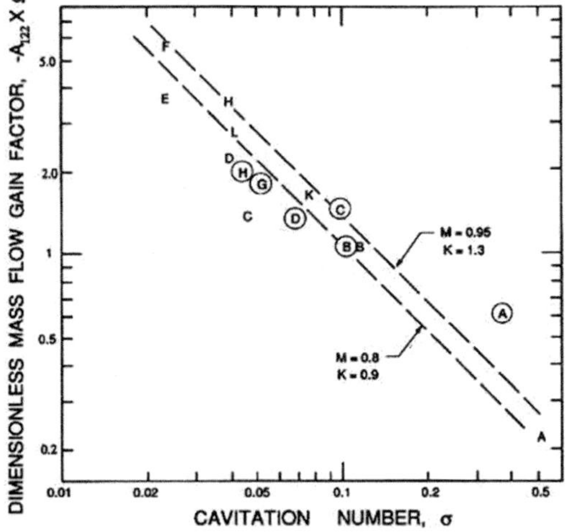

Figure 82. The mass flow gain factor, $-A_{122}$, nondimensionalized as $-A_{122}\Omega$ for the same circumstances as described in figure 80.

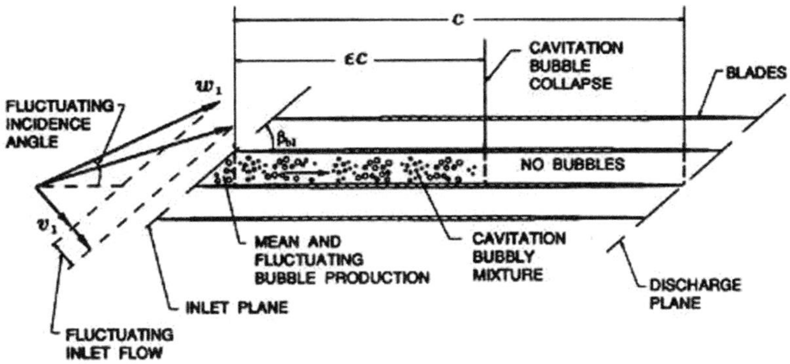

Figure 83. Schematic of the bubbly flow model for the dynamics of cavitating pumps (adapted from Brennen 1978).

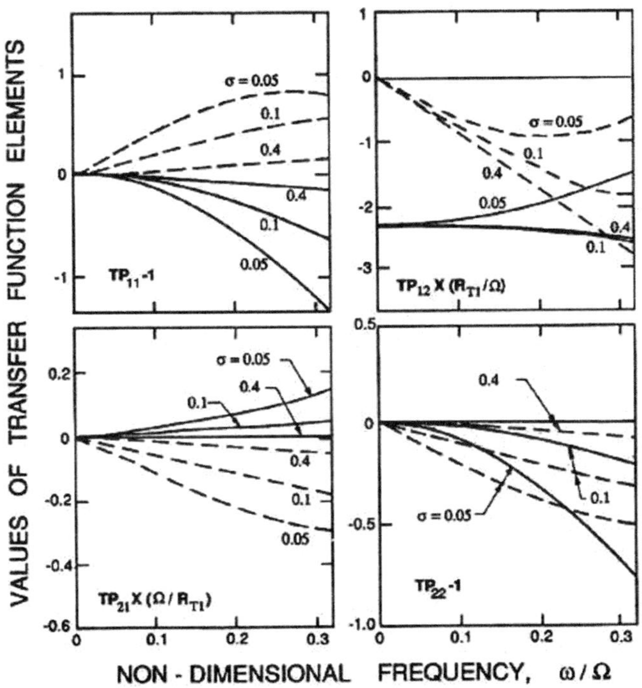

Figure 84. Transfer functions for the SSME low pressure LOX pump at $\phi_1 = 0.07$ calculated from the bubbly flow model using $K = 1.3$ and $M = 0.8$ (adapted from Brennen et al. 1982).

determine that wave speed. On the other hand, fluctuations in the inlet flow rate produce fluctuations in the angle of incidence which cause fluctuations in the rate of production of cavitation at inlet. These disturbances would then propagate down the blade passage as kinematic or concentration waves which travel at the mean mixture velocity. This process is modeled by a factor of proportionality, M, which relates the fluctuation in the angle of incidence to the fluctuations in the void fraction. Neither of the parameters, K or M, can be readily estimated analytically; they are, however, the two key features in the bubbly flow model. Moreover they respectively determine the cavitation compliance and the mass flow gain factor, two of the most important factors in the transfer function insofar as the prediction of instability is concerned.

The theory yields the following expressions for A_{111}, A_{112}, A_{121} and A_{122} at small dimensionless frequencies (Brennen 1978, 1982):

$$\begin{aligned} A_{111}\Omega &\simeq \frac{K\zeta\epsilon}{4}\left\{\cot\beta_{b1} + \phi_1/\sin^2\beta_{b1}\right\} \\ A_{112}R_{T1} &\simeq -\zeta/4\pi\sin^2\beta_{b1} \\ A_{121}\Omega^2/R_{T1} &\simeq -\pi K\zeta\epsilon/4 \\ A_{122}\Omega &\simeq -\frac{\zeta\epsilon}{4}\left\{M/\phi_1 - K\phi_1/\sin^2\beta_{b1}\right\} \end{aligned} \qquad (133)$$

where $\zeta = \ell Z_R/R_{T1}$ where ℓ is the axial length of the inducer, and Z_R is the number of blades. Evaluation of the transfer function elements can be effected by noting that the experimental observations suggest $\epsilon \approx 0.02/\sigma$. Consequently, the A_{nij} characteristics from equations (133) can be plotted against cavitation number. Typical results are shown in figures 80 to 82 for various choices of the two undetermined parameters K and M. The inertance, A_{112}, which is shown in figure 80, is independent of K and M. The calculated value of the inertance for these impellers is about 9.2; the actual value may be somewhat larger because of three-dimensional geometric effects that were not included in the calculation (Brennen et al. 1982). The parameter M only occurs in A_{122}, and it appears from figure 82 as though values of this parameter in the range $0.8 \rightarrow 0.95$ provide the best agreement with the data. Also, a value of $K \approx 1.3$ seems to generate a good match with the data of figures 81 and 82.

Finally, since $K = 1.3$ and $M = 0.8$ seem appropriate values for these impellers, we reproduce in figure 84 the complete theoretical transfer functions for various cavitation numbers. These should be directly compared with the transfer functions of figure 79. Note that the general features of the transfer functions, and their variation with cavitation number, are reproduced by the model. The most notable discrepency is in the real part of TP_{21}; this parameter is, however, usually rather unimportant in determining the stability of a hydraulic system. Most important from the point of view of stability predictions, the cavitation compliance and mass flow gain factor components of the transfer function are satisfactorily modeled.

11 Concluding Remarks

In these lectures I have tried to give not only a coherent introduction to the hydrodynamics of pumps but also a summary of recent developments, particularly in the area of unsteady flows. Indeed, in closing it is important to emphasize that much remains to be done to develop reliable methodolgies that will allow the accurate prediction of the numerous instabilities and vibration problems that plague modern high-speed turbomachines including pumps. Experimental, numerical and analytical tools for unsteady and for cavitating flows will all be needed to further this aim. One reason why the development of these tools are challenging is that one is frequently dealing with physical processes that have multiple scales both in time and in space. The problem of the multiple spatial scales involved in these highly turbulent flows is one of the more familiar sets of issues. Another example are the multiple time scales associated with cavitation where the time scale of bubble collapse and damage is much smaller than the time scale of the flow through the machine. Such scale disparities present enormous challenges for the accurate computation of these flows and frameworks for addressing these multiple scales are sorely needed.

Acknowledgements

The author wishes to thank Concepts ETI for permission to use excerpts from my book "Hydrodynamics of Pumps" for these lecture notes and all of those colleagues acknowledged in the preface of that book for their contributions to it, particularly Allan J. Acosta whose knowledge of the subject was invaluable to me. I also wish to thank NASA George Marshall Space Flight Center for their support over many years, including the support from Grant Number NAG8-1934 during the preparation of this summary.

Bibliography

Abramson, H.N. (1969). Hydroelasticity: a review of hydrofoil flutter. *Appl. Mech. Rev.*, 22, No. 2, 115–121.

Acosta, A.J. (1955). A note on partial cavitation of flat plate hydrofoils. *Calif. Inst. of Tech. Hydro. Lab. Rep. E-19.9.*

Acosta, A.J. (1958). An experimental study of cavitating inducers. *Proc. Second ONR Symp. on Naval Hydrodyn.*, ONR/ACR-38, 533–557.

Acosta, A.J. (1973). Hydrofoils and hydrofoil craft. *Ann. Rev. Fluid Mech.*, 5, 161–184.

Acosta, A.J. and Bowerman, R.D. (1957). An experimental study of centrifugal pump impellers. *Trans. ASME*, 79, 1821–1839.

Acosta, A.J. and DeLong, R.K. (1971). Experimental investigation of non-steady forces on hydrofoils oscillating in heave. *Proc. IUTAM Symp. on non-steady flow of water at high speeds, Leningrad, USSR*, 95–104.

Acosta, A.J. and Hollander, A. (1959). Remarks on cavitation in turbomachines. *Calif. Inst. of Tech. Rep. E-79.3.*

Acosta, A.J. and Parkin, B.R. (1975). Cavitation inception—a selective review. *J. Ship Res.*, 19, No. 4, 193–205.

Adamczyk, J.J. (1975). The passage of a distorted velocity field through a cascade of airfoils. *Proc. Conf. on Unsteady Phenomena in Turbomachinery, AGARD Conf. Proc. No. 177.*

Amies, G. and Greene, B. (1977). Aircraft hydraulic systems dynamic analysis. Volume IV. Frequency response (HSFR). *Wright-Patterson Air Force Base Technical Report AFAPL-TR-76-43*, IV.

Anderson, D.A., Blade, R.J. and Stevens, W. (1971). Response of a radial-bladed centrifugal pump to sinusoidal disturbances for non-cavitating flow. *NASA TN D-6556.*

Anderson, H.H. (undated) *Centrifugal pumps.* The Trade and Technical Press Ltd., Crown House, Morden, England.

Anderson, H.H. (1955). Modern developments in the use of large single-entry centrifugal pumps. *Proc. Inst. Mech. Eng.*, 169, 141–161.

Arakeri, V.H. (1979). Cavitation inception. *Proc. Indian Acad. Sci.*, C2, Part 2, 149–177.

Arndt, R.E.A. (1981). Cavitation in fluid machinery and hydraulic structures. *Ann. Rev. Fluid Mech.*, 13, 273–328.

Badowski, H.R. (1969). An explanation for instability in cavitating inducers. *Proc. 1969 ASME Cavitation Forum*, 38–40.

Badowski, H.R. (1970). Inducers for centrifugal pumps. *Worthington Canada, Ltd., Internal Report.*

Balje, O.E. (1981). *Turbomachines. A guide to design, selection and theory.* John Wiley and Sons, New York.

Batchelor, G.K. (1967). An introduction to fluid dynamics. *Cambridge Univ. Press.*

Biesheuvel, A. and van Wijngaarden, L. (1984). Two-phase flow equations for a dilute dispersion of gas bubbles in liquid. *J. Fluid Mech.*, 148, 301–318.

Birkhoff, G. and Zarantonello, E.M. (1957). *Jets, wakes and cavities.* Academic Press, NY.

Braisted, D.M. (1979). Cavitation induced instabilities associated with turbomachines. *Ph.D. Thesis, Calif. Inst. of Tech.*

Braisted, D.M. and Brennen, C.E. (1980). Auto-oscillation of cavitating inducers. In *Polyphase Flow and Transport Technology*, (ed: R.A. Bajura), ASME Publ., New York, 157–166.

Brennen, C.E. (1973). The dynamic behavior and compliance of a stream of cavitating bubbles. *ASME J. Fluids Eng.*, 95, 533–542.

Brennen, C.E. (1978). Bubbly flow model for the dynamic characteristics of cavitating pumps. *J. Fluid Mech.*, 89, Part 2, 223–240.

Brennen, C.E. (1994). *Hydrodynamics of pumps.* Concepts ETI and Oxford Univ. Press.

Brennen, C.E. (1995). *Cavitation and bubble dynamics.* Oxford Univ. Press.

Brennen, C.E. and Acosta, A.J. (1973). Theoretical, quasistatic analyses of cavitation compliance in turbopumps. *J. of Spacecraft and Rockets*, 10, No.3, 175–180.

Brennen, C.E. and Acosta, A.J. (1975). The dynamic performance of cavitating turbopumps. *Proc. Fifth Conference on Fluid Machinery, Akademiai Kiado, Budapest, Hungary*, 121–136.

Brennen, C.E. and Acosta, A.J. (1976). The dynamic transfer function for a cavitating inducer. *ASME J. Fluids Eng.*, 98, 182–191.

Brennen, C.E. and Braisted, D.M. (1980). Stability of hydraulic systems with focus on cavitating pumps. *Proc. 10th Symp. of IAHR, Tokyo*, 255–268.

Brennen, C.E., Oey, K., and Babcock, C.D. (1980). On the leading edge flutter of cavitating hydrofoils. *J. Ship Res.*, 24, No. 3, 135–146.

Brennen, C.E., Meissner, C., Lo, E.Y., and Hoffman, G.S. (1982). Scale effects in the dynamic transfer functions for cavitating inducers. *ASME J. Fluids Eng.*, 104, 428–433.

Breugelmans, F.A.E. and Sen, M. (1982). Prerotation and fluid recirculation in the suction pipe of centrifugal pumps. *Proc. 11th Int. Pump Symp., Texas A&M Univ.*, 165–180.

Brown, F.T. (1967). A unified approach to the analysis of uniform one-dimensional distributed systems. *ASME J. Basic Eng.*, 89, No. 6, 423–432.

Busemann, A. (1928). Das Förderhöhenverhältnis radialer Kreiselpumpen mit logarithmisch-spiraligen Schaufeln. *Z. angew. Math. u. Mech.*, 8, 372.

Ceccio, S.L. and Brennen, C.E. (1991). Observations of the dynamics and acoustics of travelling bubble cavitation. *J. Fluid Mech.*, 233, 633–660. Corrigenda, 240, 686.

Chahine, G.L. (1982). Cloud cavitation theory. *Proc. 14th ONR Symp. on Naval Hydrodynamics*, 165–194.

Chamieh, D. (1983). Forces on a whirling centrifugal pump-impeller. *Ph.D. Thesis, Calif. Inst. of Tech., and Div. of Eng. and App. Sci. Report No. E249.2.*

Chamieh, D.S., Acosta, A.J., Brennen, C.E., and Caughey, T.K. (1985). Experimental measurements of hydrodynamic radial forces and stiffness matrices for a centrifugal pump-impeller. *ASME J. Fluids Eng.*, 107, No. 3, 307–315.

Chivers, T.C. (1969). Cavitation in centrifugal pumps. *Proc. Inst. Mech. Eng.*, 184, Part I, No. 2, 37–68.

Constant, H. (1939). Performance of cascades of aerofoils. *Royal Aircraft Est. Note No. E3696 and ARC Rep. No. 4155.*

Cooper, P. (1967). Analysis of single- and two-phase flows in turbo-pump inducers. *ASME J. Eng. Power*, 89, 577–588.

Cumpsty, N. A. (1977). Review—a critical review of turbomachinery noise. *ASME J. Fluids Eng.*, 99, 278–293.

d'Agostino, L. and Brennen, C.E. (1983). On the acoustical dynamics of bubble clouds. *ASME Cavitation and Multiphase Flow Forum*, 72–75.

d'Agostino, L., Brennen, C.E., Acosta, A.J. (1988). Linearized dynamics of two-dimensional bubbly and cavitating flows over slender surfaces. *J. Fluid Mech.*, 192, 485–509.

d'Agostino, L. and Brennen, C.E. (1989). Linearized dynamics of spherical bubble clouds. *J. Fluid Mech.*, 199, 155–176.

Dean, R. C. (1959). On the necessity of unsteady flow in fluid machines. *ASME J. Basic Eng.*, 81, 24–28.

del Valle, J., Braisted, D.M., and Brennen, C.E. (1992). The effects of inlet flow modification on cavitating inducer performance. *ASME J. Turbomachinery*, 114, 360–365.

De Siervi, F., Viguier, H.C., Greitzer, E.M., and Tan, C.S. (1982). Mechanisms of inlet-vortex formation. *J. Fluid Mech.*, 124, 173–207.

Dixon, S.L. (1978). *Fluid mechanics, thermodynamics of turbomachinery.* Pergamon Press.

Dussourd, J. L. (1968). An investigation of pulsations in the boiler feed system of a central power station. *ASME J. Basic Eng.*, 90, 607–619.

Eckardt, D. (1976). Detailed flow investigations with a high-speed centrifugal compressor impeller. *ASME J. Fluids Eng.*, 98, 390–420.

Ek, B. (1957). *Technische Strömungslehre.* Springer-Verlag.

Emmons, H.W., Pearson, C.E., and Grant, H.P. (1955). Compressor surge and stall propagation. *Trans. ASME*, 79, 455–469.

Emmons, H.W., Kronauer, R.E., and Rockett, J.A. (1959). A survey of stall propagation—experiment and theory. *ASME J. Basic Eng.*, 81, 409–416.

Fanelli, M. (1972). Further considerations on the dynamic behaviour of hydraulic turbomachinery. *Water Power*, June 1972, 208–222.

Ferguson, T.B. (1963). *The centrifugal compressor stage.* Butterworth, London.

Fischer, K. and Thoma, D. (1932). Investigation of the flow conditions in a centrifugal pump. *Trans. ASME, Hydraulics*, 54, 141–155.

Fitzpatrick, H.M. and Strasberg, M. (1956). Hydrodynamic sources of sound. *Proc. First ONR Symp. on Naval Hydrodynamics*, 241–280.

Franz, R., Acosta, A.J., Brennen, C.E. and Caughey, T.K. (1989). The rotordynamic forces on a centrifugal pump impeller in the presence of cavitation. *Proc. ASME Symp. Pumping Machinery*, FED-81, 205–212.

Franz, R., Acosta, A.J., Brennen, C.E. and Caughey, T.K. (1990). The rotordynamic forces on a centrifugal pump impeller in the presence of cavitation. *ASME J. Fluids Eng.*, 112, 264–271.

Fung, Y.C. (1955). *An introduction to the theory of aeroelasticity.* John Wiley and Sons.

Furuya, O. and Acosta, A.J. (1973). A note on the calculation of supercavitating hydrofoils with rounded noses. *ASME J. Fluids Eng.*, 95, 222–228.

Furuya, O. (1974). Supercavitating linear cascades with rounded noses. *ASME J. Basic Eng., Series D*, 96, 35–42.

Gongwer, C. (1941). A theory of cavitation flow in centrifugal-pump impellers. *Trans. ASME*, 63, 29–40.

Greitzer, E.M. (1976). Surge and rotating stall in axial flow compressors. Part I: Theoretical compression system model. Part II: Experimental results and comparison with theory. *ASME J. Eng. for Power*, 98, 190–211.

Greitzer, E.M. (1981). The stability of pumping systems—the 1980 Freeman Scholar Lecture. *ASME J. Fluids Eng.*, 103, 193–242.

Grist, E. (1974). NPSH requirements for avoidance of unacceptable cavitation erosion in centrifugal pumps. *Proc. I.Mech.E. Conf. on Cavitation*, 153–163.

Gross, L.A. (1973). An experimental investigation of two-phase liquid oxygen pumping. *NASA TN D-7451.*

Guinard, P., Fuller, T. and Acosta, A.J. (1953). Experimental study of axial flow pump cavitation. *Calif. Inst. of Tech. Hydro. Lab. Report, E-19.3.*

Ham, N.D. (1968). Aerodynamic loading on a two-dimensional airfoil during dynamic stall. *AIAA J.*, 6, 1927–1934.

Hartmann, M.J. and Soltis, R.F. (1960). Observation of cavitation in a low hub-tip ratio axial flow pump. *Proc. Gas Turbine Power and Hydraulic Conf., ASME Paper No. 60-HYD-14.*

Henderson and Tucker. (1962). Performance investigation of some high speed pump inducers. *R.P.E. Tech. Note 214.* Reported by Janigro and Ferrini (1973) but not located by the author.

Hennyey, Z. (1962). *Linear electric circuits.* Pergamon Press.

Hergt, P. and Benner, R. (1968). Visuelle Untersuchung der Strömung in Leitrad einer Radialpumpe. *Schweiz. Banztg.*, 86, 716–720.

Hobson, D. E. and Marshall, A. (1979). Surge in centrifugal pumps. *Proc. 6th Conf. on Fluid Machinery, Budapest*, 475–483.

Holl, J.W. (1969). Limited cavitation. *Cavitation State of Knowledge*, ASME, 26–63.

Horlock, J.H. (1973). *Axial flow compressors.* Robert E. Krieger Publ. Co., New York.

Horlock, J.H. and Lakshminarayana, B. (1973). Secondary flows: theory, experiment and application in turbomachinery aerodynamics. *Ann. Rev. Fluid Mech.*, 5, 247–279.

Howard, J.H.G. and Osborne, C. (1977). A centrifugal compressor flow analysis employing a jet-wake passage model. *ASME J. Fluids Eng.*, 99, 141–147.

Howell, A.R. (1942). The present basis of axial flow compressor design: Part I—Cascade theory and performance. *ARC R and M No. 2095.*

Hydraulic Institute, New York. (1965). *Standards of the Hydraulic Institute* (11th edition).

Jaeger, C. (1963). The theory of resonance in hydro-power systems, discussion of incidents and accidents occurring in pressure systems. *ASME J. Basic Eng.*, 85, 631–640.

Jakobsen, J.K. (1964). On the mechanism of head breakdown in cavitating inducers. *ASME J. Basic Eng.*, 86, 291–304.

Jakobsen, J.K. (1971). *Liquid rocket engine turbopumps.* NASA SP 8052.

Janigro, A. and Ferrini, F. (1973). Inducer pumps. In *Recent progress in pump research*, von Karman Inst. for Fluid Dynamics, Lecture Series 61.

Jansen, W. 1964. Rotating stall in a radial vaneless diffuser. *ASME J. Basic Eng.*, 86, 750–758.

Japikse, D. (1984). *Turbomachinery diffuser design technology.* Concepts ETI, Inc., Norwich, VT.

Johnston, J.P. and Dean, R.C. (1966). Losses in vaneless diffusers of centrifugal compressors and pumps. *ASME J. Eng. for Power*, 88, 49–62.

Johnson, V.E. and Hsieh, T. (1966). The influence of the trajectories of gas nuclei on cavitation inception. *Proc. 6th ONR Symp. on Naval Hydrodynamics*, 7-1.

Kamijo, K., Shimura, T., and Watanabe, M. (1977). An experimental investigation of cavitating inducer instability. *ASME Paper 77-WA/FW-14.*

Kamijo, K., Shimura, T., and Watanabe, M. (1980). A visual observation cavitating inducer instability. *Nat. Aero. Lab. (Japan), Rept. NAL TR-598T.*

Kamijo, K., Yoshida, M., and Tsujimoto, Y. (1992). Hydraulic and mechanical performance of LE-7 LOX pump inducer. *Proc. 28th Joint Propulsion Conf., Paper AIAA-92-3133.*

Kemp, N. H. and Ohashi, H. (1975). Forces on unstaggered airfoil cascades in unsteady in-phase motion with applications to harmonic oscillation. *Proc. Symp. on Unsteady Aerodynamics, Tuscon, Ariz.*, 793–826.

Kemp, N. H. and Sears, W. R. (1955). The unsteady forces due to viscous wakes in turbomachines. *J. Aero Sci.*, 22, No.7, 478–483.

Kermeen, R.W. (1956). Water tunnel tests of NACA 4412 and Walchner Profile 7 hydrofoils in noncavitating and cavitating flows. *Calif. Inst. of Tech., Hydrodynamics Lab. Rep. 47-5.*

Knapp, R.T., Daily, J.W., and Hammitt, F.G. (1970). *Cavitation.* McGraw-Hill, New York.

König, E. (1922). Potentialströmung durch Gitter. *Z. angew. Math. u. Mech.*, 2, 422.

Lazarkiewicz, S. and Troskolanski, A.T. (1965). *Pompy wirowe. (Impeller pumps.)* Translated from Polish by D.K.Rutter. Publ. by Pergamon Press.

Lee, C.J.M. (1966). Written discussion in *Proc. Symp. on Pump Design, Testing and Operation, Natl. Eng. Lab., Scotland*, 114-115.

Lieblein, S., Schwenk, F.C., and Broderick, R.L. (1953). Diffusion factor for estimating losses and limiting blade loadings in axial-flow-compressor blade elements. *NACA RM E53D01.*

Lieblein, S. (1965). Experimental flow in two-dimensional cascades. *Aerodynamic design of axial flow compressors, NASA SP-36*, 183–226.

Lenneman, E. and Howard, J. H. G. (1970). Unsteady flow phenomena in centrifugal impeller passages. *ASME J. Eng. for Power*, 92-1, 65–72.

Lush, P.A. and Angell, B. (1984). Correlation of cavitation erosion and sound pressure level. *ASME J. Fluids Eng.*, 106, 347–351.

Makay, E. (1980). Centrifugal pump hydraulic instability. *Electric Power Res. Inst. Rep. EPRI CS-1445.*

Makay, E. and Szamody, O. (1978). Survey of feed pump outages. *Electric Power Res. Inst. Rep. FP-754.*

Mansell, C.J. (1974). Impeller cavitation damage on a pump operating below its rated discharge. *Proc. of Conf. on Cavitation, Inst. of Mech. Eng.*, 185–191.

Martin, M. (1962). Unsteady lift and moment on fully cavitating hydrofoils at zero cavitation number. *J. Ship Res.*, 6, No.1, 15–25.

McCroskey, W. J. (1977). Some current research in unsteady fluid dynamics—the 1976 Freeman Scholar Lecture. *ASME J. Fluids Eng.*, 99, 8–38.

McNulty, P.J. and Pearsall, I.S. (1979). Cavitation inception in pumps. *ASME Int. Symp. on Cavitation Inception*, 163–170.

Mikolajczak, A. A., Arnoldi, R. A., Snyder, L. E., and Stargardter, H. (1975). Advances in fan and compressor blade flutter analysis and predictions. *J. Aircraft*, 12, No.4, 325–332.

Miller, C.D. and Gross, L.A. (1967). A performance investigation of an eight-inch hubless pump inducer in water and liquid nitrogen. *NASA TN D-3807.*

Mimura, Y. (1958). The flow with wake past an oblique plate. *J. Phys. Soc. Japan*, 13, 1048–1055.

Moore, R.D. and Meng, P.R. (1970a). Thermodynamic effects of cavitation of an 80.6° helical inducer operated in hydrogen. *NASA TN D-5614.*

Moore, R.D. and Meng, P.R. (1970b). Effect of blade leading edge thickness on cavitation performance of 80.6° helical inducers in hydrogen. *NASA TN D5855.*

Murai, H. (1968). Observations of cavitation and flow patterns in an axial flow pump at low flow rates (in Japanese). *Mem. Inst. High Speed Mech., Tohoku Univ.*, 24, No.246, 315–333.

NASA. (1970). Prevention of coupled structure-propulsion instability. *NASA SP-8055.*

Natanzon, M.S., Bl'tsev, N.E., Bazhanov, V.V., and Leydervarger, M.R. (1974). Experimental investigation of cavitation-induced oscillations of helical inducers. *Fluid Mech., Soviet Res.*, 3, No.1, 38–45.

Ng, S.L. and Brennen, C.E. (1978). Experiments on the dynamic behavior of cavitating pumps. *ASME J. Fluids Eng.*, 100, No. 2, 166–176.

Pearsall, I.S. (1963). Supercavitation for pumps and turbines. *Engineering (GB)*, 196(5081), 309–311.

Ohashi, H. (1968). Analytical and experimental study of dynamic characteristics of turbopumps. *NASA TN D-4298.*

Okamura, T. and Miyashiro, H. (1978). Cavitation in centrifugal pumps operating at low capacities. *ASME Symp. on Polyphase Flow in Turbomachinery*, 243–252.

Omta, R. (1987). Oscillations of a cloud of bubbles of small and not so small amplitude. *J. Acoust. Soc. Amer.*, 82, 1018–1033.

Oshima, M. and Kawaguchi, K. (1963). Experimental study of axial and mixed flow pumps. *Proc. IAHR Symp. on Cavitation and Hydraulic Machinery, Sendai, Japan*, 397–416.

Parkin, B.R. (1952). Scale effects in cavitating flow. *Ph.D. Thesis, Calif. Inst. of Tech., Pasadena.*

Parkin, B.R. (1958). Experiments on circular-arc and flat plate hydrofoils. *J. Ship Res.*, 1, 34–56.

Parkin, B.R. (1962). Numerical data on hydrofoil reponse to non-steady motions at zero cavitation number. *J. Ship Res.*, 6, 40–42.

Paynter, H.M. (1961). *Analysis and design of engineering systems.* MIT Press.

Pearsall, I.S. (1963). Supercavitation for pumps and turbines. *Engineering (GB)*, 196(5081), 309–311.

Pearsall, I.S. (1966-67). Acoustic detection of cavitation. *Proc. Inst. Mech. Eng.*, 181, No. 3A.

Pearsall, I.S. (1978). Off-design performance of pumps. *von Karman Inst. for Fluid Dynamics, Lecture Series 1978-3.*

Peck, J.F. (1966). Written discussion in *Proc. Symp. on Pump Design, Testing and Operation, Nat. Eng. Lab., Scotland*, 256–273.

Pfleiderer, C. (1932). *Die Kreiselpumpen.* Julius Springer, Berlin.

Pipes, L.A. (1940). The matrix theory for four terminal networks. *Phil. Mag.*, 30, 370.

Pipes, L.A. (1963). *Matrix methods for engineering.* Prentice-Hall, Inc., NJ.

Platzer, M. F. (1978). Unsteady flows in turbomachines—a review of current developments. *AGARD Rept. CP-227.*

Plesset, M.S. (1949). The dynamics of cavitation bubbles. *Trans. ASME, J. Appl. Mech.*, 16, 228–231.

Rohatgi, U.S. (1978). Pump model for two-phase transient flow. In *Polyphase Flow in Turbomachinery* (eds: C.E. Brennen, P. Cooper and P.W. Runstadler, Jr.), ASME, 101–120.

Rosenmann, W. (1965). Experimental investigations of hydrodynamically induced shaft forces with a three bladed inducer. *Proc. ASME Symp. on Cavitation in Fluid Machinery*, 172–195.

Roudebush, W.H. (1965). Potential flow in two-dimensional cascades. *Aerodynamic design of axial flow compressors, NASA SP-36*, 101–149.

Roudebush, W.H. and Lieblein, S. (1965). Viscous flow in two-dimensional cascades. *Aerodynamic design of axial flow compressors, NASA SP-36*, 151–181.

Rubin, S. (1966). Longitudinal instability of liquid rockets due to propulsion feedback (Pogo). *J. Spacecraft and Rockets*, 3, No.8, 1188–1195.

Ruggeri, R.S. and Moore, R.D. (1969). Method for prediction of pump cavitation performance for various liquids, liquid temperatures, and rotative speeds. *NASA TN D-5292.*

Sabersky, R.H., Acosta, A.J. and Hauptmann, E.G. (1989). *Fluid flow (3rd edition), Chapters 12 and 13.* Macmillan Publ. Co.

Sack, L.E. and Nottage, H.B. (1965). System oscillations associated with cavitating inducers. *ASME J. Basic Eng.*, 87, 917–924.

Salemann, V. (1959). Cavitation and NPSH requirements of various liquids. *ASME J. Basic Eng.*, 81, 167–180.

Samoylovich, G.S. (1962). On the calculation of the unsteady flow around an array of arbitrary profiles vibrating with arbitrary phase shift. *Prikladnaya Matematika i Mekhanika*, No.4.

Schorr, B. and Reddy, K.C. (1971). Inviscid flow through cascades in oscillatory and distorted flow. *AIAA J.*, 9, 2043–2050.

Silberman, E. (1959). Experimental studies of supercavitating flow about simple two-dimensional bodies in a jet. *J. Fluid Mech.*, 5, 337–354.

Silberman, E. and Song, C.S. (1961). Instability of ventilated cavities. *J. Ship Res.*, 5, 13–33.

Sisto, F. (1953). Stall-flutter in cascades. *J. Aero. Sci.*, 20, 598–604.

Sisto, F. (1967). Linearized theory of non-stationary cascades at fully stalled or supercavitating conditions. *Zeitschrift fur Angewandte Mathematik und Mechanik*, 8, 531–542.

Sisto, F. (1977). A review of the fluid mechanics of aeroelasticity in turbomachines. *ASME J. Fluids Eng.*, 99, 40–44.

Sloteman, D.P., Cooper, P., and Dussourd, J.L. (1984). Control of backflow at the inlets of centrifugal pumps and inducers. *Proc. Int. Pump Symp., Texas A&M Univ.*, 9–22.

Song, C.S. (1962). Pulsation of ventilated cavities. *J. Ship Res.*, 5, 8–20

Sparks, C.R. and Wachel, J.C. (1976). Pulsations in liquid pumps and piping systems. *Proc. 5th Turbomachinery Symp.*, 55–61.

Spraker, W.A. (1965). The effect of fluid properties on cavitation in centrifugal pumps. *ASME J. Eng. Power*, 87, 309–318.

Stahl, H.A. and Stepanoff, A.J. (1956). Thermodynamic aspects of cavitation in centrifugal pumps. *Trans. ASME*, 78, 1691–1693.

Stanitz, J.D. (1952). Some theoretical aerodynamic investigations of impellers in radial- and mixed- flow centrifugal compressors. *Trans. ASME*, 74, 473–497.

Stepanoff, A.J. (1948). *Centrifugal and axial flow pumps.* John Wiley & Sons, Inc.

Stepanoff, A.J. (1961). Cavitation in centrifugal pumps with liquids other than water. *ASME J. Eng. Power*, 83, 79–90.

Stepanoff, A.J. (1964). Cavitation properties of liquids. *ASME J. Eng. Power*, 86, 195–200.

Stodola, A. (1927). *Steam and gas turbines. Volumes I and II.* McGraw-Hill, New York.

Strecker, F. and Feldtkeller, R. (1929). Grundlagen der Theorie des allgemeinen Vierpols. *Elektrische Nachrichtentechnik*, 6, 93.

Streeter, V.L. and Wylie, E.B. (1967). *Hydraulic transients.* McGraw-Hill.

Streeter, V.L. and Wylie, E.B. (1974). Waterhammer and surge control. *Ann. Rev. Fluid Mech.*, 6, 57–73.

Stripling, L.B. and Acosta, A.J. (1962). Cavitation in turbopumps - Part I. *ASME J. Basic Eng.*, 84, 326–338.

Stripling, L.B. (1962). Cavitation in turbopumps - Part II. *ASME J. Basic Eng.*, 84, 339–350.

Strub, R.A. (1963). Pressure fluctuations and fatigue stresses in storage pumps and pump-turbines. *ASME Paper No. 63-AHGT-11.*

Sturge, D.P. and Cumpsty, N.A. (1975). Two-dimensional method for calculating separated flow in a centrifugal impeller. *ASME J. Fluids Eng.*, 97, 581–579.

Tsujimoto, Y., Imaichi, K., Tomohiro, T., and Gatoo, M. (1986). A two-dimensional analysis of unsteady torque on mixed flow impellers. *ASME J. Fluids Eng.*, 108, No. 1, 26–33.

Tsujimoto, Y., Kamijo, K., and Yoshida, Y. (1992). A theoretical analysis of rotating cavitation in inducers. *ASME Cavitation and Multiphase Flow Forum*, FED 135, 159–166.

Tsukamoto, H. and Ohashi, H. (1982). Transient characteristics of centrifugal turbomachines. *ASME J. Fluids Eng.*, 104, No. 1, 6–14.

Tulin, M.P. (1953). Steady two-dimensional cavity flows about slender bodies. *David Taylor Model Basin Rep. 834.*

Tulin, M.P. (1964). Supercavitating flows - small perturbation theory. *J. Ship Res.*, 7, No. 3, 16–37.

Tyler, J.M. and Sofrin, T.G. (1962). Axial compressor noise studies. *Soc. Automotive Eng.*, 70, 309–332.

Vaage, R.D., Fidler, L.E., and Zehnle, R.A. (1972). Investigation of characteristics of feed system instabilities. *Final Rept. MCR-72-107, Martin Marietta Corp., Denver, Col.*

van der Braembussche, R. (1982). Rotating stall in vaneless diffusers of centrifugal compressors. *von Karman Inst. for Fluid Dyn., Technical Note 145.*

Verdon, J.M. (1985). Linearized unsteady aerodynamic theory. *United Technologies Research Center Report R85-151774-1.*

Wade, R.B. and Acosta, A.J. (1966). Experimental observations on the flow past a planoconvex hydrofoil. *ASME J. Basic Eng.*, 88, 273–283.

Wade, R.B. and Acosta, A.J. (1967). Investigation of cavitating cascades. *ASME J. Basic Eng., Series D*, 89, 693–706.

Whitehead, D. (1960). Force and moment coefficients for vibrating airfoils in cascade. *ARC R&M 3254, London.*

Wiesner, F.J. (1967). A review of slip factors for centrifugal impellers. *ASME J. Eng. for Power*, 89, 558–576.

Wijdieks, J. (1965). Greep op het ongrijpbare—II. Hydraulische aspecten bij het ontwerpen van pompinstallaties. *Delft Hydraulics Laboratory Publ. 43.*

Wislicenus, G.F. (1947). *Fluid mechanics of turbomachinery.* McGraw-Hill, New York.

Wood, G.M. (1963). Visual cavitation studies of mixed flow pump impellers. *ASME J. Basic Eng.*, Mar. 1963, 17–28.

Woods, L.C. (1955). On unsteady flow through a cascade of airfoils. *Proc. Roy. Soc. A*, 228, 50–65.

Woods, L.C. (1957). Aerodynamic forces on an oscillating aerofoil fitted with a spoiler. *Proc. Roy. Soc. A*, 239, 328–337.

Woods, L.C. (1961). *The theory of subsonic plane flow.* Cambridge Univ. Press.

Worster, R.C. (1963). The flow in volutes and its effect on centrifugal pump performance. *Proc. Inst. of Mech. Eng.*, 177, No. 31, 843–875.

Wu, T.Y. (1956). A free streamline theory for two-dimensional fully cavitated hydrofoils. *J. Math. Phys.*, 35, 236–265.

Wu, T.Y. (1962). A wake model for free streamline flow theory, Part 1. Fully and partially developed wake flows and cavity flows past an oblique flat plate. *J. Fluid Mech.*, 13, 161–181.

Wu, T.Y. and Wang, D.P. (1964). A wake model for free streamline flow theory, Part 2. Cavity flows past obstacles of arbitrary profile. *J. Fluid Mech.*, 18, 65–93.

Wu, T.Y. (1972). Cavity and wake flows. *Ann. Rev. Fluid Mech.*, 4, 243–284.

Yamamoto, K. (1991). Instability in a cavitating centrifugal pump. *JSME Int. J., Ser. II*, 34, 9–17.

Yoshida, Y., Murakami, Y., Tsurusaki, T., and Tsujimoto, Y. (1991). Rotating stalls in centrifugal impeller/vaned diffuser systems. *Proc. First ASME/JSME Joint Fluids Eng. Conf.*, FED-107, 125–130.

Young, W.E., Murphy, R., and Reddecliff, J.M. (1972). Study of cavitating inducer instabilities. *Pratt and Whitney Aircraft, Florida Research and Development Center, Rept. PWA FR-5131.*

Cavitation Instabilities in Turbopump Inducers

Yoshinobu Tsujimoto[1]

[1] Engineering Science, Osaka University, Osaka, Japan

Abstract. This article describes about various types of cavitation instabilities including rotating cavitation and alternate blade cavitation. In section 1, general characteristics of cavitating flow through turbopump inducers are reviewed. In section 2, rotating cavitation in 3- and 4-bladed inducers are discussed focusing on the effect of the number of blades. Section 3 is intended to show the relationship between the operating condition and various types of cavity oscillations in an inducer.

1 Cavitating Flow Through Inducers

In the first detailed report on cavitating flow through inducers, Acosta (1958) mentions also to cavitation instabilities. In this section general characteristics of cavitating flow through inducers are discussed based on this pioneering work.

1.1 Non-Cavitating and Cavitating Performance of Inducers

Figure 1.1 shows the non-cavitating total pressure rise coefficient $\psi = \Delta p_t / (\rho U_t^2)$ of herical non-swept inducers with the blade angles $\beta_o = 12°$, 9°, and 6°, at the tip, plotted against the flow coefficient $\phi = \bar{v}_x / U_t$, where Δp_t is the total pressure rise measured at the mid passage position, ρ the fluid density, U_t the impeller tip velocity, and $\bar{v}_x$ the mean axial velocity. The solidity at the tip is 2.5. The 12° inducer has 4 blades and the others 3 blades. Figure 1.2 may be consulted for the definition of impeller geometry.

The nominal incidence-free flow coefficients at the tip $\phi_{sf} = tan\, \beta_o$ are 0.212, 0.158, and 0.105 for the cases with $\beta_o = 12°$, 9°, and 6°, respectively. The flow coefficients with zero pressure rise are 75-85% of the incidence-free flow coefficients. The maximum efficiency occurs at 65-75% of the zero pressure rise flow coefficient. The nominal incidence angles at the tip and at the maximum efficiency ϕ_{opt} evaluated from $\beta_o - tan^{-1} \phi_{opt}$ are, 5.16°, 4.43°, and 2.57° for the cases with $\beta_o = 12°$, 9°, and 6°, respectively. Unlike other turbomachines, inducers are used with a certain incidence angle from this reason and more importantly to avoid premature head breakdown caused by the cavitation on the pressure side of the blades.

The performance curves have negative slope throughout the flow range measured. This is because the incidence angle does not exceed the small blade angle even with zero flow and hence the blade stall never occurs for inducers with a small blade angle.

1.2 Cavitation Performance

Figure 1.3 shows the suction performance of the impellers. Here, the cavitation number k is defined as $k = (p_1 - p_v)/(\rho w_1^2)$ where p_1 is the inlet pressure, p_v the vapor pressure, and w_1 the inlet relative velocity at the tip. Although not very clear, the breakdown cavitation number is smaller for smaller blade angle. It was found that for practically all the flow rates and impellers tested (6°, 9°, and 12°) the minimum cavitation number reached before breakdown was less than two times the cavitation number possible in a given cascade with a given angle of attack, $k \approx \alpha(\beta - \alpha)$ where α and β are local angle of attack and blade angle respectively, obtained from momentum considerations based on the fact that there is no net force parallel to the plate. This relation has a maximum of $\beta^2/4$ and is zero at $\alpha = 0$ and $\alpha = \beta$. The first result follows from the assumed zero thickness of the blades, and the latter, only occurs when there is no flow through the blade row. This formula also shows that smaller blade angles are better for obtaining lower cavitation numbers.

1.3 Development of Cavitation

The development of cavitation for the 12° inducer is shown in Figs.1.4-1.7 for various values of flow coefficient ϕ. At higher flow coefficient $\phi = 0.14$ shown in Fig.1.4, alternate blade cavitation is observed at smaller cavitation number of $k = 0.06$, with larger cavity on the second blade from the bottom (upstream). This arrangement is stable and it does not always occur on the same blades. It is described: "At lower flow rates, the alternate blade cavitation appears to propagate from blade to blade in much the same way as propagating stall in a cascade." This would be the first observation of rotating cavitation but no unsteady measurements were made. It is also described "The frequency of propagation depends upon the cavitation number, being high at high k's and decreasing to zero frequency just before cavitation breakdown." This character agrees with that of rotating cavitation described in the following sections. The boundary of the tip cavity is rather smooth at this higher flow rate.

At $\phi = 0.12$ (Fig.1.5) where the efficiency becomes the maximum, the boundary of the tip cavity becomes less smooth, especially for $k = 0.09$ and $k = 0.04$. This is caused by the backflow vortices, that are clearer at smaller flow rates. With $k = 0.09$ and $k = 0.04$, the upstream boundary of the tip cavity is in a plane nearly perpendicular to the inducer axis including the leading edge of the blades. At $k = 0.04$ where the cavity starts to extend into the blade passage, Fig.1.3 shows that the head starts to decrease. Near the cavitation breakdown (sudden decrease of head) with $k = 0.023$ and $k = 0.020$, the cavity extends into the blade passage and the cavity boundary becomes closer to the direction of the blade surface, suggesting higher axial flow velocity near the tip. This is caused by a change of the flow pattern due to cavitation.

As the flow rate is further decreased (Fig.1.6 for $\phi = 0.10$ and Fig.1.7 $\phi = 0.08$), the backflow vortex cavitation becomes clearer. The picture with $\phi = 0.08$ and $k = 0.13$ clearly shows that the vortex extends from the blade surface to the inlet pipe wall, surrounding the blade tip cavity. The backflow region is smaller for the cases of $k = 0.07$ and $k = 0.03$, as compared with the cases of $k = 0.15$ and $k = 0.13$ suggesting that the strength of backflow is decreased by the increase of cavitation.

1.4 Various Modes of Cavitating Flow and Head Breakdown

Figure 1.8 shows various modes of cavitating flow. For oscillating cavitation, it is described: " In this regime blade forces can be quite high and the various mechanical parts of the pump assembly can be easily excited to resonance." Thus the problems we are encountering now have been predicted nearly 50 years ago!! In addition, several methods for the suppression of the oscillating cavitation were examined and it was found that increasing the tip clearance and leading edge sweep has favorable effects for the suppression of oscillating cavitations. These methods for suppression are now being used as described in another lecture of the present course.

Thus, the existence of cavitation instabilities has been known from the early stage of inducer research but the detailed unsteady observations are now being made associated with the problems in large-scale rocket engines.

2 Rotating Cavitation in 3- and 4-Bladed Inducers

Although the existence of rotating cavitation and alternate blade cavitation was suggested in the pioneering paper described in section 1, detailed characteristics were not clarified. The number of blades affects the range of rotating cavitation through the occurrence of alternate blade cavitation. So, the character of rotating cavitation is explained in the present section for the cases of 3- and 4-bladed inducers.

2.1 Rotating Cavitation in 3-Bladed Inducers

The first detailed report of rotating cavitation was made by Rosenmann (1965) on a three bladed inducer with the blade angle of 8.99° and the solidity of 1.99 at the tip, with the design flow coefficient of $\phi_d = 0.111$ corresponding to a tip incidence angle of 2.66 degrees. The observation was made by the unsteady force measurement on the rotor. Figure 2.1 shows the results obtained at the design flow coefficient. Here, $\tau = (p_1 - p_v)/(\rho U_t^2/2)$ is the cavitation number, $\psi_H = \Delta p_t/(\rho U_t^2)$ the head coefficient, $\psi_p = \psi_H/\eta$ the power coefficient with the hydraulic efficiency η, C_s the force vector rotation coefficient defined as (rotational speed of radial force vector)/(shaft rotational speed) which corresponds to the propagation velocity ratio defined as (rotational speed of cavitating region)/(shaft rotational speed). C_R is the radial force vector coefficient normalized by the axial force evaluated as the product of the pressure difference across the inducer and the cross sectional blade area at inducer exit. We observe the following important characteristics of rotating cavitation:

(1) The rotating cavitation is observed at the design flow rate. This is quite different from the rotating stall which occurs at partial flow rate.
(2) The force vector rotation coefficient C_s is larger than 1, which means that the cavitating region rotates faster than the impeller. This is also different from rotating stall in which the stalled region rotates slower than the impeller.
(3) As the cavitation number τ is decreased, the radial force C_R suddenly increases at a certain cavitation number τ. There, the rotation coefficient is $C_s \approx 1.05 > 1$ and then it decreases to 1 with further decrease in τ.

(4) In a certain range of cavitation number ($0.035 < \tau < 0.050$) the radial force rotates fixed to the rotor and sometimes wanders at random. This is called attached asymmetric cavitation. The radial force decreases when the head starts to decrease steeply (this is called "head breakdown"). This means that rotating cavitation occurs in a range of cavitation number just above the head breakdown cavitation number. The small head decrease under the occurrence of rotating cavitation and the attached asymmetric cavitation is a result of those cavitation instabilities.

The first visual observation of rotating cavitation was made by Kamijo et al. (1980) on a three bladed inducer with the inlet blade angle 10° and the solidity 2.5 at the tip. Figure 2.2 shows the pictures from a high-speed film arranged to show the size of the cavity on each blade on every turn. The blade passes in the order of blade number 1, 2 and 3. The time proceeds from left to right. Here we focus on the shorter cavity. It moves from blade number 2 to 1 then 3 and returns to the original blade 2, in 4 turns of the impeller. Thus the cavity moves in the direction of impeller rotation and makes complete one turn relative to the impeller, while the impeller makes 4 turns. This means that the propagation velocity ratio is (4+1)/4=1.25.

2.2 Rotating Cavitation and Alternate Blade Cavitation in a 4-Bladed Inducer

To study the cavitation instabilities in a 4-bladed inducer, water and liquid hydrogen tests were made with VULCAIN liquid hydrogen turbopump inducer (Goirand et al., 1992). Figure 2.3 shows the sketch of the cavity and the spectrum of pressure fluctuation at nominal flow coefficient obtained from a water test. It is reported that no significant dependence on the rotational speed was found in the water tests. In this figure, τ and τ_0 are the cavitation number and the reference cavitation number, f and f_0 are the frequency of the pressure fluctuation and the rotational frequency of the impeller, F_r and $F_{r,0}$ are the radial forces under cavitating and non-cavitating conditions, respectively.

(1) At sufficiently large cavitation number with $\tau/\tau_0 = 1.477$, equal cavity appears on each blade and the frequency of the pressure fluctuation is $4f_0$, the blade passing frequency.
(2) At $\tau/\tau_0 = 1.055$, alternate blade cavitation appears with stable and balanced pattern, with the pressure fluctuation of $f/f_0 = 2.0$.
(3) At $\tau/\tau_0 = 0.738$, rotating cavitation with unbalanced geometry appears. The frequency of pressure fluctuation is slightly higher than the rotational frequency. Radial force is 3 to 5 times that under non-cavitating condition.
(4) At $\tau/\tau_0 = 0.423$, either of the following two modes occurs. One is the unbalanced pattern fixed to the impeller with the pressure fluctuation of $f/f_0 = 1.0$. The other is balanced equal cavitation.
(5) At $\tau/\tau_0 = 0.336$, stable and balanced flow with fully developed cavitation occurs at the beginning of inducer head drop.

The frequency ratio f/f_0 of the rotating cavitation is summarized in Fig.2.4. The results of liquid hydrogen (LH2) tests with three rotational speeds $\Omega = 0.77\Omega_0, 1.0\Omega_0, 1.08\Omega_0$ (Ω_0 is the reference speed) and the results of 3-bladed inducer by Kamijo et al. (1980) are also shown. The range of frequency and its dependence on the cavitation number are the same for all cases. If we compare the results of water tests with three (Kamijo's) and four (Vulcain LH2) bladed inducers, we find that the onset cavitation number is shifted to lower values for the case with the four bladed inducer. This is caused by the alternate blade cavitation which occurs at a larger cavitation number for the case with 4-bladed inducer. In the LH2 test, the occurrence region is shifted to still smaller cavitation number, caused by the thermal effect.

3 Various Cavitation Instabilities in a 3-Bladed Inducer

In the preceding sections we focused on rotating cavitation and alternate blade cavitation. In this section, various types of cavity oscillations are shown in a wide range of flow rate and cavitation number (Tsujimoto et al. (1997)).

3.1 Experimental Apparatus

The pump loop used is shown in Fig.3.1. The baseline experiments were made with an inlet pipe A with an inner diameter 200mm and without the outlet accumulator B. The system parameter was changed by replacing the inlet pipe A with one with a diameter 150mm or by adding an accumulator B with a volume of air $1.5\times10^{-3}\,m^3$ under the test condition. The base pressure (and hence the cavitation number) was adjusted by using a vacuum pump connected to the pressure control tank with about $0.65\,m^3$ of gas/vapor.

Figure 3.2 shows the test section and the performance curve. The impeller is a scale model of LOX turbopump inducer for HII rocket, with three blades $Z_i = 3$, outer diameter 149.8 mm, blade angles at the tip 7.5° at the inlet and 9.0° at the outlet; the housing is made of transparent acrylic resin, with inner diameter 150.8 mm (constant tip clearance of 0.5mm). The inlet pressure fluctuation was measured at 27.5mm upstream of the inducer leading edge, at two circumferential locations with various separation angles θ. The number of cells was determined from a plot of the phase difference of those pressure signals. The rotational frequency of the impeller was maintained $N = 4{,}000 \pm 2rpm$. The static pressure performance is also shown in Fig.3.2. The performance curve has a negative slope throughout the flow range and no conventional surge nor rotating stall is expected to occur. The design point of the inducer is ($\phi = 0.078$, $\psi_s = 0.130$). Since the flow meter was not available at the time of unsteady pressure measurements, the operating condition is shown by using the value of the static pressure coefficient ψ_s at sufficiently large cavitation number.

3.2 Cell Number Identification

Figure 3.3(a) shows an example of cascade plot of inlet pressure spectra for various cavitation numbers σ, for a static pressure coefficient $\psi_s = 0.123$ and a rotational frequency $f_N = 4000/60 \approx 66.7Hz$. Hence the blade passing frequency is $3f_N = 200Hz$. We observe the frequency components $i-v$ and $i'-v'$. If we represent the frequencies of i and i' components by f_i and $f_{i'}$, we observe the following relationship

$$f_i + f_{i'} = 3f_N \tag{3.1}$$

within the accuracy of $\pm1.5Hz$. Hence, it is quite possible that either one of i or i' components is substantial and the other appears as a result of a nonlinear interaction with the blade passing component.

Figure 3.3(b) shows the phase difference $Arg(\tilde{p}_\theta / \tilde{p}_0)$ plotted against the angular separation θ of the pressure taps (see Fig.3.2), where the pressure fluctuation is represented by $p_\theta = \tilde{p}_\theta \exp(2\pi jft)$. By definition, the total amount of the continuous change of $Arg(\tilde{p}_\theta / \tilde{p}_0)$ corresponding to the change in θ of 2π, is 2π times the number of cell n. If $Arg(\tilde{p}_\theta / \tilde{p}_0)$ decreases/increases with an increase in θ, it shows that the pressure pattern rotates in the direction of/opposite to the impeller rotation. We represent the direction by the plus/minus signs on n for co-/counter rotating patterns. From the plots similar to Fig.3.3 (b), the value of n' for the i' component is also determined. Then we can obtain the following relationship consistent to the results of nonlinear interaction considerations.

$$n + n' = Z_i = 3 \tag{3.2}$$

The propagation velocity ratio, defined as the ratio of the rotational velocity of pressure pattern to the impeller rotational frequency is given by

$$f/(nf_N) \tag{3.3}$$

The number of cell n and the propagation velocity ratio $f/(nf_N)$ thus determined are shown in Fig.3.3(b). We cannot distinguish the "substantial" component from the pressure measurements alone. However, from the flow visualization and other studies described later, the components $i - v$ appear to be the substantial components.

3.3 Map of Oscillating Cavitation

Figure 3.4 presents the cascade plot of inlet pressure power spectra for various values of static pressure coefficient ψ_s. The design point is $\psi_s = 0.130$. The number of cells and the propagation velocity ratio for each component are summarized in Table 3.1. The ranges of occurrence of each component are shown in the suction performance plots of Fig.3.5.

3.4 Interpretation of Oscillating Cavitation

Cavitation in backflow vortices. Figure 3.6(a) presents a picture from high-speed video under the condition with the component i at $\psi_s = 0.130$ and $\sigma = 0.015$. We observe five clouds of cavitation extending upstream from the inducer inlet. They are supposed to be formed in the backflow vortices as observed by Acosta (1958). The cavitation clouds slowly rotate at an angular velocity close to that corresponding to $f/(nf_N) = 0.16$. Hence, the component i is considered to be caused by the backflow vortices. Figure 3.6(b) shows the picture at a larger cavitation number $\sigma = 0.07$ where the component v is found. We find a system of cavitation in the tip leakage flow, which basically rotates attached to the blades, and that in backflow vortices formed on the boundary of the backflow region, rotating much slower than the impeller, as sketched in Fig. 3.6(c). As shown in Fig 3.6(b), the backflow vortices occur rather irregularly and it was difficult to determine the its number and speed definitely from the video picture. However, they were found to be consistent with the values $n = 5$ and $f/(nf_N) = 0.21$ determined from the pressure measurements. The passage of the backflow vortices is considered to be the cause of the component v. Although the structure similar to that as sketched in Fig. 3.6(c) is observed at other conditions, distinct pressure fluctuation could be found only in the regions of i and v shown in Fig.3.5.

Forward traveling rotating cavitation. Figure 3.7 shows the fluctuation of cavity length l_c on three blades of the impeller under the condition with the component iv. Forward propagation of cavity pattern is clearly shown and hence the component iv is the normal rotating cavitation. It should be noted that the maximum cavity length is slightly larger than the circumferential blade spacing H and the mean value of l_c / H is about 0.75. As shown in Fig. 3.5, rotating cavitation iv appears in two separate regions with ψ_s larger/smaller than the design value. In the region with higher static pressure coefficient, the length of blade cavity is smaller and the cavitating region extends more upstream. With the attached cavitation component iii, we observe one shorter cavity and two longer cavities fixed to the three blades of the impeller. Hence, the component iii is caused by unequal cavities attached to the rotor. With the component ii, the change in the size of the cavities is not so clear as with the components iii and iv. However, the phase plots in Fig. 3.3(b) for ii is quite similar to that for iii and iv. This resemblance suggests that the component ii with $n=1$ and $f/(nf_N)=0.9$ is substantial. If so, this could be the "backward traveling mode" predicted by the 1-D stability analysis (Tsujimoto et al. (1993)) presented in another lecture of this series.

Surge mode oscillation. For the components vi, vii, and vii', the pressure fluctuations are in phase at all circumferential locations. They are called herein "surge mode oscillation" after "surge" in gas handling turbomachines. The fluctuation of the cavity length is plotted in Fig. 3.8 under the surge mode oscillation vi. The major differences between the component vi and the components vii and vii' are that, for the component vi, the amplitude is larger and the frequency is fairly constant (18 Hz) as shown in Fig.3.4. Figure 3.9 shows the effects of rotational frequency f_N at $\psi_s = 0.08$. A detailed examination shows that the frequencies of iv and vii' are proportional to f_N, while that of vi is fixed at 18 Hz. At $\psi_s = 0.08$, it was also found that $f_{iv} - f_N$ is equal to $f_{vii'}$ within the measurement error ($\pm 1.5Hz$). Hence, the component vii' is caused by a nonlinear interaction of component iv (rotating cavitation) with the rotational frequency component. The difference, $f_{iv} - f_N$, corresponds to the frequency of cavity oscillation observed on a blade. Figure 3.9 shows that the component vi occurs when the frequency of vii' approaches 18 Hz. This suggests that component vi is a result of the resonance of a certain vibration mode with 18 Hz and the cavity oscillation due to rotating cavitation. Various attempts have been made to identify the "vibration mode," but no clear mode could be found.

Although the frequency of vii is also close to $f_{iv} - f_N$, a meaningful difference (larger than $\pm 1.5Hz$, up to $\pm 5Hz$) was observed. In addition, the component vii can appear even without the rotating cavitation component iv. Figure 3.10 shows the effects of the rotational speed, N, on the frequency of the component vii. The frequency is nearly proportional to N at lower frequencies but the rate of increase decreases at higher frequencies. The component vii is considered to be a substantial one and herein called "cavitation surge", since the proportionality of the frequency to the rotational speed is an important characteristics of cavitation surge. No "resonance" (or, a component equivalent to vi) was observed even when f_{vii} approached 18 Hz. This suggests that the component vi is not a simple “resonance”.

The modes of pressure fluctuations iv, vi, and vii in the inlet and outlet pipes are shown in Fig.3.11. For the surge mode oscillations vi and vii, the phase is nearly constant throughout the pipes and the amplitude decreases linearly as the tank is approached from the inducer. The pressure fluctuation amplitude of the rotating cavitation component iv is significantly smaller at the inducer outlet. It is quite surprising that the rotating cavitation component iv can be observed as far downstream as 33 impeller diameters from the impeller.

3.5 Effects of Piping System

In order to examine the effects of piping system, experiments were carried out with the following two additional configurations (see Fig. 3.1):

-Diameter of the inlet pipe A is decreased from 200mm to 150mm (Case I).

-An accumulator B filled with air is added (CaseII).

The frequency of each component is plotted in Fig. 3.12. The surge mode oscillation vi disappeared when we added the accumulator B. We observe no significant change in the frequencies except for the cavitation surge vii. This shows that all of the rotating cavitation types $i-v$ are quite independent on the system. The frequency of cavitation surge vii is decreased by the decrease of the inlet pipe diameter (increase of effective inlet pipe length) and increased by the addition of the tank B (decrease of effective outlet pipe length). Although the frequency of cavitation surge is system dependent as shown above, Brennen (1994) proposes an empirical relation between f/f_N and σ for cavitation surge. This relation is shown by the dashed curve in Fig.3.12. The present result agrees qualitatively with the empirical relation. The frequency of the surge mode oscillation vi was not altered by the change of the inlet pipe but it disappeared if we added the accumulator to the outlet line. It disappeared also when a small amount of air was introduced to the outlet line. These observations suggest that the component vi is closely related with the outlet line but it is not fully understood yet.

4 Conclusion

Characteristics of cavitation instabilities in turbopump inducers are summarized as follows:

(1) Cavitation instabilities can be found also at design flow rate.
(2) Cavitation instabilities occur at a higher cavitation number than the breakdown cavitation number.
(3) Alternate blade cavitation occurs only for inducers with even number of blades.
(4) The propagation velocity ratio of normal rotating cavitation is larger than 1 and gets smaller as the cavitation number is decreased.
(5) In some cases attached asymmetrical cavitation occurs at a cavitation number smaller than that of rotating cavitation.
(6) The frequency of cavitation surge is proportional to rotational frequency of the impeller.
(7) All cavitation instabilities cease once the head breakdown starts at lower cavitation number.

Bibliography

Acosta, A.J., (1958). An Experimental Study of Cavitating Inducers. *Second Symposium on Naval Hydrodynamics, Hydrodynamoc Noise, Cavity Flow.* August 25-29, Washington D.C., ACR-38, 533-557.

Brennen, C.E., (1994). *Hydrodynamics of Pumps.* Concepts ETI and Oxford University Press.

Goirand, B., Mertz, A-L., Jousellin, F., and Rebattet, C., (1992). Experimental Investigations of Radial Loads Induced by Partial Cavitation with a Liquid Hydrogen Inducer. *ImechE*, C453/056, 263-269.

Rosenmann, W., (1965). Experimental Investigations of Hydrodynamically Induced Shaft Forces With a Three Bladed Inducer. *Symposium on Cavitation in Fluid Machinery, ASME Winter Annual Meeting,* 172-195.

Kamijo, K., Shimura, T., and Watanabe, M., (1980). A Visual Observation of Cavitating Inducer Instability. *Technical report of National Aerospace Laboratory,* TR-598T.

Tsujimoto,Y, Yoshida, Y., Maekawa, Y., Watanabe, S., and Hashimoto, T., (1997). Observations of Oscillating Cavitation of an Inducer. *ASME Journal of Fluids Engineering,* Vol.119, No.4, December, 775-781.

Tsujimoto, Y., Kamijo, K., and Yoshida,Y., (1993). A Theoretical Analysis of Rotating Cavitation in Inducers. *ASME Journal of Fluids Engineering*, Vol.115, No.1, March, 135-141.

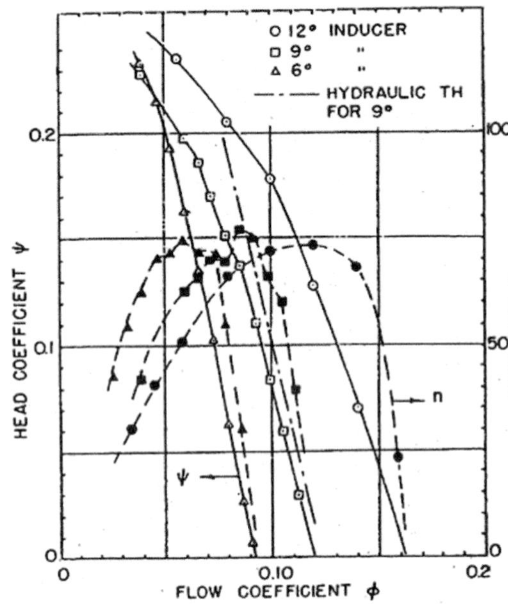

Figure 1.1 Non cavitating performance

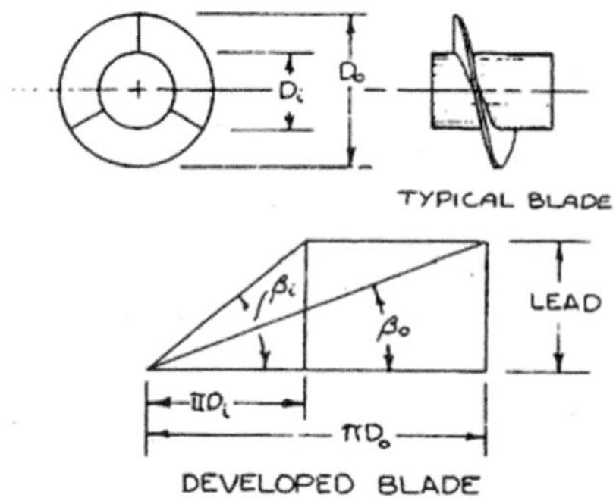

Figure 1.2 Definition of impeller geometry

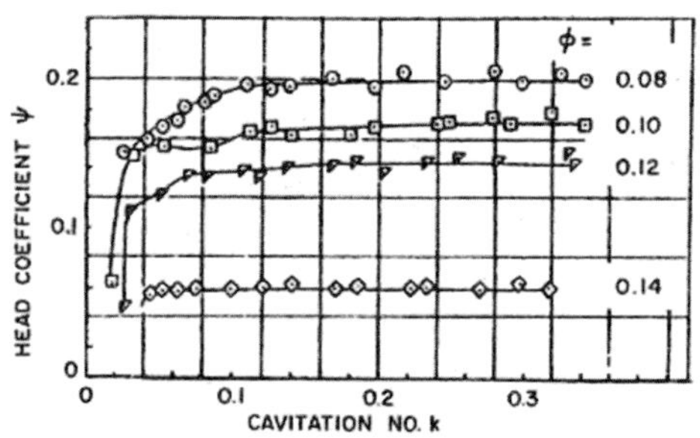

(a) 12° inducer

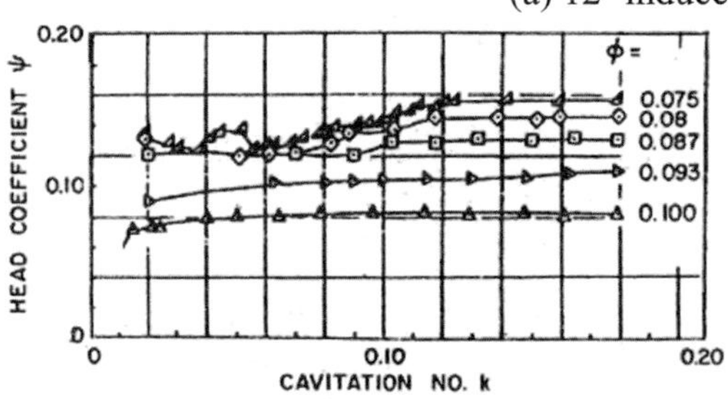

(b) 9° inducer

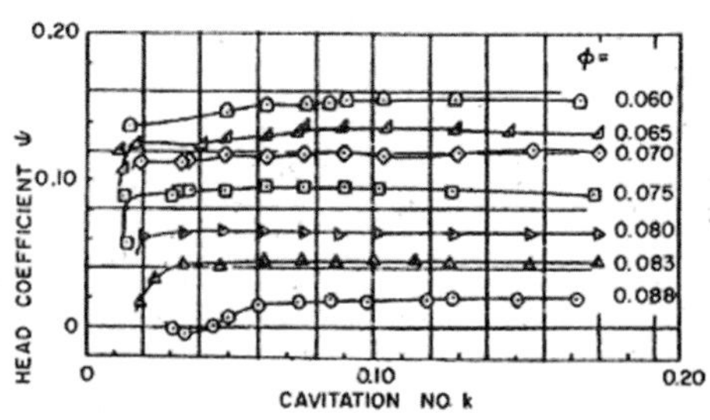

(c) 6° inducer

Figure 1.3 Suction performance

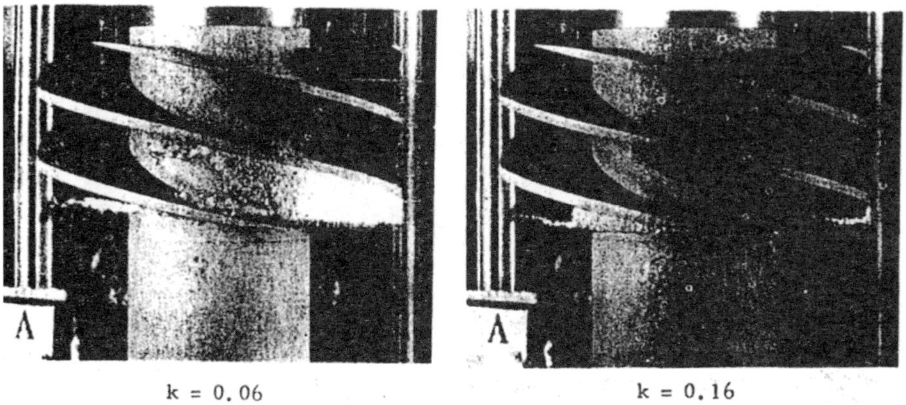

Figure 1.4 Cavitation on the 12° inducer at a flow coefficient of $\phi = 0.14$ showing the occurrence of alternate blade cavitation

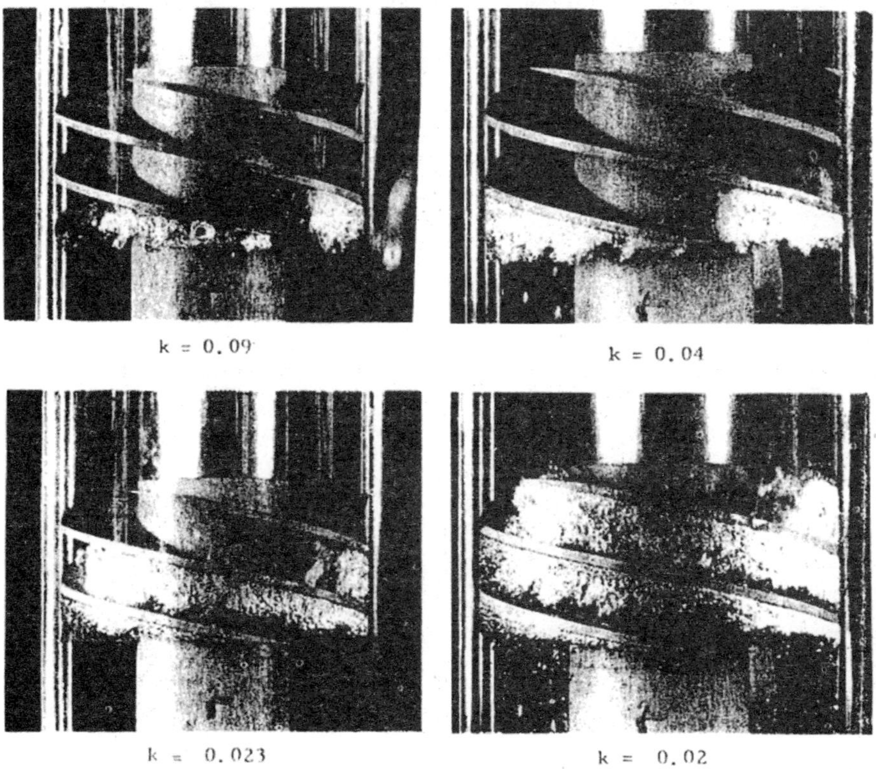

Figure 1.5 Development of cavitation in a 12° helical inducer for a flow coefficient $\phi = 0.12$

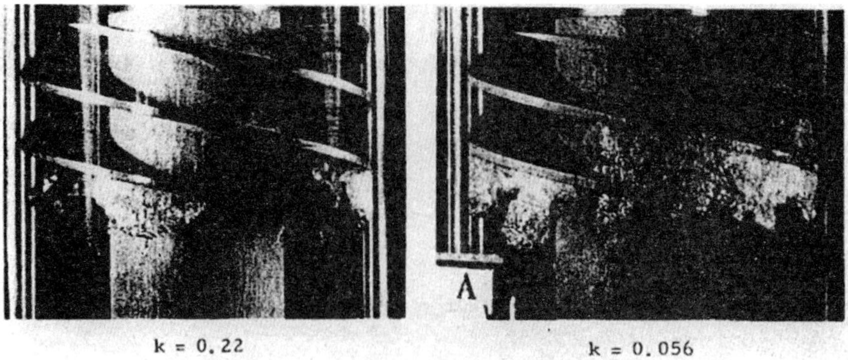

Figure 1.6 Cavitation on the 12° inducer at a flow coefficient of $\phi = 0.10$

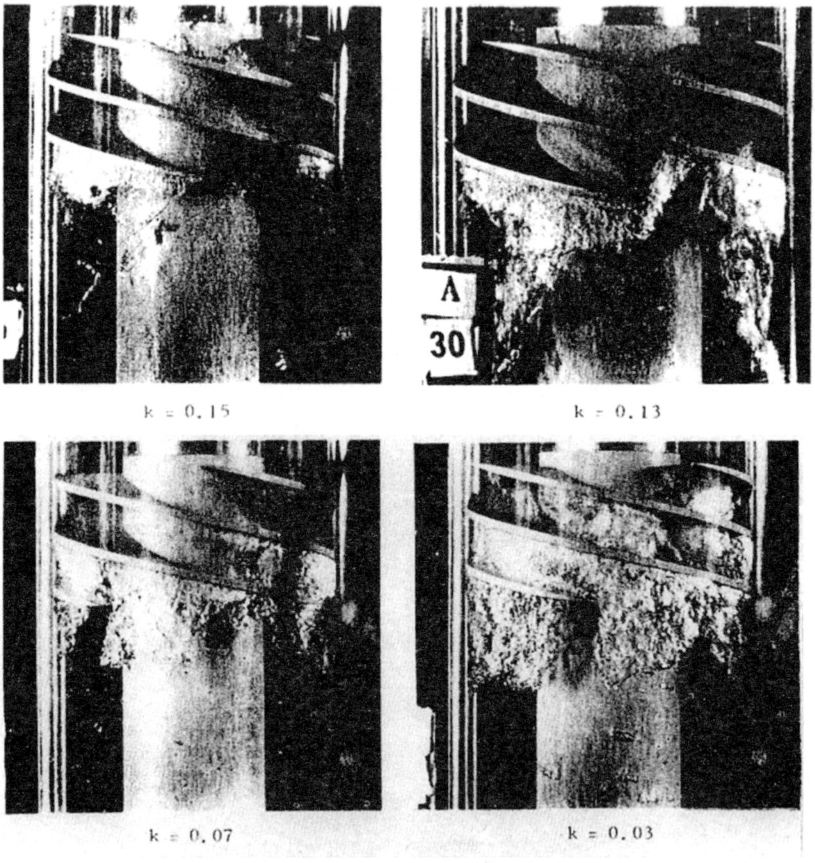

Figure 1.7 Cavitation development on the 12° inducer at a flow coefficient $\phi = 0.08$

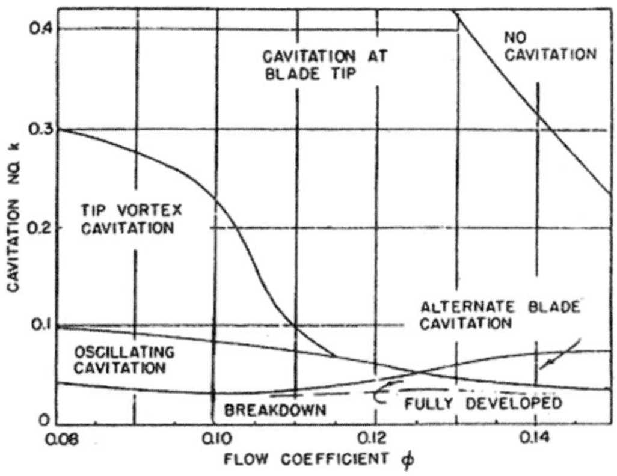

Figure 1.8 Various modes of cavitating flow in a 12° herical inducer as a function of cavitation number and flow coefficient

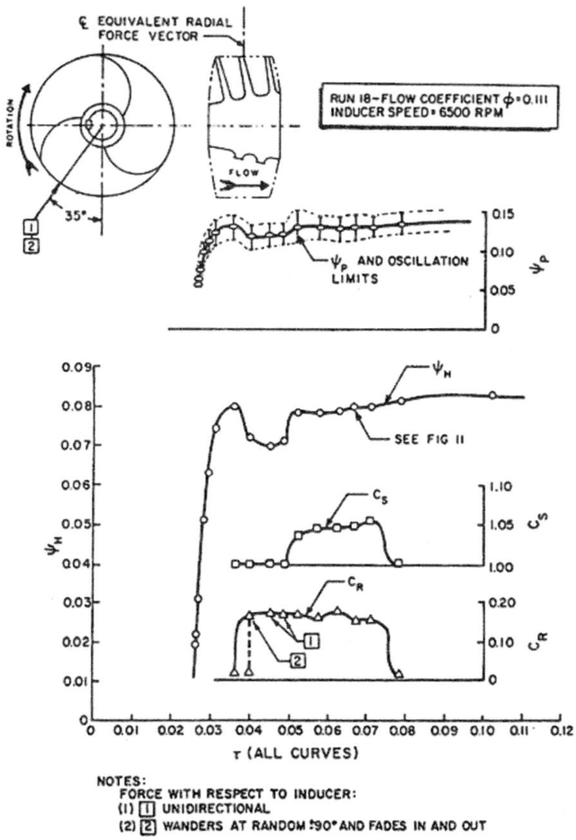

Figure 2.1 Radial forces, head coefficient vs cavitation number τ

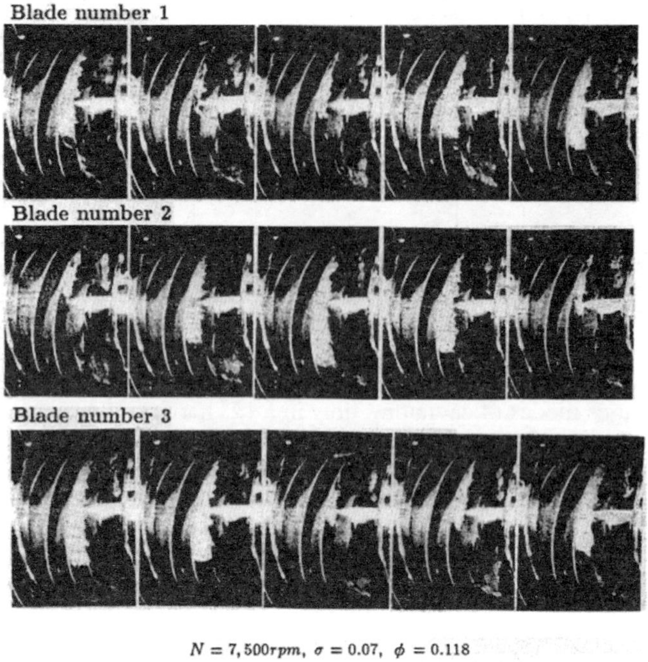

Figure 2.2 Rotating cavitation. Pictures are arranged to show the cavity size on each blade. The blades pass in the order of Blade number 1, 2, and 3. Time proceeds from left to right.

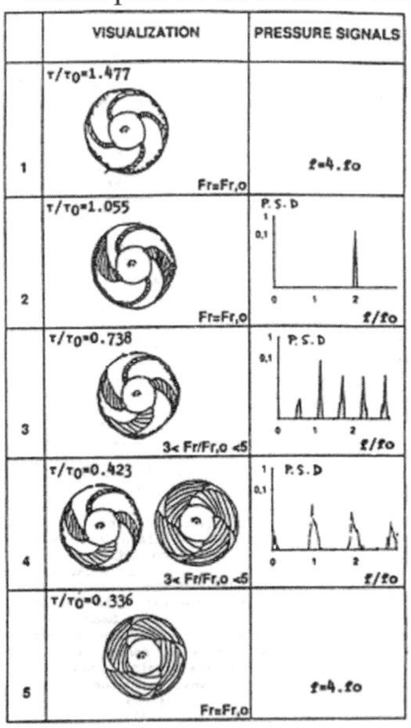

Figure 2.3 Cavity pattern at nominal flow rate

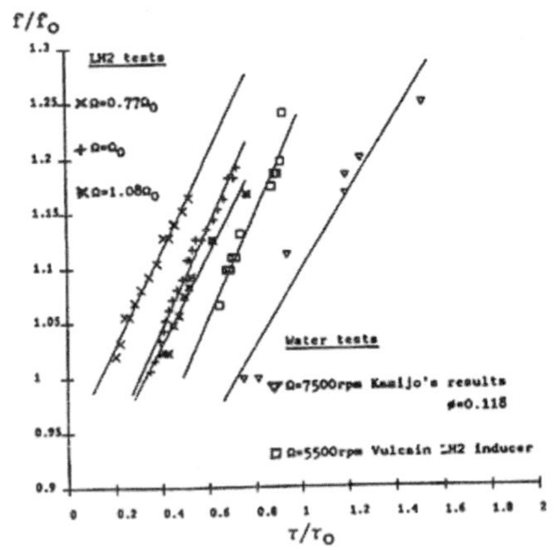

Figure 2.4 Propagation velocity ratio f / f_0 against cavitation number τ / τ_0

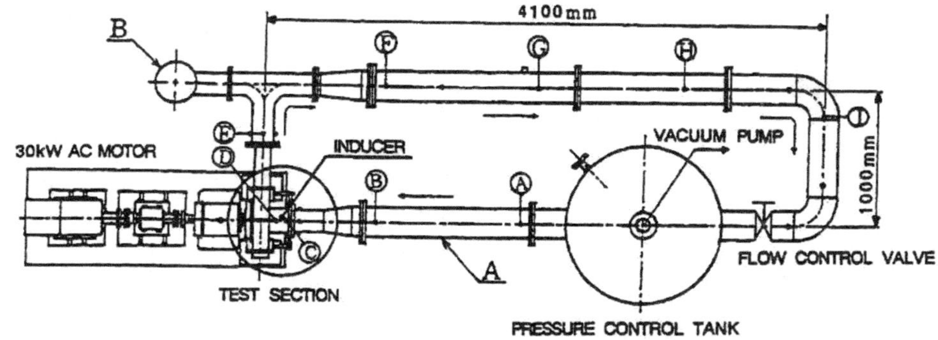

Figure 3.1 Test loop

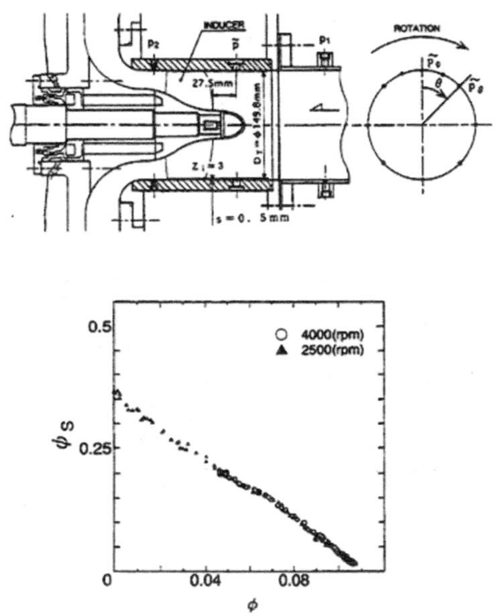

Figure 3.2 Inducer crosssection and inlet pressure measurement locations, and static pressure performance

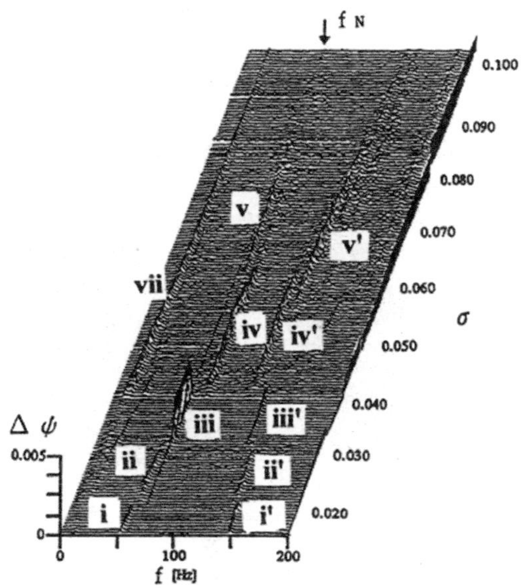

Figure 3.3 (a) Spectra of inlet pressure fluctuation for $\psi_S = 0.123 \pm 0.002$

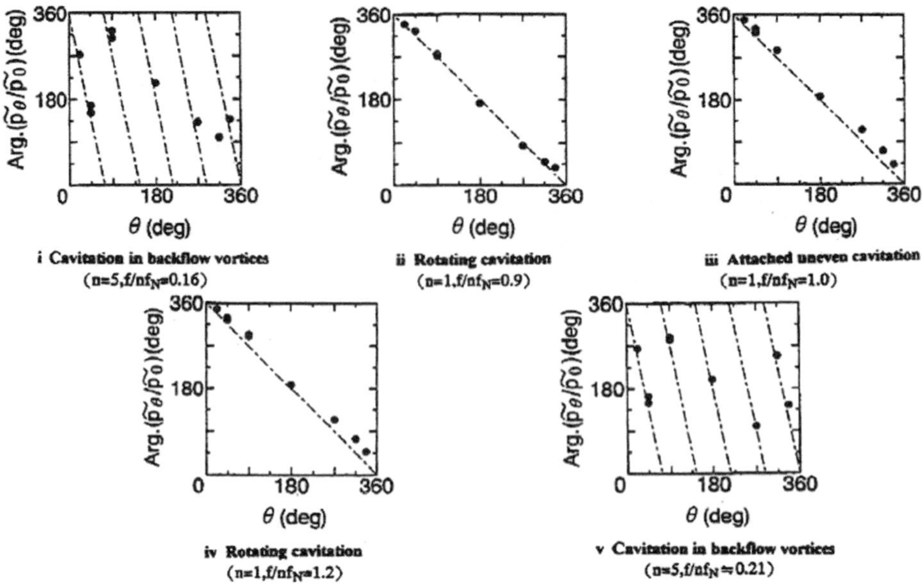

Figure 3.3(b) Phase difference $Arg(\widetilde{p}_\theta / \widetilde{p}_0)$ plotted against the angular separation θ of the ressure taps

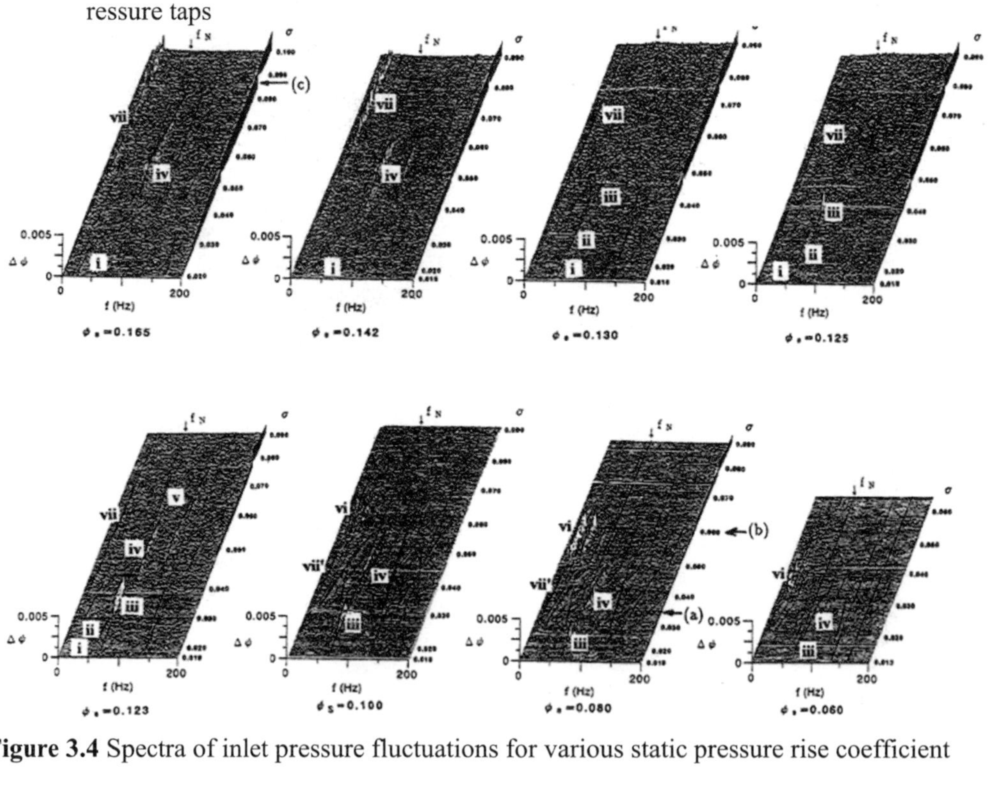

Figure 3.4 Spectra of inlet pressure fluctuations for various static pressure rise coefficient

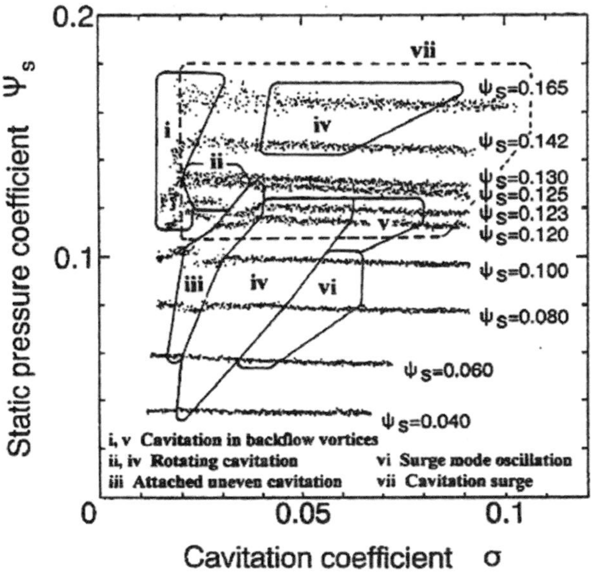

Figure 3.5 Suction performance and a map of various oscillating cavitation types

		Number of cells n	Rotational velocity ratio f/nf_N
i	Cavitation in backflow vortices	5	0.16
ii	Rotating cavitation	1	≈0.9
iii	Attached uneven cavitation	1	1.0
iv	Rotating cavitation	1	1.1~1.3
v	Cavitation in backflow vortices	5	0.21
vi	Surge mode oscillation	0	≈18Hz/ (3300~5000 rpm)
vii	Cavitation surge	0	8~19Hz/4000 rpm, proportional to rpm

Table 3.1 Number of cells and propagation velocity ratio

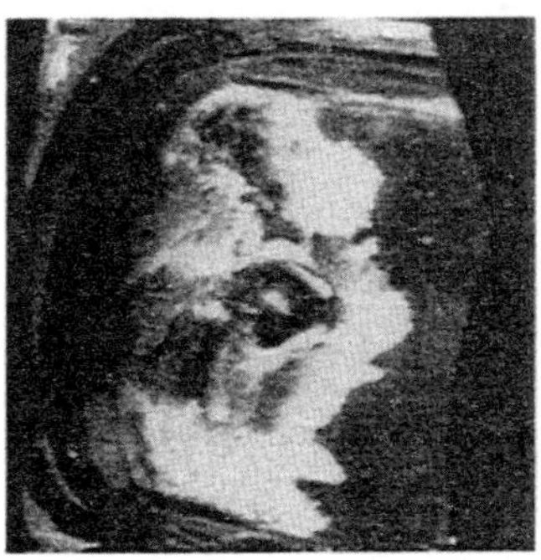

Figure 3.6(a) Cavitation in backflow vortices, component i, at smaller flow rate.

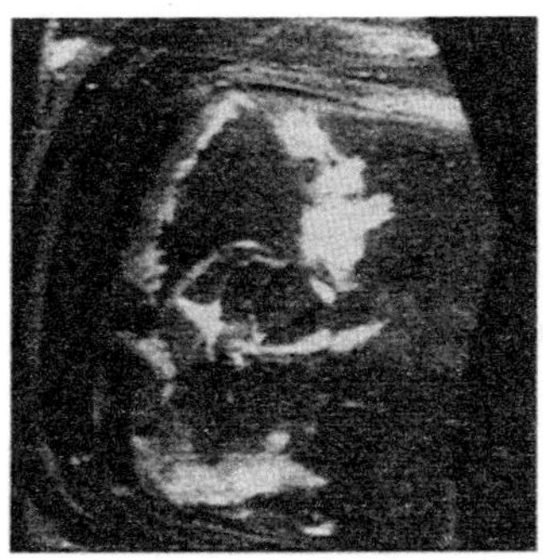

Figure 3.6(b) Cavitation in backflow vortices, component v, at larger flow rate

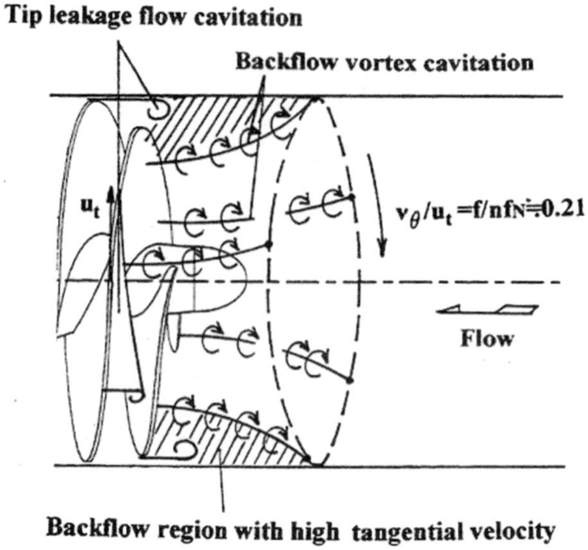

Figure 3.6(c) Tip leakage flow cavitation and backflow vortex cavitation

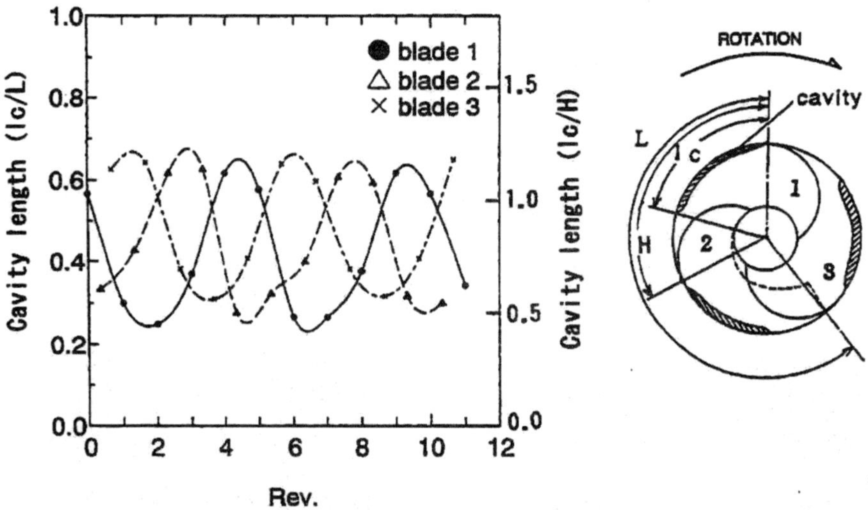

Figure 3.7 Oscillation of cavity length under rotating cavitation (Component *iv* , $\psi_s = 0.08$, $\sigma = 0.041$)

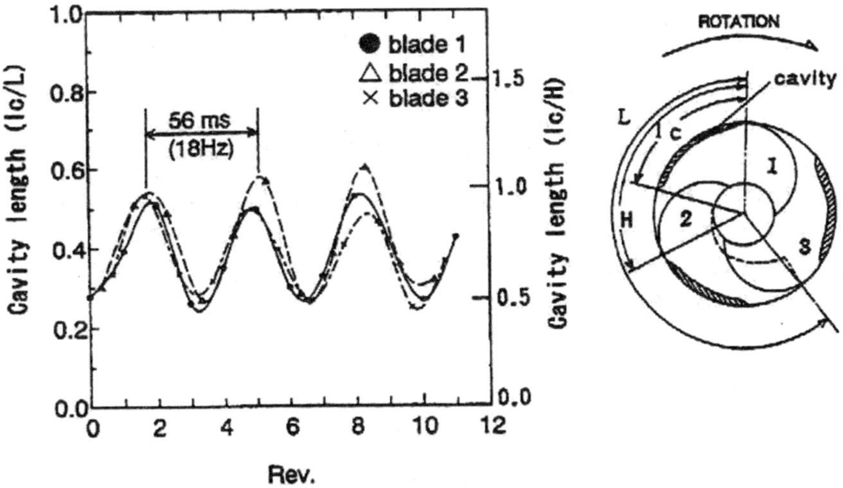

Figure 3.8 Oscillation of cavity length, under surge mode oscillation (Component *vi*, $\psi_S = 0.08$, $\sigma = 0.054$)

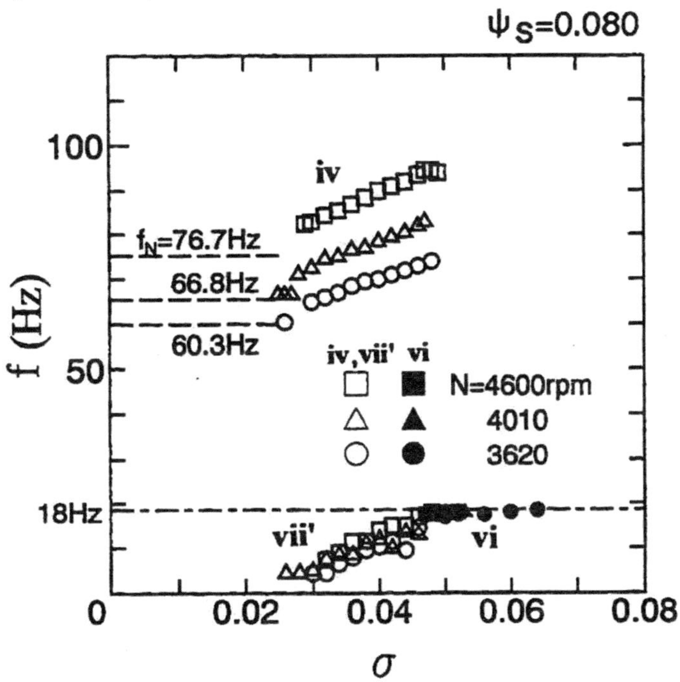

Figure 3.9 Effects of rotational speed *N* on the frequencies of rotating cavitation *iv*, surge mode oscillation *vi*, and cavitation surge *vii'*.

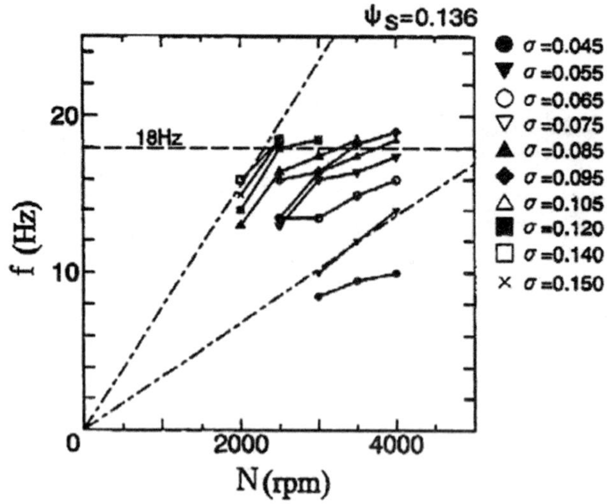

Figure 3.10 Effects of rotational speed N, on the frequencies of cavitation surge *vii* . The dash dot lines show proportionality relations between N and f

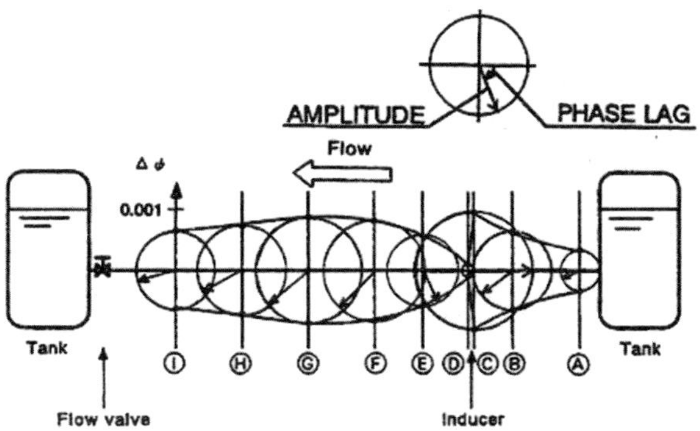

(a) Rotating cavitation *iv* ($\psi_S = 0.080$, $\sigma = 0.035$, point (a) in Fig.3.4)

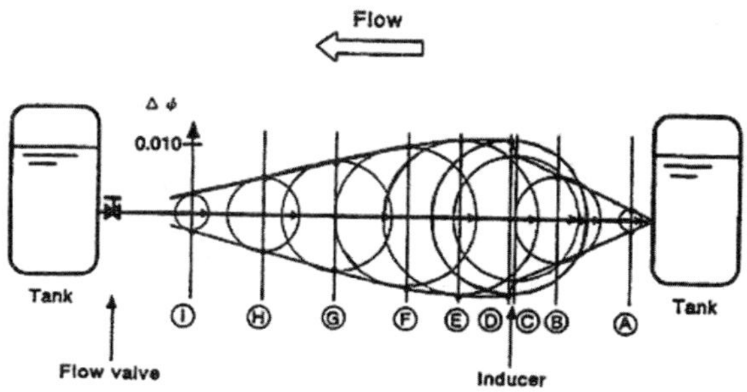

(b) Surge mode oscillation *vi* ($\psi_S = 0.080$, $\sigma = 0.060$, point (b) in Fig.3.4)

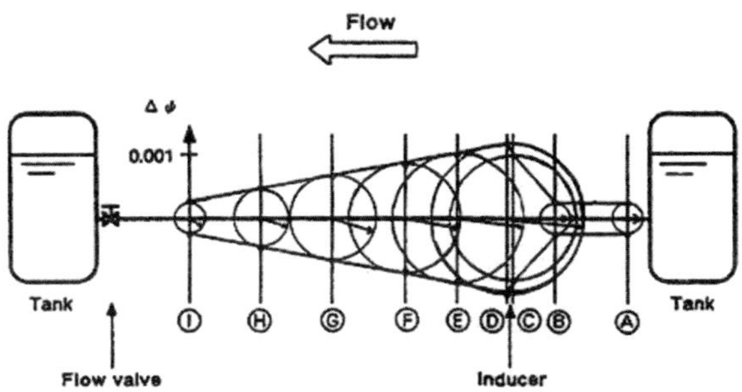

(c) Cavitation surge *vii* ($\psi_S = 0.165$, $\sigma = 0.087$, point (c) in Fig.3.4)

Figure 3.11 Modes of pressure fluctuations *iv*, *vi*, and *vii* represented by the amplitude and the phase at each axial locations shown in Fig.3.1

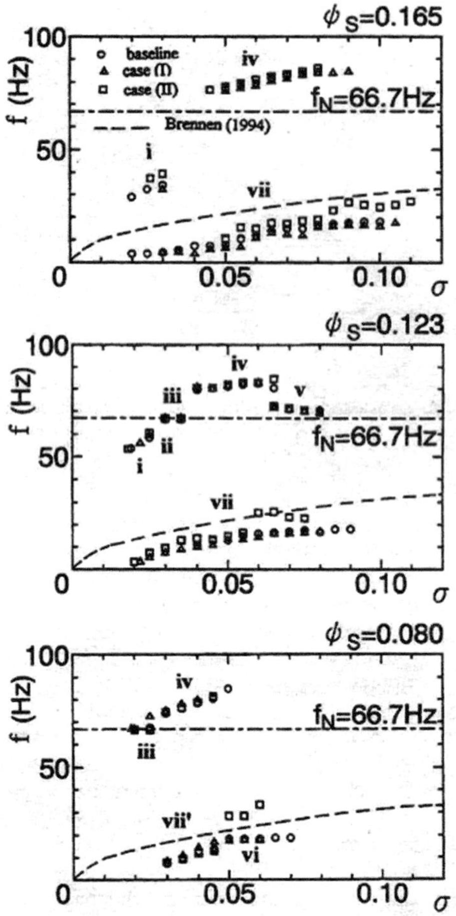

Figure 3.12 Effect of the piping system on the frequencies of oscillating cavitation

Stability Analysis of Cavitating Flows Through Inducers

Yoshinobu Tsujimoto[1]

[1] Engineering Science, Osaka University, Osaka, Japan

Abstract. Three types of stability analysis of cavitating flows through inducers are presented. The first is a one-dimensional analysis in which the characteristics of cavitation are represented by two factors that should be evaluated by other appropriate methods. This method has been used for the analysis of rotating cavitation and cavitation surge, as well as more general instabilities, surge and rotating stall to show the similarity and difference of the characteristics and the mechanisms of those instabilities. The second is a stability analysis of two-dimensional cavitating flow using a closed blade surface cavity model. This model has been used to clarify various types of cavitation instabilities. The third one was developed to understand a cavitation instability associated with the degradation of pressure performance due to cavitation.

1 One Dimensional Stability Analysis

Cavitation instabilities in inducers can be sorted into two types, cavitation surge and rotating cavitation, although they include various higher order modes. Cavitation surge is a system instability in which the flow rate of the hydraulic system fluctuates with in-phase cavity fluctuations on each blade. The effects of various system parameters on cavitation surge are examined in detail by Young (1972). Rotating cavitation is a local cavitation instability in which the cavities propagate from blade to blade in the same way as rotating stall. In the most typical mode, the cavitating region rotates faster than the impeller for rotating cavitation, while the stalled region rotates slower than the impeller in rotating stall.

Rotating cavitation, as well as cavitation surge, can be treated by assuming that the cavity volume V_c is a function of the cavitation number $\sigma_1 = (p_1 - p_v)/(\rho W_1^2/2)$, (p_1 : inlet pressure, p_v : vapor pressure, ρ : liquid density, W_1 : inlet relative velocity) and the local incidence angle to the blade α_1 (Tsujimoto et al., 1993). The cavity volume is normalized by using the blade spacing h and is represented by a:

$$a(\sigma_1, \alpha_1) = V_c/(h^2 \times 1) \tag{1}$$

Then, the mass flow gain factor M and the cavitation compliance K are defined as

$$M = \frac{\partial a}{\partial \alpha_1} \quad \text{and} \quad K = -\frac{\partial a}{\partial \sigma_1} \tag{2}$$

Since the cavity volume will increase if the incidence angle is increased or the inlet pressure is decreased, both mass flow gain factor and cavitation compliance should have positive values. These parameters are introduced by Brennen and Acosta (1976) and evaluated from quasi-steady calculations of blade surface cavitation. An extensive series of experiments were carried out and the results are reported in Ng and Brennen (1978) and Brennen et al. (1982). These are still the only reliable data and are recently assessed to include the dependence on the frequency (Rubin (2004)). A bubbly flow model was proposed by Brennen (1982) which can predict not only the values of M and K but also of all the transfer matrix elements correlating the pressure and mass flow fluctuations at the pump inlet and exit. Otsuka et al. (1996) carried out unsteady calculations of blade surface cavitation and represented M and K as complex functions of the frequency.

One-dimensional linear stability analyses are possible for surge, rotating stall, cavitation surge, and rotating cavitation (Tsujimoto et al. (2001)). We consider a system composed of an inlet conduit, an impeller cascade with infinite number of blades, a downstream tank, and an exit valve, as shown in Fig.1.1. The velocity triangle at the inlet of the impeller is shown in Fig.1.2. For one-dimensional instabilities of surge and cavitation surge, an axial velocity disturbance in a finite length upstream pipe is considered. For two-dimensional instabilities of rotating stall and rotating cavitation, a two-dimensional sinusoidal potential flow disturbance is assumed in the upstream of the impeller. In both cases the pressure fluctuation at the inlet of the impeller can be correlated with the axial flow velocity disturbance by applying the momentum equation to the inlet flow. This is why one-dimensional stability analysis is possible also for two-dimensional instabilities such as rotating stall and rotating cavitation. The pressure increase in the impeller is obtained by assuming that the flow in the impeller is perfectly guided by the blades. To simulate surge and rotating stall, two types of loss are assumed: an incidence loss proportional to the loss coefficient ς_s and the square of the incidence velocity ΔV at the inlet, and a through flow loss proportional to the loss coefficient ς_Q and the square of the flow velocity W_s through the impeller flow channel.

The effect of cavitation is taken into account in the continuity relation across the impeller by representing the cavitation characteristics by M and K. If we represent the axial velocity disturbance at the inlet and the outlet of the impeller by δu_1 and δu_2, respectively, and represent the cavity volume per blade by V_c, the continuity equation across the impeller can be represented by

$$h(\delta u_2 - \delta u_1) = (\partial / \partial t)\delta V_c \tag{3}$$

If we consider that the cavity volume is a function of the cavitation number σ_1 and the incidence angle α_1, the cavity volume fluctuation can be expressed as

$$\delta V_c = h^2 (M\delta\alpha_1 - K\delta\sigma_1) \tag{4}$$

with the mass flow gain factor M and the cavitation compliance K defined in Eq. (2). The effect of cavitation on the pressure performance of the impeller is neglected since most of cavitation instabilities occur in a range where the pressure performance is not affected by the existence of cavitation.

1.1 Surge, Cavitation Surge, Rotating Stall, and Rotating Cavitation

To obtain simple expressions of the onset condition and the frequency of surge, cavitation surge, rotating stall and rotation cavitation, various simplifying assumptions are made for each instability. For cavitation surge and rotating cavitation, it is assumed that the downstream flow rate fluctuation does not occur. This is a good approximation for typical inducers with a smaller blade angle. For surge, it is assumed that the flow from the rotor is discharged to a surge tank followed by an exit valve. For rotating stall, it is assumed that the flow from the impeller is discharged directly to a space of constant pressure. By writing down the relations connecting the flow disturbances in the upstream and the downstream of the impeller, we obtain a set of linear equations in terms of the amplitudes of fluctuations. From the coefficient matrix of the linear equations, we obtain a polynomial characteristic equation in terms of a complex frequency whose real part represents the frequency and the imaginary part the damping rate of possible instability mode.

The onset conditions and the frequencies obtained by solving the characteristic equations are summarized in Table 1.1, where $\psi_{ts} = (p_2 - p_{t1})/(\rho U_T^2)$ is the inlet total to outlet static pressure coefficient of the impeller, $\phi = U/U_T$ is the flow coefficient, with the mean axial velocity U and the circumferential velocity of the impeller U_T, β^* is the mean blade angle measured from axial direction, l is the chordlength of the blades, L is the length of the inlet conduit, R is the resistance of the exit valve, $B = \sqrt{\rho C/L}U_T$ is called Greitzer's B factor, with the compliance of a tank placed downstream of the impeller $C = V/(\rho a^2 f)$ for liquids or $C = A/(\rho g f)$ for gas, V the volume of the tank, a the speed of sound, f the cross-sectional area of the inlet pipe, A the free surface area of the surge tank, and g the gravitational acceleration constant. L_u is an impeller loss coefficient and β_2^* the blade outlet angle. s is the circumferential wavelength of the disturbance, $\overline{\beta_1}$ the mean flow angle at the inlet and ϕ^* the incidence free flow coefficient. V_p is the propagation velocity of rotating stall and rotating cavitation observed in a stationary frame, and n is the frequency of surge and cavitation surge.

The results shown in Table 1.1 show the following characteristics.

- Both surge and rotating stall occur at small flow rates with a positive slope of $\psi_{ts} - \phi$ performance. Under this condition, the head produced by the rotor will increase if the flow rate is increased, which accelerate the flow and results in further increase of the flow rate. This positive feedback is the cause of surge and rotating stall.
- Both cavitation surge and rotating cavitation occur when $M > 2(1+\sigma_1)\phi K$.
- The frequency of surge is basically identical to $n = 1/(2\pi\sqrt{\rho CL})$, the natural frequency of a Helmholtz resonator composed of a tank with the compliance C and an inlet pipe with the length L, and does not depend on the impeller speed.
- The frequency of cavitation surge is proportional to the impeller speed U_T and inversely proportional to the square root of L and K. This shows that cavitation surge is an oscillation of the upstream fluid associated with the compliance of cavitation in the impeller.
- The propagation velocity ratio defined as V_p/U_T for rotating stall is less than 1, suggesting that the stalled region rotates slower than the impeller.
- Two modes of rotating cavitation are predicted. One of them rotates faster than the impeller and this mode is generally observed in experiments (Tsujimoto et al. (1997)). The other mode rotates slower than the impeller and occasionally observed as a mode rotating in the opposite direction of the impeller. This will be discussed in the following section.

The criterion $M > 2(1+\sigma_1)\phi K$ for the onset of cavitation surge and rotating cavitation clearly shows the importance of the positive mass flow gain factor for which the cavity volume decreases as the flow rate increases. The mechanism of the instability can be explained as follows. When the flow rate is increased, the angle of attack to a rotor blade is decreased. If the value of mass flow gain factor is positive, the cavity volume will also decrease from the definition of mass flow gain factor. If the cavity volume is decreased, the inflow to the rotor will increase to fill up the space occupied by the cavity volume decreased. Thus, the increase of flow rate results in further increase of flow rate. This mechanism of instability depends totally on the continuity relation and is not associated with impeller performance. Actually, both rotating cavitation and cavitation surge occur at a higher inlet pressure where the performance degradation due to cavitation is insignificant.

On the contrary, positive cavitation compliance has an effect to suppress instabilities: when the inlet flow rate is increased, the inlet pressure will decrease due to the Bernoulli effects and the cavity volume will increase if $K > 0$, resulting in the decrease of the inlet flow rate. So, $K > 0$ provides negative feedback and stabilizes the system. The onset condition of cavitation surge and rotating cavitation does not depend on the steady $\psi_{ts} - \phi$ performance and they may occur even at the design flow rate. This makes the cavitation instabilities more serious than surge and rotating stall that occur at off design points. The mechanism of instabilities in turbomachinery is summarized in Fig.1.3.

1.2 Relation Between Rotating Stall and Rotating Cavitation

In order to illustrate the relation between rotating stall and rotating cavitation, a stability analysis was made under the assumption that the two-dimensional flow extends to upstream and downstream infinity (Tsujimoto, 1993). In the downstream, the flow is composed of a potential flow disturbance and a vortical flow disturbance due to the vorticity shed from the impeller caused by the unsteadiness of the flow. The mass continuity and pressure rise relations across the cascade result in a third order characteristic equation of the non-dimensional frequency k^* whose real part k_R^* gives the propagation velocity ratio V_p / U_T and the imaginary part k_I^* the damping rate:

$$(k^* - k_1^*)(k^* - k_2^*)(k^* - k_3^*) = 0 \tag{5}$$

Examination of three roots of Eq. (5). Figure 1.4 shows the roots for an inducer tested by Kamijo et al. (1980) for which the first detailed visual observation of rotating cavitation was made. The static performance of the inducer is shown in Fig.1.5, from which the values of the loss coefficients ζ_Q and ζ_S are determined. In this figure, ϕ^* and ψ^* are the flow and pressure coefficients normalized by using the inducer tip speed. ψ_{th}^* is the Euler's head at the mean radius, ψ_t^* is the total head and ψ_{ts}^* is the inlet total to outlet static pressure coefficient.

Figure 1.4 (a)-(j) show the three roots, $k_i^* = (k_{Ri}^*, k_{Ii}^*)$, $i = 1,3$, of the characteristic equation (5), with the value of parameters which is different from the standard values shown in (a). They are assigned to k_1^*, k_2^* and k_3^* following the criteria ($k_{R1}^* > 1$, $k_{R2}^* < 0$ and $k_{R3}^* < 1$). The values in the lower line of k_1^* and k_2^* in (a) are the values for the rotating cavitation obtained from the simplified analysis discussed in the preceding section, and the values in the lower line of k_3^* in (a)-(g) are the values for rotating stall by the simplified analysis.

As shown in Fig. 1.4(a), the values of k_1^*, k_2^* and k_3^* are close to the values determined from the simplified analysis. This suggests that k_1^* and k_2^* represent the rotating cavitation and k_3* the rotating stall, and that they can be approximately treated by the simplified method outlined in the last section. For the standard case (a), the imaginary parts of k_1*, k_2* and k_3* are all negative, showing that both rotating cavitation and rotating stall can occur simultaneously. The fact that the root k_3*, which represents rotating stall estimated under the effect of cavitation, is close to the non-cavitating rotating stall solution, means that the rotating stall is not affected largely by the cavitation. As shown in (b)-(g), the rotating stall is damped ($k_{I3}* > 0$) when $\phi*$ increases or the shock loss is neglected. On the other hand, the values of k_1* and k_2* are almost independent on the values of $\phi*$, ζ_Q, ζ_S and σ, so long as the values of M and K are kept constant. Rotating cavitations are amplified even with a negative slope of $\psi_{ts}*$ at larger $\phi*$ or with $\zeta_S = 0$, which is quite different from the case of rotating stall.

From these numerical results, we can conclude that rotating cavitation and rotating stall are, practically, mutually independent and completely different phenomena, in the sense that their causes are different and that they behave differently, although both can be treated and deduced from the same characteristic equation (5).

Table 1.2 shows the relative amplitudes of the pressure and axial velocity fluctuations at the inlet and outlet of the cascade, for the case of Fig.1.4 (a) and corresponding to $\delta p_1 / \rho U_1^2 = 1$. For each case δp_2 is much smaller than δp_1. For rotating cavitation (k_1* and k_2*), δu_2 is small compared with δu_1, showing that the fluctuation at the inlet is almost absorbed by the change of cavity volume. This is caused by the fact that the blade angle β_2* is close to $\pi / 2$, and supports the experimentally obtained conclusion (Kamijo et al. (1980)) that "rotating cavitation is related mainly to the inlet flow conditions." On the other hand, for rotating stall (k_3*), δu_2 is nearly equal to δu_1, with a small effect of cavity volume change.

As shown above, direct effects of $\phi*$ and σ on k_1* and k_2* are small. It can be shown that k_1* and k_2* are mainly dependent on M and K. Since M and K are functions of $\phi*$ and σ, rotating cavitations are affected by $\phi*$ and σ through the values of M and K.

Rotating cavitation. Contour maps of k_1* and k_2* in the M - K plane are shown in Figs.1.6 and 1.7. The values of parameters not specified in the figures are the same as those in Fig.1.4(a). The solid lines are obtained from Eq.(5), while the broken lines are determined from the simplified analysis ("Eqs.(15), (17) " in the figure show the results of the simplified analysis). The difference between these results is small, showing that the simplified analysis simulates rotating cavitation very well. The neutral stability curve is shown by the solid line of $k_I* = 0$, which is close to the criterion obtained by the simplified analysis. The rotating cavitation grows in the region with $k_I* < 0$, beneath the neutral stability curve.

In order to make comparisons with experimental results, calculations were made also for $\sigma = 0.06$ and 0.02. It was found that the contour maps are almost unchanged. Hence, the ranges of M and K for three values of σ are shown in the figures, estimated from Brennen et al. (1982). For k_1* shown in Fig.1.6, the propagation velocity ratio for $\sigma = 0.02$ is $k_R* = 1.1 - 1.4$, which is close to the experimental value of $k_R* = 1.16$ (Kamijo et al. (1980)).

As we reduce the cavitation number, the estimated ranges of M and K shifts to the location with smaller propagation velocity ratio k_R*. The experiment shows a similar tendency. Figure 1.8 shows the supersynchronous shaft vibration of LE-7 LOX turbopump caused by the k_1* rotating cavitation. The reduction of the supersynchronous frequency with time is caused by the reduction of the inlet pressure with time, showing the above-mentioned tendency.

Since $k_{R2}*<0$, the characteristic root k_2* corresponds to a mode of rotating cavitation which propagates in the direction opposite that of the impeller rotation. This backward rotating cavitation was found later by Hashimoto et al. (1997). The propagation velocity ratio V_p/U_T observed was -1.36 at $\sigma=0.072$, which agrees with the result in Fig.1.7 $k_R*\approx -1.25$ for $\sigma=0.06$.

Usually, only forward rotating cavitation corresponding to k_1* is observed and the observation of the backward rotating cavitation corresponding to k_2* is limited to a few cases. This contradicts to the results of Figs.1.6 and 1.7, in which the amplifying rate $-k_I*$ is much larger for the backward rotating cavitation k_2*.

2 Two-Dimensional Flow Stability Analysis with a Closed Cavity Model

The analysis in the preceding section is basically one-dimensional and only the effect of total cavity volume fluctuation is included. The stability analysis of two-dimensional cavitating flow using a closed model of blade surface cavitation is presented in this section.

2.1 Method of Stability Analysis

We consider a cascade as shown in Fig. 2.1 (Horiguchi et al. (2000)). For simplicity, we assume that the downstream conduit length is infinite and no velocity fluctuation occurs there. The upstream conduit length is assumed to be finite, L, in the x-direction and connected to a space with constant (static = total) pressure at the inlet AB.

We assume small disturbances with time dependence $e^{j\omega t}$ where $\omega=\omega_R+j\omega_I$ is the complex frequency with ω_R the frequency and ω_I the damping rate, to be determined from the analysis. The velocity disturbance is represented by a source distribution $q(s_1)$ on the cavity region, vortex distributions $\gamma_1(s_1)$ and $\gamma_2(s_2)$ on the blades, and the free vortex distribution $\gamma_t(\xi)$ downstream of the blades, shed from the blades associated with the blade circulation fluctuation. We divide the strength of those singularities and the cavity length into steady and unsteady components, and represent the velocity with steady uniform velocity ($U,U\alpha$), the steady disturbance (u_s,v_s), and the unsteady disturbance ($\tilde{u},\tilde{v}$). We assume that $\alpha<<1$, $|\tilde{u}|,|\tilde{v}|<<|u_s|,|v_s|<<U$ and neglect higher order small terms.

The boundary conditions are:

(1) The pressure on the cavity should equal vapor pressure.
(2) The normal velocity on the wetted blade surface should vanish.
(3) The cavity should close at the (moving) cavity trailing edge (closed cavity model).
(4) The pressure difference across the blades should vanish at the blade trailing edge (Kutta's condition).

(5) Upstream and downstream conditions: the total pressure along AB is assumed to be constant and the downstream velocity fluctuation is assumed to be zero.

By specifying the strength of the singularity distributions at discrete points on the coordinates fixed to the fluctuating cavity as unknowns, the boundary conditions can be represented by a set of linear equations in terms of those unknowns. If the unknowns are separated into steady and unsteady components, the steady boundary conditions result in a set of non-homogeneous linear equations. This set of equations can be used to show that the steady cavity length l_s normalized by the blade spacing h, l_s/h is a function of $\sigma/2\alpha$. On the other hand, the unsteady component of the boundary conditions results in a set of homogeneous linear equations. For non-trivial solutions, the determinant of the coefficient matrix should be zero. Since the coefficient matrix is a function of the steady cavity length and complex frequency, the complex frequency $\omega = \omega_R + j\omega_I$ is determined from this relation as a function of the steady cavity length l_s/h, or equivalently of $\sigma/2\alpha$. This shows that the frequency ω_R and the damping rate ω_I, as well as possible modes of instability, depend only on the steady cavity length l_s/h, or equivalently on $\sigma/2\alpha$, once the geometry and other flow conditions are given.

2.2 Results of Stability Analysis

The steady cavity length obtained by assuming equal cavity on each blade is plotted in the upper part of Fig.2.2a (Horiguti et al. (2000)), for a cascade with the stagger $\beta = 80$ deg and the chord-pitch ratio $C/h = 2.0$, typical for turbopump inducers. In this calculation, a periodicity over 4 blades is assumed and hence it corresponds to the case of a 4-bladed inducer. It is well known that alternate blade cavitation, in which the cavity length differs alternately, may occur for inducers with an even number of blades. The cavity lengths of alternate blade cavitation are shown in the upper part of Fig.2.2b. Alternate blade cavitation starts to develop when the cavity length, l_s, of equal cavitation exceeds 65% of the blade spacing, h. Figure 2.3 shows the flow field around alternate blade and equal length cavitations. Near the trailing edge of cavities, we can observe a region where the flow is inclined towards the suction surface. In this region the incidence angle to the neighboring blade on the suction side is smaller. This region starts to interact with the leading edge of the next blade when the cavity length becomes about 65% of the blade spacing. If the cavity length on one blade becomes longer than 65% of the blade spacing, the incidence angle to the next blade on the suction side becomes smaller and hence the cavity length on the next blade will decrease. This is the mechanism of the development of alternate blade cavitation.

Strouhal numbers $St = \omega_R l_s / 2\pi U$ of various amplifying modes are shown in the lower part of Fig.2.2(a) and (b) for equal length cavitation and alternate blade cavitation. The symbol $\theta_{n,n+1}$ shows the phase advance of the disturbance on the upper blade (n+1) with respect to that on the lower blade (n) by one pitch, which is obtained as a result of the stability analysis. Here we focus on Mode I. For Mode I, the frequency is zero and the phase difference $\theta_{n,n+1}$ is 180 deg, corresponding to exponential transitions between equal and alternate blade cavitation. This mode appears for equal cavitation longer than 65% of the blade spacing, h, which shows that longer equal cavitation is statically unstable to a disturbance corresponding to the transition to alternate blade cavitation. Alternate blade cavitation does not have this mode and hence it is statically stable.

We now return to Fig.2.2(a). Mode II is a surge mode oscillation without interblade phase difference: $\theta_{n,n+1} = 0$. It was found that the frequency of this mode correlates with $1/\sqrt{L}$ where L is the length of the upstream conduit. So this mode represents normal cavitation surge. Mode II is system dependent while all other modes are system independent. Mode IX is also a surge mode oscillation with no interblade phase difference but has higher frequency. This mode is herein called "higher order surge mode oscillation". Figure 2.4 shows the shape of cavity oscillations for these modes. The cavity volume fluctuation of Mode IX is much smaller than that of conventional cavitation surge, Mode II. For this reason the frequency does not depend on the inlet conduit length. In addition, the frequency of this mode does not depend on the geometry of the cascade and this mode occurs also for single isorated hydrofoils (Watanabe et al.1998). This mode starts to appear at much larger values of $\sigma / 2\alpha$ than other modes.

Modes III-VI are various modes of rotating cavitation with various interblade phase differences. Observed from a stationary frame, the disturbance of Mode III rotates around the rotor with an angular velocity higher than the impeller speed. This is conventional rotating cavitation. Mode IV represents one-cell rotating cavitation propagating in the opposite direction of the impeller rotation and is called "backward rotating cavitation". Mode V represents 2-cell rotating cavitation. Mode VI is one-cell forward rotating cavitation with a larger propagating speed than Mode III and this mode is called "higher-order rotating cavitation". All modes except for Mode IX start to occur when the cavity length exceeds 65% of the spacing. So, those modes might be caused by the interaction of the local flow near the cavity trailing edge with the leading edge of the opposing blade, as for alternate blade cavitation. Mode IX occurs for much shorter cavities and no physical explanation has been given so far.

By the two-dimensional stability analysis, various types of higher order modes are predicted in addition to cavitation surge, forward and backward propagating modes of rotating cavitation. These higher order modes are experimentally observed less frequently compared to cavitation surge and forward rotating cavitation but they do occur (Tsujimoto et al., 2004). Since the frequencies of those higher order modes are high enough, resonance with blade bending mode vibration is possible (Tsujimoto et al. (2002)). So, it is important to confirm that those instabilities are adequately suppressed by testing the inducer under all conditions encountered in real flight. It is also important to identify the reason why they occur in some cases and not in others.

Although the critical values of $\sigma / 2\alpha$ do not agree with experiments, the critical value in terms of cavity length $l_s = 0.65h$ agrees with experiments reasonably if applied at the tip. The one-dimensional criterion $M > 2(1+\sigma_1)\phi K$ is satisfied under the condition $l_s = 0.65h$. So, the two-dimensional analysis gives more useful guideline for the prediction of cavitation instabilities although the one-dimensional criterion is more useful in considering the effects of various types of cavitation.

3 Two-Dimensional Analysis of Cavitation Instabilities with a Waked Cavity Model

3.1 Experimental Observation of Rotating Choke

During the firing test of the LE-7A LOX/LH2 engine of the HIIA rocket, a large amplitude rotor vibration of the fuel turbopump occurred at a frequency of about 350Hz which is about one-half of the shaft rotational speed Ω =700Hz. The vibration occurred when the pump inlet pressure was reduced. This vibration caused the failure of the bolts fastening the bearing cartridge. To investigate into this phenomenon, a series of tests was conducted at full speed with liquid hydrogen. The test results are reported by Shimura et al. (2002).

Figure 3.1 shows the suction performance and the amplitude of the shaft vibration for two groups of flow rates around $Q/Q_d = 0.98$ and $Q/Q_d = 0.95$. For $Q/Q_d = 0.95$, the amplitude becomes larger at the cavitation number where the inducer head decreases rapidly. The head is kept nearly constant, above and below the cavitation number. The head decrease is not caused by the instability since the head decrease is not recovered at smaller cavitation number where the amplitude of vibration is smaller again. At a cavitation number higher than the breakdown cavitation number, σ_{high}, the head is larger for smaller flow rate and the performance curve has a negative slope in the flow rate vs. head plane. On the other hand, at a lower cavitation number than the breakdown, σ_{low}, the head is smaller for smaller flow rate and the performance curve has a positive slope. These relations are sketched in Fig.3.2. For compressors and fans, it is well known that the positive slope of the performance curve can cause surge and rotating stall at a smaller flow rate.

Without cavitation, inducers with smaller blade angle from tangent never stall and the performance curve has positive slope for all flow rate. On the other hand, the head starts to decrease when the cavity extends into the flow channel between blades. It was shown by Stripling and Acosta (1962) that the head decrease due to cavitation can be explained by the mixing loss downstream of the cavity terminus. This is called "choke (by cavitation)". The cavity thickness is larger for smaller flow rate and hence the head decrease due to cavitation is larger, as sketched in Fig.3.3. If the head decrease due to cavitation associated with the reduction of flow rate is larger than the increase of Euler's head, the performance curve will have a positive slope. So, it is quite possible that the positive slope at smaller cavitation number is caused by the choke.

Figure 3.4 shows the spectrum of the inlet pressure fluctuation and the phase difference of the signals from two pressure transducers located apart by 144 degrees circumferentially (Shimura et al. (2002)). The rotational frequency of the impeller is 723Hz. The phase difference at the spectrum peak, 366Hz, is 147 degrees, which is close to the geometrical separation of 144 degrees. This suggests that a disturbance with one cell is rotating around the rotor at 366Hz, with the rotational speed ratio 366/723=0.506. Since the head decrease due to cavitation is caused by choke, it is appropriate to call the instability "rotating choke". All reported cavitation instabilities occur in a range of cavitation number where degradation of performance due to cavitation is limited. So, this is a new type of cavitation instability caused by the positive slope of pressure performance due to cavitation choke. We should note here that the positive slope of suction performance has a stabilizing effect, as shown by Young et al. (1972) for surge mode oscillation.

3.2 Theoretical Analysis of Rotating Choke

Since the instability is associated with the head breakdown due to cavitation, the analytical model should be able to simulate the head breakdown. The closed cavity model cannot predict the head decrease due to cavitation and the results of stability analysis shown in Fig.2.2 does not include such instabilities associated with choke. To simulate rotating choke, a cavity model with a cavity wake is applied.

The viscous/inviscid interaction approach developed for cavitation-free flows is used. According to this approach the flow is divided into an internal turbulent flow behind the cavity and an external inviscid flow in which the cavity is closed on the body of viscous layer displacement (Semenov et al. (2004)). In the cavity wake three specific regions are considered. The first is the transitional mixing region where the velocity profile across the wake is under formation and the density changes smoothly due to flow re-circulation. This region is needed to provide a good agreement of suction performance with experiment (Semenov et al. (2003)). The second is the near wake in which the velocity profile changes according to the external pressure gradient along the wake. Von Karman's integral equation is used and it is assumed that the coefficient of turbulent mixing is constant (Gogish et al. (1979)). The third is the far wake which starts from the trailing edge of blades and adjusts itself on the downstream flow. The conditions of interaction between viscous and inviscid flows in the near wake make it possible to find the shape of the boundary of the inviscid flow and to calculate both steady and unsteady flows. The normal velocity of the outer flow on the boundary is equated with that determined from the change of the wake thickness.

The head decrease due to cavitation is associated with the increase of relative velocity caused by the displacement effect of the cavity wake. The stability analysis is similar to that with closed cavity model (Horiguchi et al. (2000)) described in section 2. The problems for the inviscid and viscous flow are reduced to a system of linear equations by using linear interpolation of source and vortex and finite difference approximation of the ordinary differential equations for the viscous wake. The solution of the eigenvalue problem makes it possible to determine the frequencies and the stability of the various modes of cavitation instabilities.

Figure 3.6 shows the suction performance predicted by the model. On the suction performance curves, the occurrence of rotating cavitation is shown by open symbols, while the occurrence of rotating choke is shown by closed symbols. Other types of instabilities such as shown in Fig. 2.2 are also predicted by the present model but they are not shown here. We note that the rotating cavitation mainly occurs in the range where the performance degradation is insignificant, while rotating choke occurs where the head is decreased due to cavitation.

The performance curve in the flow coefficient vs. head plane is shown in Fig.3.7, where the occurrences of rotating cavitation and rotating choke are also shown by open and closed symbols, respectively. The same observations as with Fig. 3.6 are also evident in this figure. Rotating choke occurs typically in the region with positive slope of the performance curve, which is quite different from rotating cavitation. This clearly shows that rotating choke is caused by the positive slope of the performance curve due to choke.

Figure 3.8 shows the propagation velocity ratio of rotating cavitation and rotating choke. Rotating choke is found only at a region with smaller σ. Rotating cavitation rotates faster than impeller but rotating choke rotates about a half of the impeller speed. Rotating stall in impellers with longer blade rotates with a speed only slightly smaller than the impeller speed, caused by the inertia effect of the fluids in the impeller (Tsujimoto, et al. (1993), Tsujimoto et al. (2001)). Lower speed of the rotating choke may be caused by the decrease of the inertia effect due to cavitation.

Figure 3.9 shows the cavity shapes under rotating cavitation and rotating choke at their onset points (the largest cavitation number). For rotating choke the mixing region extends to the throat of the blade channel. On the other hand, rotating cavitation starts to occur with much shorter cavity. With a closed cavity model, it has been shown that various types of cavitation instability start to occur when the cavity length reaches about 65% of the spacing as mentioned in section 2. Figure 3.10 shows the steady cavity length l_c of the closed cavity model and the total length of cavity and mixing region $l_c + l_m$ of the waked cavity model, normalized by the blade spacing h, plotted against $\sigma/(2\alpha)$. The regions of rotating cavitation and rotating choke occurrence are also shown. It is clearly shown that rotating cavitation starts to occur when the cavity length (closed cavity model) or the total length of cavity and mixing region (waked cavity model) exceeds 65% of the blade spacing, h. This suggests that the total length of cavity plus mixing region plays an important role in the present model. On the other hand, rotating choke starts to occur when the total length of cavity and mixing region exceeds 150% of the spacing.

Thus, the present model can simulate two important characteristics of rotating choke, the onset condition and frequency. However, predicted suction performance shown in Fig.3.6 is significantly different from the experimental curve as shown in Fig.3.1. This may be caused by non-linear and/or three-dimensional effects. Positive slope and hence rotating choke are unusual and not observed for many inducers. Further research is needed for the complete understanding of rotating choke.

4 Conclusion

The one dimensional stability analysis shows that the onset condition of cavitation surge and rotating cavitation can be represented by $M > 2(1+\sigma_1)\phi K$ and is independent on the pressure rise performance of the impeller. This shows that cavitation instabilities may occur at the design flow coefficient. The frequency of cavitation surge is proportional to the rotational frequency of the rotor. Rotating cavitation has a mode which rotates faster than impeller rotation.

The two dimensional cavitating flow analysis using a closed cavity model shows that the cavitation instability depends on the steady cavity length l_s / h, or equivalently on $\sigma / 2\alpha$. Various modes of cavitation instabilities start to occur when the steady cavity length becomes larger than 65% of the blade spacing, caused by the interaction of the local flow near the cavity closure with the leading edge of the opposing blade. In addition to alternate blade cavitation, cavitation surge and rotating cavitation, and various higher order modes are predicted.

Rotating choke which occurs caused by the positive slope of the performance due to the blockage effect of cavitation could be predicted by using a waked cavity model. It starts to occur when the cavity-wake system extends into the blade passage, and the cavitating region rotates about 50% of the impeller speed.

Although real inducer flows are far more complicated with three-dimensional cavities such as tip leakage and backflow cavitation, the results of the two-dimensional flow stability analysis predicts cavitation instabilities surprisingly well if we apply them at the blade tip. However, in real engineering, the problem is to identify whether or not the predicted modes actually occurs or not under a certain geometry. To satisfy this requirement, we need to integrate the effects of the 3-D cavitations in the stability analysis.

Bibliography

Brennen, C.E., and Acosta, A.J., (1976). The Dynamic Transfer Function for a Cavitating Inducer. *ASME Journal of Fluids Engineering*, Vol.98, No.2, 182-191.

Brennen, C.E., (1982). Bubbly Flow Model for the Dynamic Characteristics of Cavitating Pumps. *J. Fluid Mech.,* Vol.89, Part 2, 223-240.

Brennen, C.E., Meissner, C., Lo, E.Y., and Hoffmann, G.S., (1982). Scale Effects in the Dynamic Transfer Functions for Cavitating Inducers. *ASME Journal of Fluids Engineering*, Vol.104, No.4, 428-433.

Gogish, L.V. and Stepanov, G.Yu., (1979). *Turbulent Separated Flows* (in Russian), Moscow, Nauka.

Hashimoto, T., Yoshida, M., Kamijyo, K. and Tsujimoto, Y., (1997). Experimental Study on Rotating Cavitation of Rocket Propellant Pump Inducers. *AIAA Journal of Propulsion and Power*, Vol.13, No.4, 488-494.

Horiguchi, H., Watanabe, S., and Tsujimoto, Y., (2000). A Linear Stability Analysis of Cavitation in a Finite Blade Count Impeller. *ASME Journal of Fluids Engineering*, Vol.122, No.4, 798-805.

Horiguchi,H., Watanabe, S., Tsujimoto, Y., and Aoki,M., (2000). Theoretical analysis of Alternate Blade Cavitation in Inducers. *ASME Journal of Fluids Engineering*, Vol.122, No.1, 156-163.

Kamijo, K., Shimura, T., and Watanabe, M., (1980). A Visual Observation of Cavitating Inducer Instability. *Technical Report of National Aerospace Laboratory*, TR-598T.

Ng, S.L., and Brennen, C.E., (1978). Experiments on the Dynamic Behavior of Cavitating Pumps. *ASME Journal of Fluids Engineering*, Vol.100, No.2, 166-176.

Otsuka, S., Tsujimoto, Y., Kamijo, K., and Furuya, O., (1996). Frequency Dependence of Mass Flow Gain Factor and Cavitation Compliance of Cavitating Inducers. *ASME Journal of Fluids Engineering*, Vol.118, No.2, 400-408.

Rubin, S., (2004). An Interpretation of Transfer Function Data for a Cavitating Pump. *40th AIAA/ASME/SAE/ASEE Joint Propulsion Conference,* 11-14 July, Fort Launderdale, Florida, AIAA-2004-4025.

Semenov, Y., and Tsujimoto, Y., (2003). A Cavity Wake ModelBased on the Viscous/Inviscid Interaction Approach and Its Application to Nonsymmetric Cavity Flows in Inducers. *ASME Journal of Fluids Engineering*, Vol.125, No.5, 758-766.

Semenov,Y., Fujii,A., and Tsujimoto,Y., (2004). Rotating Choke in Cavitating Turbopump Inducer. *ASME Journal of Fluids Engineering*, Vol.126. No.1, 87-93.

Shimura, T., Yoshida, M., Kamijo, K., Uchiumi, M., Yasutomi, Y., (2002). Cavitation Induced Vibration Caused by Rotating-stall-type Phenomenon in LH2 Turbopump. *Proceedings of the 9th of International Symposium on Transport Phenomena and Dynamics of Rotating Machinery*. Honolulu, Hawaii, February 10-14.

Stripling, L.B. and Acosta A.,J., (1962). Cavitation in Turbopump – Part 1. *ASME Journal of Fluids Engineering*, Vol.84, No.3, 326-338.

Tsujimoto, Y., Kamijio, K. and Yoshida, Y., (1993). A Theoretical Analysis of Rotating Cavitation in Inducers. *ASME Journal of Fluids Engineering*, Vol.115, No.1, 135-141.

Tsujimoto, Y., Yoshida, Y., Maekawa, Y, Watanabe, S., and Hashimoto, T., (1997). Observations of Oscillating Cavitation of an Inducer. *ASME Journal of Fluids*

Engineering,Vol.119, No.4, December, 775-781.

Tsujimoto,Y., Kamijo,K., and Brennen, C., (2001). Unified Treatment of Cavitation Instabilities of Turbomachines. *AIAA Journal of Propulsion and Power*, Vol.17, No.3, 636-643.

Tsujimoto, Y., and Semenov,Y.,(2002). New Types of Cavitation Instabilities in Inducers. *Proceedings of the 4th International Symposium on Launcher Technology*, 3-6 December, Liege, Belgium.

Tsujimoto, Y., Horiguchi,H., and Fujii, A., (2004). Non-Standard Cavitation Instabilities in Inducers. *Proceedings of the 10th International Symposium on Heat Transfer and Dynamics of Rotating Machinery*, March 7-11, Honolulu, Hawaii.

Watanabe, S., Tsujimoto, Y., Franc, J.P., and Michel, J.M., (1998). Linear Analysis of Cavitation Instabilities. *Proc. Third International Symposium on Cavitation,* April, Grenoble, France, 347-352.

Young, W.E. et al., (1972). Study of Cavitating Inducer Instabilities. *Final Report, NACA-CR-123939*, 1972-8.

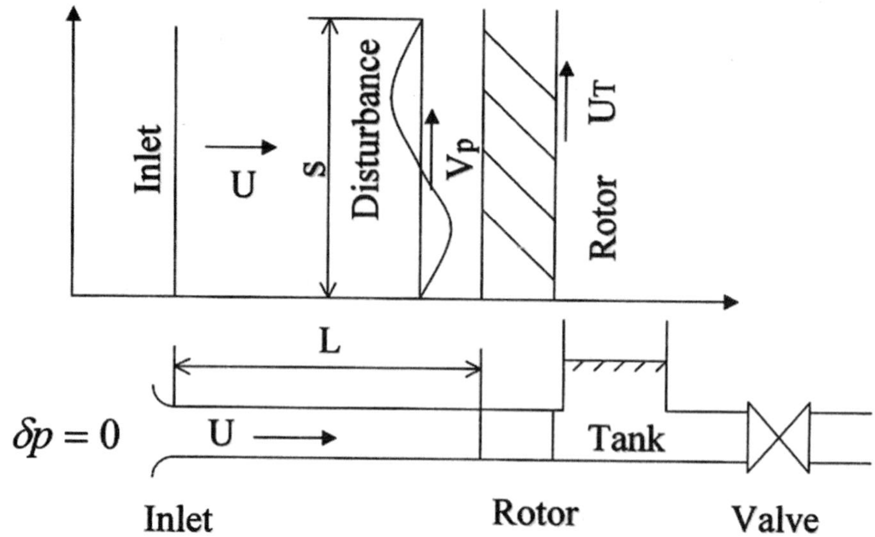

Figure 1.1 The hydraulic system for the analysis of instabilities in turbomachinery

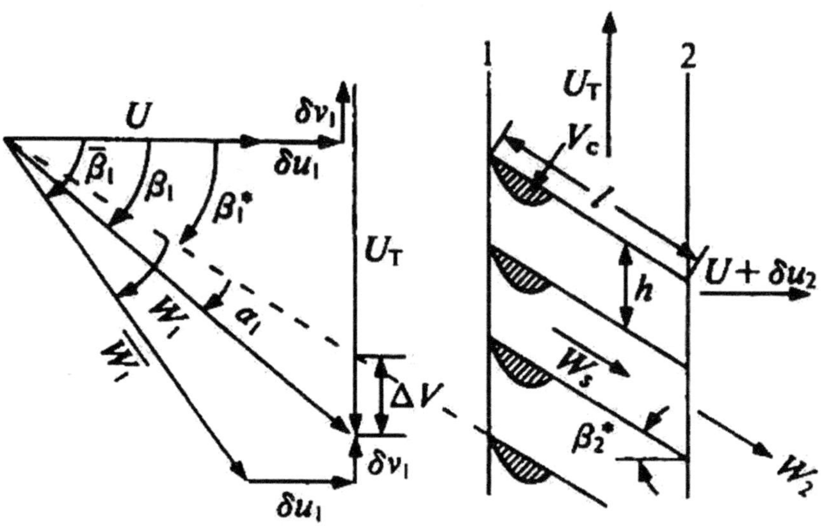

Figure 1.2 Velocity triangle at the inlet

Instability	Onset condition	Frequency
Surge	$\frac{\partial \psi_{ts}}{\partial \phi} > \frac{1+(1/\cos\beta^*)(l/L)}{B^2\phi R}$	$n = \frac{1}{2\pi}\frac{1}{\sqrt{\rho CL}}\sqrt{\frac{1+(1/R)\left[L_u+\left(1/\cos^2\beta_2^*\right)\right]}{[1+(1/\cos\beta^*)l/s]}}$
Rotating stall	$\frac{\partial \psi_{ts}}{\partial \phi} > 0$	$\frac{V_p}{U_T} = 1 - \frac{2\zeta_s[1-(\phi/\phi^*)]}{1+(2\pi/\cos\beta_1)l/s]} < 1$
Cavitation surge	$M > 2(1+\sigma)\phi K$	$n = \frac{U_T}{2\pi}\frac{1}{\sin\bar{\beta}_1}\frac{1}{\sqrt{2KLh}}$
Rotating cavitation	$M > 2(1+\sigma)\phi K$	$V_p/U_T > 1, \quad V_p/U_T < 0$

Table 1.1 The onset condition and the frequency of instabilities in turmomachinery

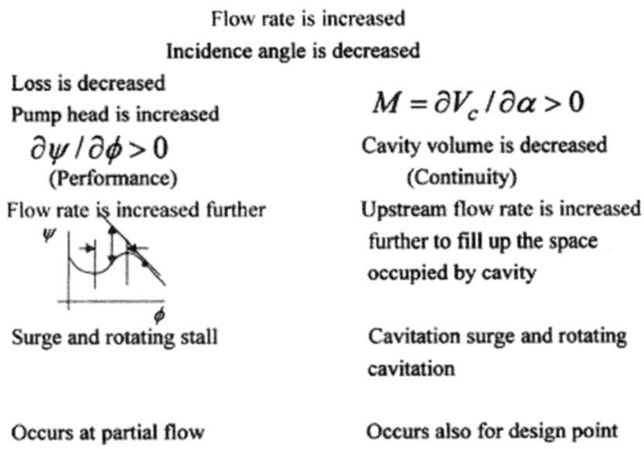

Figure 1.3 Mechanisms of instabilities in turbomachinery

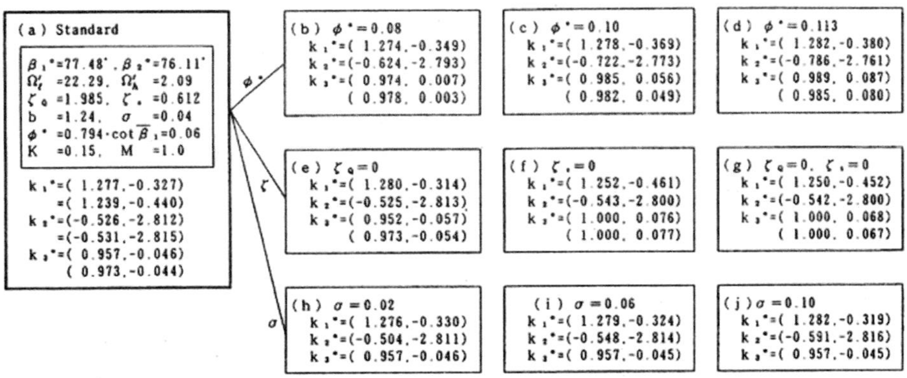

Figure 1.4 Solutions of Eq.(5) showing rotating cavitation (k_1*, k_2 *) and rotating stall k_3 *

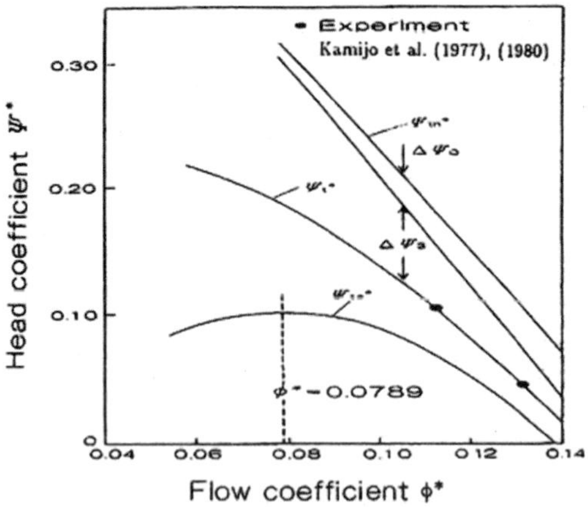

Figure 1.5 Static performance of the inducer

		$k^* = k_1^*$	$k^* = k_2^*$	$k^* = k_3^*$
$\lvert \delta p_2/\rho U_2^2 \rvert$	=	(0.0591)	(0.00836)	(0.0497)
$\lvert \delta u_1/U_1 \rvert$	=	(0.0563)	(0.00257)	(0.0783)
$\lvert \delta u_2/U_2 \rvert$	=	(0.0129)	(0.00020)	(0.0713)

Table 1.2 Amplitudes of pressure and velocity fluctuations

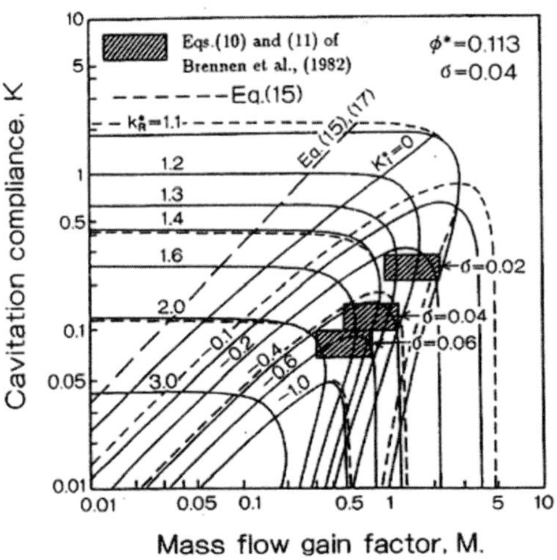

Figure 1.6 Forward rotating cavitation k_1 *

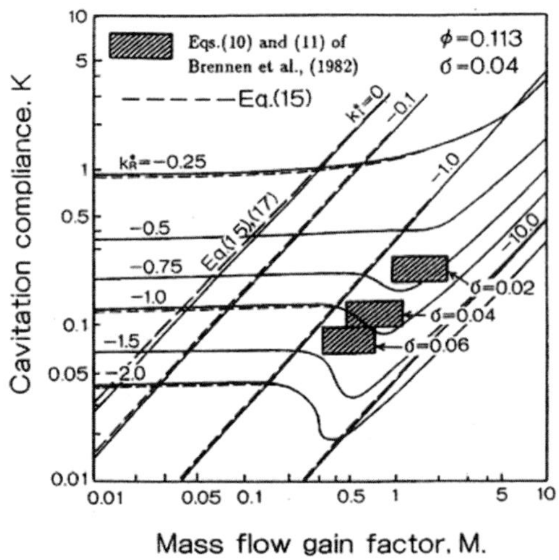

Figure 1.7 Backward rotating cavitation k_2 *

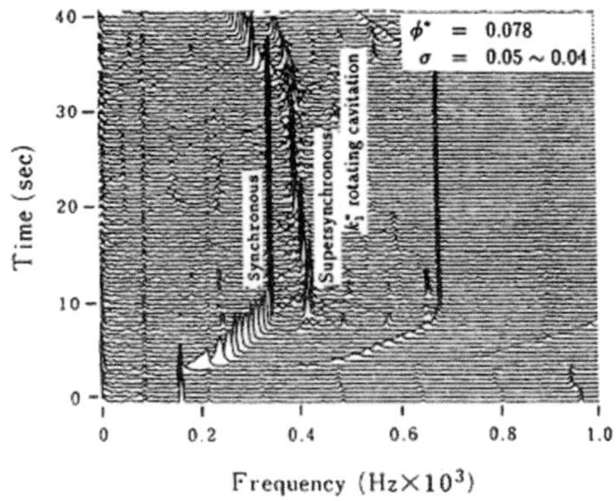

Figure 1.8 Supersynchronous shaft vibration caused by forward rotating cavitation

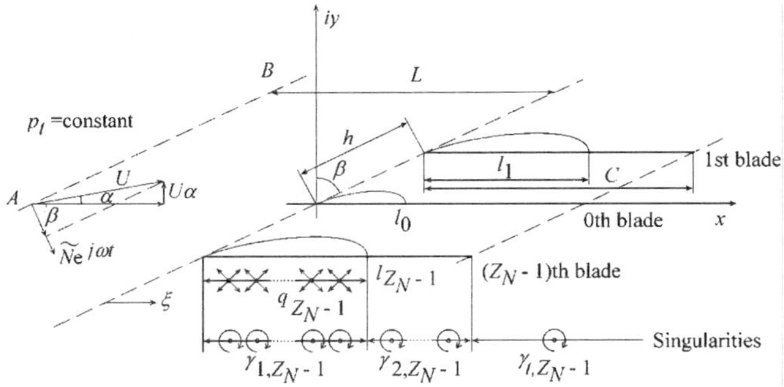

Figure 2.1 2D stability analysis of cavitating flow through a cascade

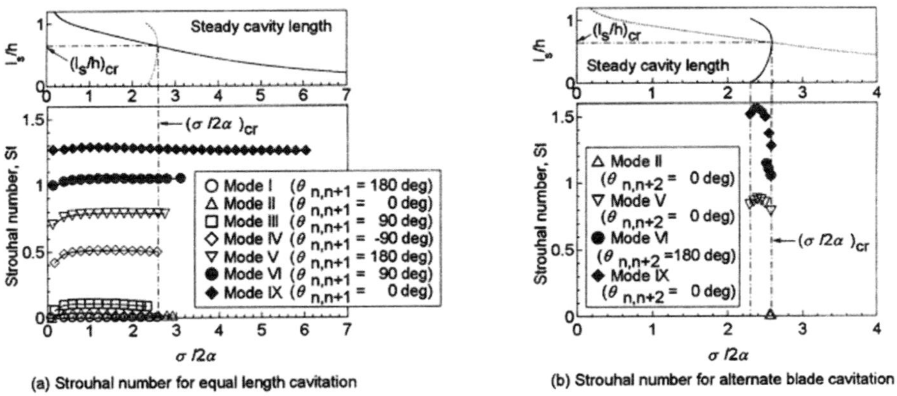

Figure 2.2 Steady cavity length (upper figures) and Strouhal number (lower figures) of various modes of cavitation instabilities, for a 4-bladed inducer with the solidity $C/h = 2.0$, stagger $\beta = 80^o$ and the inlet duct length $L/C = 1000$.

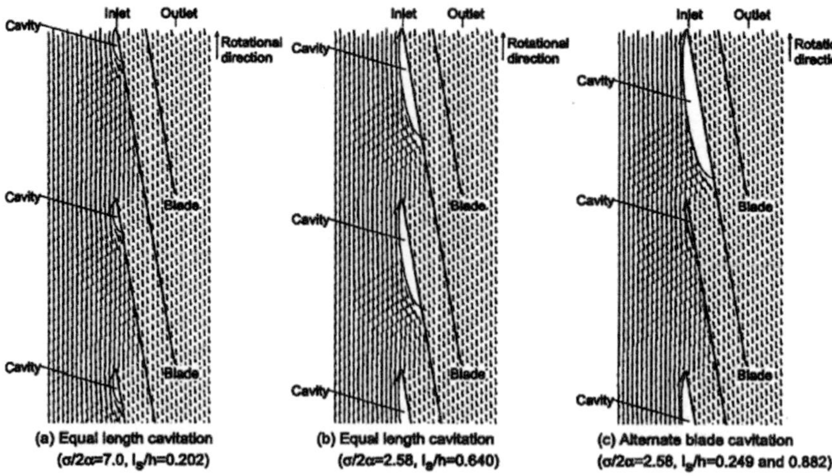

Figure 2.3 Alternate blade and equal cavitation in a cascade with the solidity $C/h = 2.0$, stagger $\beta = 80^o$ and the inlet duct length $L/C = 1000$. The incidence angle is $\alpha = 4^o$.

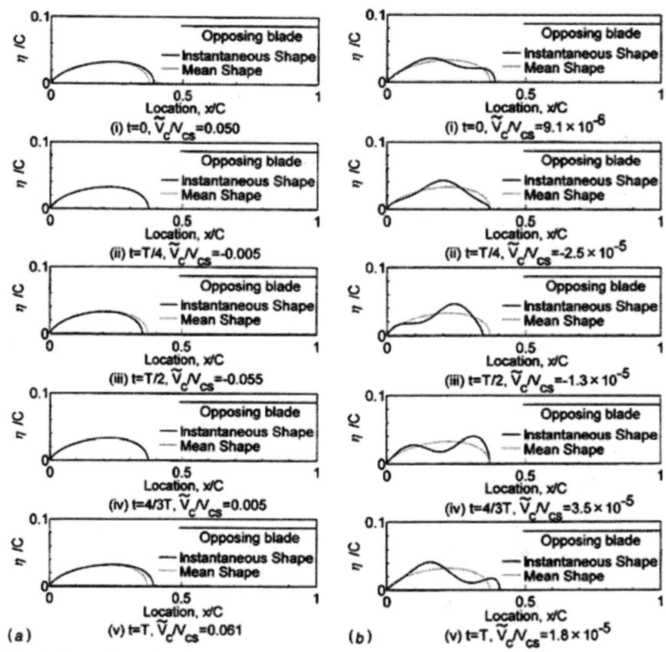

Figure 2.4 Oscillating cavity shape under (a) cavitation surge (Mode II) and (b) higher order surge mode oscillation (Mode IX). $\sigma/2\alpha = 2.0$ and $\alpha = 4.0^o$.

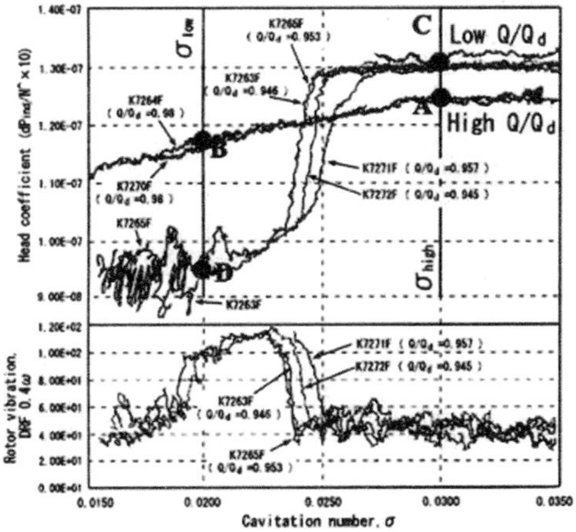

Figure 3.1 Suction performance and shaft vibration amplitude, with two groups of flow rate.

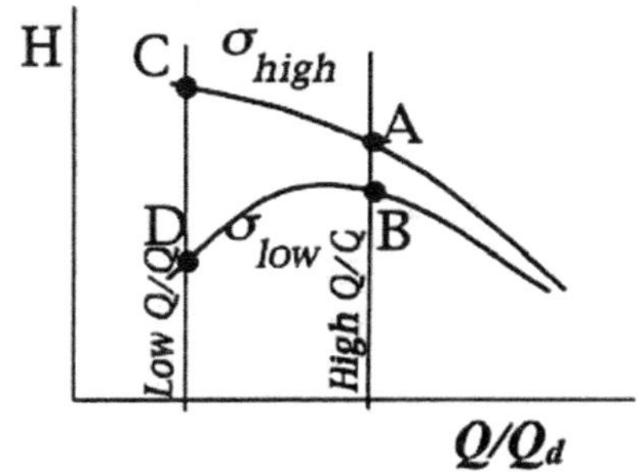

Figure 3.2 Sketch of head-flow rate curve at higher and lower cavitation number

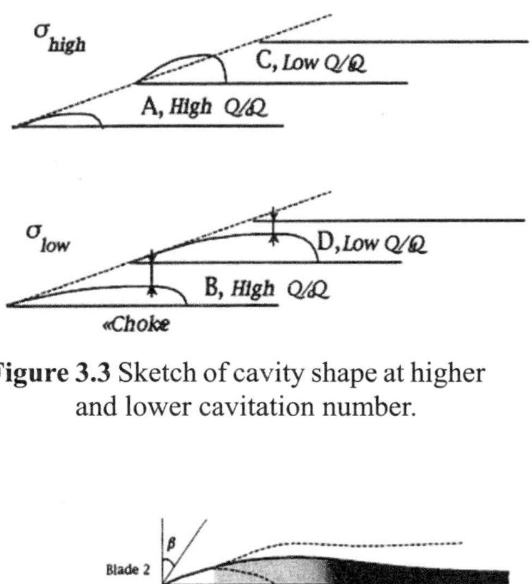

Figure 3.3 Sketch of cavity shape at higher and lower cavitation number.

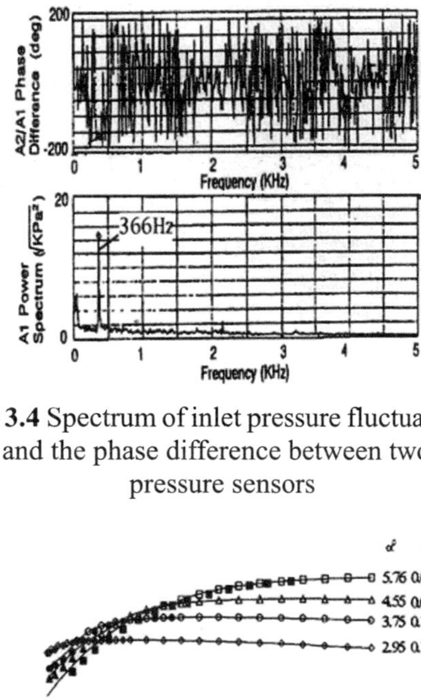

Figure 3.4 Spectrum of inlet pressure fluctuation and the phase difference between two pressure sensors

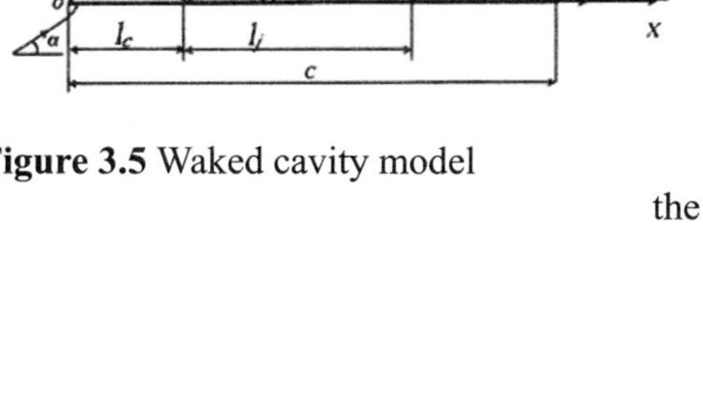

Figure 3.5 Waked cavity model

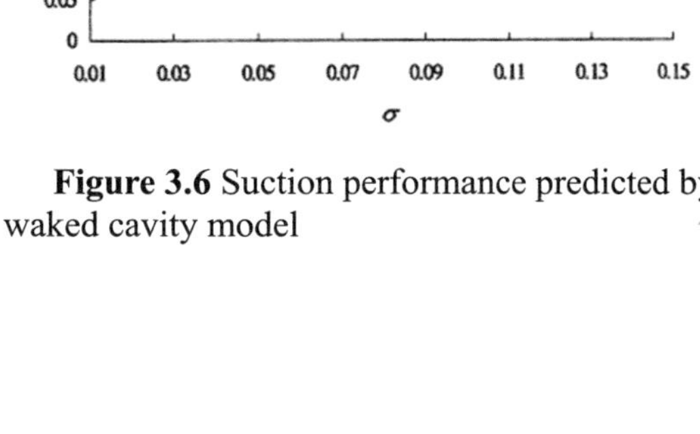

Figure 3.6 Suction performance predicted by the waked cavity model

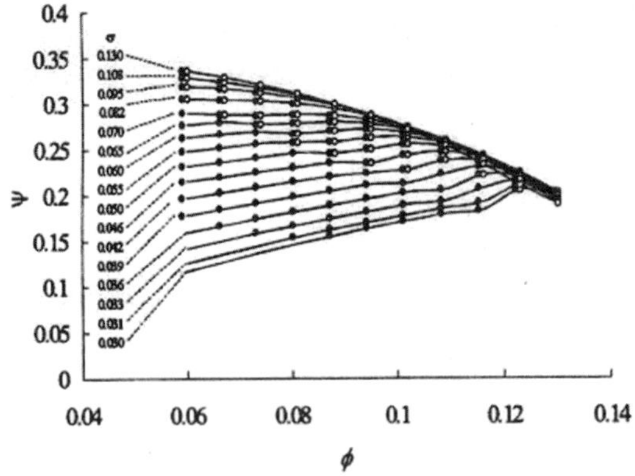

Figure 3.7 Head coefficient-flow coefficient curves for various cavitation number.

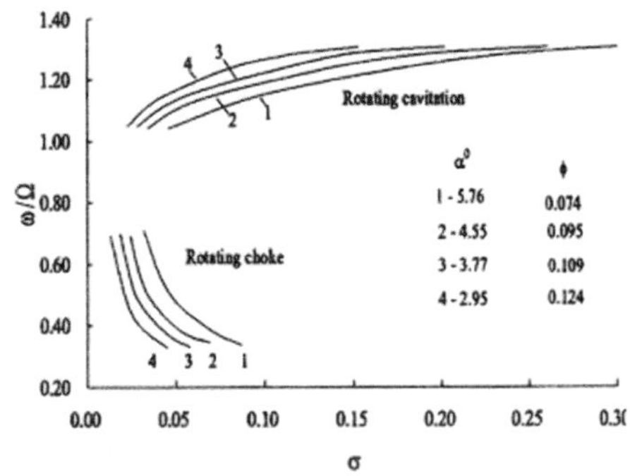

Figure 3.8 Propagation velocity ratio of rotating cavitation and rotating choke

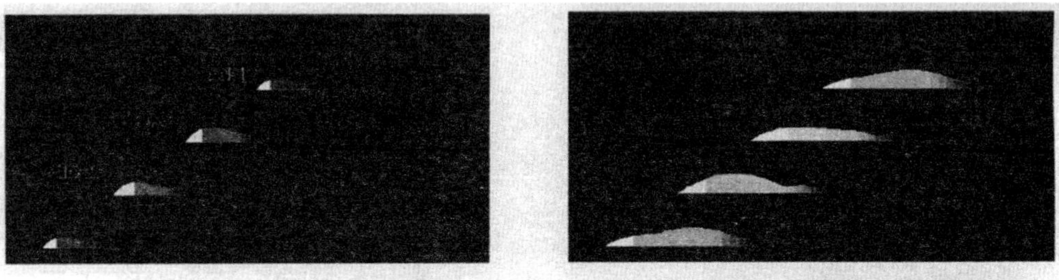

a) rotating cavitation ($\sigma = 0.26$) *b) rotating choke* ($\sigma = 0.077$)

Figure 3.9 Cavity shapes at the onset point of rotating cavitation and rotating choke at $\phi = 0.095$. Cavity is shown by the white area. The density in the mixing region and the wake is shown by the darkness. Vertical dimension is magnified with a factor 3.

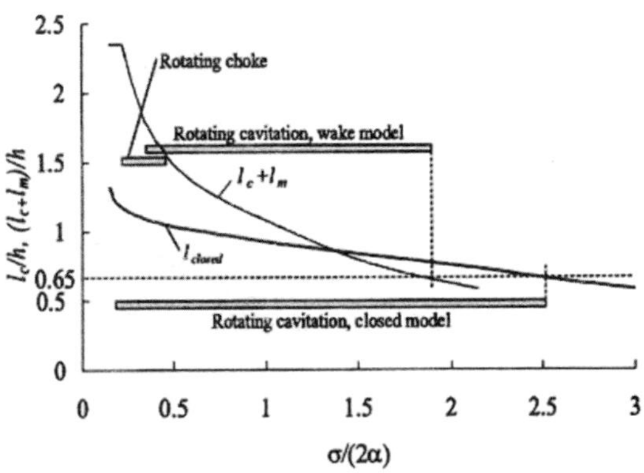

Figure 3.10 Cavity length of closed cavity model l_{closed} , and the total length of cavity and mixing region $l_c + l_m$ of the waked cavity model. The gray bars show the range of rotating cavitation and rotating choke. ($\beta = 80^o$, $C/h = 2.35$, $\alpha = 4^o$)

Suppression of Cavitation Instabilities

Yoshinobu Tsujimoto[1]

[1] Engineering Science, Osaka University, Osaka, Japan

Abstract. Avoidance of cavitation instabilities is essential for the design of reliable inducers. Two possible methods are examined in this article, i.e., the leading edge sweep, and the casing enlargement at the inlet. Although they do have certain effects for the avoidance, it is becoming clear that in certain cases these methods are not sufficient to completely suppress cavitation instabilities. So, the suppression of cavitation instabilities is still an important problem to be solved.

1 Leading Edge Sweep

It has been shown that various types of cavitation instabilities occur if the steady cavity length becomes larger than about 65% of the blade spacing. Since the cavity length is a function of $\sigma/2\alpha$ and the cavity length increases as we decrease the value of $\sigma/2\alpha$, using smallest possible incidence angle α is desirable to avoid cavitation instabilities. However, if we use too small α, the allowance for higher flow rate will decrease, since the head will decrease rapidly at higher flow due to the development of cavity on the pressure surface of the blades.

It has been empirically known that the leading edge sweep has favorable effects on cavitation. It is shown by considering the flow components normal to the leading edge that the steady cavity length can be correlated with a parameter corresponding to $\sigma/2\alpha$ (Acosta et al (2001)). If we combine this result with the fact that various types of cavitation appear when the steady cavity length becomes larger than about 65% of the blade spacing, we can explain why we can delay the occurrence of cavitation instabilities. This will be discussed in the present section.

1.1 Geometrical Relations

We consider a cascade shown in Fig.1.1; several views are shown which are needed for clarity. The upper or plan view of the cascade shows straight uncambered blades s apart along the cascade axis (the plane normal to the inducer axis). The blades are inclined at blade angle β with respect to this axis. The leading edges of these blades are shown in the meridional view inclined at angle δ from what would be a radial line in a real inducer. Let us select two points on one blade, O, and A. Corresponding points in the meridional plane lie along $l-l$ spanning a vertical height η_s (in radial direction). These points and at the next blade O', A' are observed in the true view of the blade leading edge defined by cut $B-B$. The bold line OA is the true view of the leading edge. This line makes an angle λ to the plane of the inducer axis, namely, the projection of $O-O'$. The adjacent blade in this true view is $O'-A'$. The plane normal to

$OA, O'A'$ and to the plane of the paper is the *cross flow plane*. Note that the line $O'A'$ is hidden from view by the first blade. Now in the true view plane, BB, project a normal to OA from O' ending at P, a point on the leading edge of blade B_1. Point P also appears in the plan view on the leading edge of blade B_1. Imagine now we progress from P normal to the leading edge along blade B_1 in the cross flow plane, CC, until we are underneath the normal to the next blade, B_2, at point Q. From Q we move to point O' on blade B_2 seen in the plan view. The cross flow plane in the plan view may be identified by the points P, Q, O' (a portion of it is shown, cross-hatched for clarity).

We will be concerned with the flow velocities and the effective cascade geometry in this cross flow plane. Before completing the definitions of the cross-flow cascade geometry, let us consider the velocity components: The velocity approaching the real inducer is presumed to be purely axial; relative to the rotating inducer the flow speed is V_1 and is inclined to the blades with incidence angle α shown in the plan view of Fig.1.1. There is a component of V_1 that is normal to the cascade blade, V_n, and a component tangential to the blade surface, V_t, also shown in Fig.1.1. The tangential component is resolved into component V_c normal to the true view of the leading edge, and V_p parallel to it. We see in the cross flow plane (Fig.1.1, section CC) *component* V_c approaching a cascade but one characterized by a new spacing, s_e, and a new blade angle, β_e. The normal distance between the blades B_1, B_2 shown in the cross-flow and plan view is of course the same, i.e.,

$$s \sin\beta = s_e \sin\beta_e \text{ or } \beta_e = \sin^{-1}((s/s_e)\sin\beta) \tag{1}$$

Then sweep angle λ is constructed in the true view plane from

$$tan\,\lambda = \frac{\eta_s}{(\eta_s\ tan\delta / sin\,\beta)},\quad \lambda = tan^{-1}\left(\frac{sin\,\beta}{tan\,\delta}\right) \tag{2}$$

The effective spacing is determined from its projection $P-O'$ in the true view plane and the normal distance above to get

$$s_e = \sqrt{(s\cos\beta\, sin\,\lambda)^2 + (s\, sin\,\beta)^2} \tag{3}$$

Note that $s_e \le s$. The true thickness of the blades is t; the ratio t/s is an important geometric parameter governing cavitation. Clearly then the *effective* thickness-spacing ratio $(t/s)_{eff}$ is

$$\left(\frac{t}{s}\right)_{eff} = \left(\frac{t}{s}\right)\left(\frac{s}{s_e}\right) \ge \left(\frac{t}{s}\right) \tag{4}$$

Thus the cross-flow geometry is *blunter* than the normal flow. For a leading edge in the shape of a wedge of included angle θ, it follows that

$$\theta_e = tan^{-1}\left(\frac{tan\,\theta}{sin\,\lambda}\right) \tag{5}$$

If, as is the case with many small commercial inducers, leading edges are machined on a lathe, $\theta = \beta$. Finally, there is the flow incidence angle in the cross flow plane for which

$$\alpha_c = tan^{-1}(V_n / V_c),\quad \alpha_c = tan^{-1}(tan\,\alpha / sin\,\lambda) \tag{6}$$

1.2 Cavitation Scaling

Let the pressure in the cavity be p_v and the velocity there be V_k. Then in the usual way from Bernoulli equation, i.e.,

$$\frac{p_{t1}}{\rho} = \frac{p_1}{\rho} + \frac{1}{2}V_1^2 = \frac{p_v}{\rho} + \frac{1}{2}V_k^2 \tag{7}$$

we define the cavitation number

$$k = (p_1 - p_v)/(\rho V_1^2/2) \tag{8}$$

and so we can say

$$V_k = V_1\sqrt{1+k} \tag{9}$$

in the physical plane. From the velocity triangles in Fig.1.1, we have the relations

$$V_1^2 = V_n^2 + V_t^2, V_t^2 = V_c^2 + V_p^2, V_k^2 = V_{kc}^2 + V_p^2, \text{ and } V_{ctotal}^2 = V_n^2 + V_c^2 \tag{10}$$

If we use these relations, Eq.(7) can be rewritten as

$$\frac{p_1}{\rho} + \frac{1}{2}V_{ctotal}^2 = \frac{p_v}{\rho} + \frac{1}{2}V_{kc}^2 \tag{11}$$

which corresponds to Bernoulli equation in the cross flow plane. We can define a cross flow cavitation number as

$$k_c = (p_1 - p_v)/(\rho V_{c\,total}^{\ 2}/2) \tag{12}$$

And in the cross flow plane we have

$$V_{kc} = V_{c\,total}\sqrt{1+k_c} \tag{13}$$

If we can determine the cavitating flow in the cross flow plane with the cavitation number k_c, the flow in the physical plane with the cavitation number k can be obtained by simply adding the velocity component V_p parallel to the leading edge. From Eqs.(8), (12) and (10), we obtain a simple relation

$$k_c = k/(\cos^2\alpha \, sin^2\lambda + sin^2\alpha) \tag{14}$$

Note that if $\lambda \to \pi/2$ (no sweep), $k \to k_c$. For typically small incidence angles $\alpha \ll 1$,

$$k_c = (k/sin^2\lambda)(1 - O(\alpha^2)) \approx k/sin^2\lambda \tag{15}$$

which is the relation we will use. The equivalent formula for a swept isolated wing quoted by Ihara et al. is (in our notation)

$$k_c = k/sin^2\lambda \tag{16}$$

the slight difference arises because our cross-flow velocity is parallel to the chord.

The effect of sweep appears through two mechanisms. One is through the change of cascade geometry in the cross flow plane as shown by Eqs. (3)-(6) obtained in the preceding section. The other is through the cavitation scaling as shown by Eq. (16). It has been shown by two-dimensional linear analysis such as Acosta (1955) that the cavity length, and hence the cavity development is a function of $k/2\alpha$. If we combine Eq. (6) and Eq. (16), we obtain

$$k_c/2\alpha_c = \frac{k/2\alpha}{sin\,\lambda} \tag{17}$$

If we increase the leading edge sweep, λ is decreased. Then $k_c/2\alpha_c$ is increased and hence the cavity length $l/s = (l_c/s_c)(\cos\beta/\cos\beta_c)$ is decreased. This can be the major reason why the cavitation performance is increased by simply sweeping the leading edge.

1.3 Comparison of Steady Cavity Length with Experiments

In order to study the effect of leading edge sweep, systematic experiments were carried out at Osaka University under the support of SNECMA. Five inducers quoted here have helical blades with the same camber line with straight part near the leading edge. Forward and backward swept inducers were produced by cutting back the straight part of an unswept inducer so that the inlet blade angle is not changed. Thus, all of the inducers have the same inlet and outlet blade angles. The geometries of the inducers are shown in Fig.1.2 and the dimensions in Table 1.1. The basic design is the same as for the LE-7 LOX turbopump inducer except that 3 blades are employed for LE-7. Inducer 0 is without sweep and has a straight radial leading edge. Inducers B50 and B90 are produced by cutting back the leading edge. Inducers F30 and F50 are produced by offsetting the leading edge upstream by 35 deg and then given the forward sweep by cutting back the inner part of the leading edge. The tip/leading edge corner is rounded with the radius 4 mm. The leading edge curve is obtained by shifting the circumferential location of an involute curve with base radius 26 mm (this produces sweep with 85.3 deg) proportionally to the amount of sweep. The blade thickness is 2 mm and the suction surface near the leading edge is filed to a wedge angle 2.75 deg with the leading edge radius of 0.2mm.

Figure 1.3 shows the non-cavitating characteristics of the inducers. For forward swept inducers F30 and F50, the head is smaller than that for Inducer 0 at smaller flow coefficient $\phi < 0.06$ but the head is not largely affected by the sweep in a wide range $0.06 < \phi < 0.1$ around the design point $\phi = 0.078$.

Figure 1.4 shows the plot of cavity length l/s at the tip against the cavitation number k for three inducers 0, B50 and F 30. The nominal incidence angle at the tip, $\alpha = \beta - \tan^{-1}\phi$ is also shown in the figure. For all inducers and all incidence angles shown, alternate blade cavitation (in which cavity length differs alternately) starts to develop when the cavity length exceeds about 65% of the spacing. The cavitation becomes unsteady for the condition with k smaller than that with the data point. In most cases unsteady cavitation starts to occur when the length of the shorter cavity exceeds 65% of the spacing. As expected, the cavity develops faster for the cases with larger incidence angle α. These results clearly show the favorable effects of sweep and agree fairly well with the theoretical findings.

Figure 1.5 shows the plot of cavity length against $k/2\alpha$, where α is the nominal incidence angle at the tip. As expected from linearized analysis, the development of cavity is nearly the same for all the incidence angles. The comparisons among three inducers clearly show that the development of steady cavity is significantly delayed by giving both forward and backward sweep.

Neglecting all the difference of the cascade geometry in the cross flow plane, the cavity length l/s is replotted against $k_c/2\alpha_c$ in Fig.1.6. Nominal values at the tip have been used for the plot. We find that the alternate blade cavitation starts to develop at $k_c/2\alpha_c \approx 0.9$ and it shifts to unsteady cavitation at $k_c/2\alpha_c \approx 0.4$, for all the inducers.

The present results show that the delay of cavity development can be explained by the cross flow effect. The secondary flow caused by the leakage from the pressure side near the tip should be quite different for forward and backward sweep. However, the delay of cavity development is quite the same for forward and backward sweep as shown in Fig.1.5 and it can be explained by the cross flow effects as shown in Fig.1.6. This fact shows that the cross flow effect is more important than the secondary flow effects. It has been shown that the various kinds of unsteady cavitation depend only on the steady cavity length l/s or equivalently on $k_c/2\alpha_c$. In this respect the present correlation with $k_c / 2\alpha_c$ explains not only the steady cavity development but also the onset of unsteady cavitation for $k_c / 2\alpha_c < 0.4$.

1.4 Effects of Sweep on Cavitation Instabilities

In the last section it was shown that the development of steady cavitation is significantly delayed by the cross flow effects. In this section the effect of sweep on cavitation instabilities and also the limitation of the $k_c / 2\alpha_c$ correlation are discussed (Tsujimoto et al., 2001, Yoshida et al., 2001). Experiments were carried out for five inducers as shown in Table 1.1 and Fig.1.2, with the rotational speed of 4,000 rpm (66.7Hz). The blade angles are the same for all the inducers but the solidity at the tip differs as shown in Table 1.1. Figure 1.7 shows the velocity fluctuations at 5 mm upstream of the leading edge, at 7 radial locations, for Inducer 0, B90 and F50, measured by LDV. The circumferential locations are shown by θ measured from the leading edge. Figure 1.8 shows the circumferentially averaged distributions of axial and tangential velocities at $\phi = \phi_d = 0.078$ and at $\phi = 0.070$. At the design point $\phi = \phi_d = 0.078$ we observe a backflow near the tip for Inducer 0 and B90, but not with Inducer F50. At the reduced flow rate $\phi = 0.070$ we observe the backflow for all inducers but the backflow region with $\overline{V_z} / U_1 < 0$ is smaller for F50. The region with swirl, $\overline{V_\theta} / U_1 > 0$, becomes larger in the order of F50, 0, B90. This shows that the backward sweep enhances the backflow but the forward sweep has the effect to suppress the backflow.

Figure 1.9 shows the sketches of cavitation for the five inducers with equal length and alternate blade cavitation. First, we consider the equal length cavitation. For non-swept inducer 0, we observe a blade surface cavitation whose length is larger at the tip. For the backward swept inducers B50 and B90, we observe cavitation only near the tip. For forward swept inducers F30 and F50, we observe blade surface cavitation and the cavity length becomes maximum around the midspan. With alternate blade cavitation, the cavities near the tip merges into two separate cavities. The blade surface cavity for F30 and F50 appears alternately, typical for alternate blade cvitation. Thus the cavity development differs from inducer to inducer and also differs at different radial location but we will take the cavity length at the tip for characterizing the cavity development.

Figure 1.10 shows the specta of inlet pressure fluctuations at $\phi = 0.080$. We observe rotating cavitation (and cavitation surge for B50 and F30) at a cavitation number smaller than that with alternate blade cavitation, except for B90. They are herein called “unsteady cavitation”. Note that no unsteady cavitation was found for B90. The region of cavitation numbers for various types of cavitation is summarized in Fig. 1.11 in terms of the sweep angle λ at the tip, for $\phi = 0.080$. The boundary cavitation numbers decrease as we decrease λ (increase sweep), to $\sin\lambda = 0.5$, but further decrease in λ does not cause further decrease in σ. This shows that the cross flow effects represented by $k_c / 2\alpha_c$ appear only for the cases of smaller sweep, with λ larger than 35 deg. The cavity length at the tip l/h is also shown in Fig. 1.11. We find that the alternate blade cavitation starts to occur with $l/h \approx 0.65 - 0.70$. The unsteady cavitation occurs when the length of shorter cavity l_s / h of alternate blade cavitation becomes 55-70 % of the blade spacing.

Figure 1.12 shows the spectra of inlet pressure fluctuations at a reduced flow coefficient $\phi = 0.070$. For B90, the spectrum is noisy and no distinct cavitation instabilities could be identified except for the cavitation surge at very small cavitation number. This is because the boundary of the backflow region is near the location of the pressure sensor and it detects irregular passage of backflow vortex cavitation. On the other hand, we can identify various components for F50. First, we should note that we can identify forward rotating cavitation at the cavitation number larger than the alternate blade cavitation onset. In addition to this, we observe a backward rotating cavitation. The backward rotation was confirmed by the plot of the phase of the pressure fluctuation at various circumferential locations. Detailed examination with a wavelet transform suggested that the direction of propagation is changing between forward to backward, about every 10 turns of the impeller. The propagation velocity of the forward mode is about 1.65, which is significantly higher than normal rotating cavitation.

Figure 1.13 shows the suction performance and the regions of various cavitation instabilities for the five inducers. Note that the breakdown cavitation number is decreased from $\sigma = 0.03$ for Inducer 0 to $\sigma = 0.02$ for B50 and B90. Further improvement was observed for forward swept inducers, especially for F50: no breakdown was observed at the minimum cavitation number $\sigma = 0.015$.

We find that the region of σ for alternate blade cavitation and also unsteady cavitations is shifted to significantly smaller σ for B50 and F30. However, further improvement was not found if we increase the sweep further (B90 and F50). With smaller flow rate, rotating cavitation was found at larger cavitation number than for alternate blade cavitation for forward swept inducers. In this respect, forward sweep is not recommended although the breakdown cavitation number is significantly decreased by the sweep.

2 Housing Enlargement at the Inlet

2.1 Results with LE-7 LOX Turbopump

The method of casing enlargement at the inlet was first applied by Kamijo et al. (1993), to suppress rotating cavitation in LE-7 LOX turbopump inducer. The major design parameters are shown in Table 2.1. A comparatively large sweepback was necessary in order to decrease the stress at the hub near the leading edge. Five kinds of inducer housing with dimensions shown in Fig.2.1 and Table 2.2 were tested in order to find a method to suppress the rotating cavitation observed with Housing A.

Suction performance and rotating cavitation. Figure 2.2 shows the inducer suction performance. With Housing A and C, no significant difference occurs for $\sigma > 0.05$. However, the head decreases slightly in the region $\sigma = 0.02 - 0.05$ with Housing A. Figure 2.3 shows the spectrum of impeller displacement for the case with Housing A. A super synchronous shaft vibration is observed in the range corresponding to $\sigma = 0.027 - 0.05$, and the amplitude of synchronous vibration is increased in the range corresponding to $\sigma = 0.02 - 0.027$. These ranges agree with the range of head degradation shown in Fig. 2.2. In particular, the head decrease is larger at $\sigma = 0.027$, where the shift between super synchronous and synchronous vibration occurs. Figure 2.4 shows the ratio of supersynchronous vibration frequency f_{NN} to the impeller rotational frequency f_N. This ratio decreases as we decrease the cavitation number. From the frequency ratio observed, it was concluded that the supersynchronous vibration is caused by rotating cavitation.

Suppression. It was conjectured that rotating cavitation might be closely related to the tip leakage and backflow vortex cavitation, judging from the visual observations. Some efforts were made to influence the tip vortex cavitation. The influence of tip clearance on the supersynchronous shaft vibration was investigated by using the inducer housings D and E. The tip clearances for housings D and E are 0.5 and 1.0mm, respectively. Increasing the tip clearance was fairly effective in decreasing the amplitude of the supersynchronous shaft vibration as shown in Fig.2.5, but it could not completely extinguish the vibration. Acosta (1958) reported that increasing the tip clearance helped prevent oscillating cavitation to some extent. Figure 2.6 presents the spectrum of impeller displacement tested using the housing C with larger diameter at the inlet. Note that the housing diameter D_1 is increased by only twice the tip clearance $2C_2 = D_2 - D_1$ from D_2. The supersynchronous shaft vibration is suppressed completely. This device was successfully applied to the flight model.

In order to identify the possible cause of suppression, the values of mass flow gain factor was measured with housings Case 1 (similar to Housing A) and Case 2 (similar to Housing C) (Shimura (1993)). The result is shown in Fig. 2.7, in a map showing the results of one-dimensional stability analysis of rotating cavitation (Tsujimoto et al., 1993). It was found that the mass flow gain factor is decreased and the cavitation compliance is increased by modifying the casing to Case 2 (similar to Housing C), both contributing to the suppression of rotating cavitation. However, it is still not clear why such changes occurred by a small modification of the housing.

2.2 Results with an Inducer for LE7A LH2 Turbopump

In order to examine the effects of housing enlargement at the inlet on another geometry of inducer, a series of tests were conducted on the original design of LE7A LH2 turbopump (Fujii et al., 2004). It suffered also from rotating choke and it was eventually replaced with a new design. The dimensions of the inducer are shown in Table 2.3. It has a smaller blade angle at the inlet and hence smaller flow coefficient as compared with the LE7 LOX inducer mentioned in the preceding section. The geometry of the impeller is shown in Fig. 2.8. Eight housing geometries as shown in Fig.2.9 are tested. The “tightness” of the casing is decreased in the order Casing 0-Casing 7. The effects of tip clearance can be examined by compareing the results for Casing 0 and 3. Figure 2.10 shows typical spectrum of inlet pressure fluctuation, at the design flow coefficient $\phi = \phi_d = 0.067$ at 3000 rpm, for Casing 0, 3 and 6. Except for Casing 6, rotating cavitation denoted by R.C. first appears followed by attached asymmetric cavitation denoted by A.C. For Casing 6, no rotating cavitation occurred. Figure 2.11 shows the regions of rotating cavitation, attached asymmetric cavitation and cavitation surge with various casing geometries at three typical flow rates smaller than design ($\phi = 0.060$), at design ($\phi = \phi_d = 0.067$), and larger than design ($\phi = 0.072$) flow coefficients. These results show that,

(1) Enlarging the tip clearance is effective in reducing the onset cavitation number of rotating cavitation, for all flow rates (from comparison of Casing 0 and 3).
(2) The casing enlargement at the inlet has favorable effects at higher flow coefficient but adverse effects at smaller flow coefficient (from comparisons of Casing 3,4, and 5).
(3) The overlapping of the enlarged part with the blade has favorable effects at higher flow rates. It appears suddenly at a certain flow coefficient as we increase the flow rate (from Casing 5 and 6)
(4) At the larger flow coefficient $\phi = 0.072$, the onset cavitation number of rotating stall decreases as we decrease the “tightness” of the casing (in the order Casing 0, 1, ...)
(5) At the larger flow coefficient $\phi = 0.072$, the transition cavitation number between rotating

cavitation and attached asymmetric cavitation is increased as we move the enlargement point closer to the impeller (Casing 0, 1, and 2), or decreasing the amount of enlargement (Casing 5, 4, and 3).

(6) At the larger flow coefficient $\phi = 0.072$, there can be seen a tendency that the transition point between cavitation surge and attached asymmetric cavitation is increased as we reduce the "tightness" of the casing (Casing 0, 1, 2, …,7).

Figure 2.12 shows the maximum length of the backflow vortex cavitation for Casing 0, 3 and 6. At larger cavitation number $\sigma > 0.04$, the vortex length is not largely affected by the cavitation number, and the backflow vortex cavitation gets longer in the order of Casing 0, 3, and 6. This may suggest that the backflow or backflow vortex cavitation has an effect to suppress cavitation instabilities. However, we have larger backflow at smaller flow rates but the range of cavitation instabilities is larger for smaller flow rates. So, it is difficult to explain the effect of casing modification only from the backflow and backflow cavitation. When attached asymmetric cavitation appears, the backflow vortex cavitation suddenly becomes shorter. This suggests an abrupt change of radial flow pattern.

Figure 2.13 shows the incidence angle distribution evaluated from the velocity measurements at $z / D = -0.057$, for the Casing 0 and Casing 6, under non-cavitating condition. We find that the incidence angle is smaller for the Casing 6, caused by the increase of axial velocity due to the blockage effect of backflow and the tangential velocity imparted by swirling backflow. If the angle of attack gets smaller, the length of the cavity is expected to get smaller. A stability analysis of 2-D cavitating flow (Horiguchi et al. (2000)) suggests that the cavitation instabilities occur when the cavity length gets larger than about 65% of the blade spacing. So, the decrease of angle of attack may be the cause of suppression. However, we should note that the cavitation instabilities cannot be suppressed completely by the decrease of the incidence angle, as is clear from the fact that we do have cavitation instabilities also at larger flow rate.

From the comparison of the results shown in this and the preceding sections, the effectiveness of casing enlargement differs significantly depending on the design of the inducer. The mechanisms of suppression is not very clear as compared to the case of leading edge sweep. However, in certain cases it has a significant effect with a small modification and it worth to be examined.

3 Conclusion

The effects of leading edge sweep and the housing enlargement at the inlet on cavitation instabilities are discussed. The major effect of leading edge sweep can be explained by the cross flow effects, when the amount of sweep is not very large. With larger sweep, three dimensional flow effects including the effect of backflow would become more important. The effect of housing enlargement depends on the design of the impeller: larger effect can be expected for impellers with larger blade angle. However, the mechanism of suppression is not clear as yet. In both of the methods treated here, the backflow at the inlet is increased. So, it is needed to clarify the relation between the backflow and cavitation instabilities.

There are certain cases when the cavitation instabilities cannot be suppressed completely by using the methods introduced here. So, we need to find out additional methods of suppression, as well as to clarify the mechanisms of suppression by the casing enlargement at the inlet.

Bibliography

Acosta, A.J., (1955). A Note on Partial Cavitation of Flat Plate Hydrofoils. *Caltech Hydro Lab. Report* No. E-19.9, Oct.

Acosta, A.J., Tsujimoto,Y., Yoshida,Y., Azuma,S., and Cooper, P., (2001). Effects of Leading Edge Sweep on the Cavitating Characteristics of Inducer Pumps. *International Journal of Rotating Machinery*, Vol.7, No.6, 397-404.

Fujii, A., Azuma, S., Yoshida, Y., Tsujimoto, Y., Uchiumi, M., Warashina, S., (2004). Effects of Inlet Casing Geometries on Unsteady Cavitation in an Inducer. *Transactions of JSME,* Ser.b, Vol. 70, No. 694, 1459-1466.

Goirand, B., Mertz, A.L., Joussellin, F., and Rebattet, C., (1992). Experimental Investigations of Radial Loads Induced by Partial Cavitation with a Liquid Hydrogen Inducer. *ImechE*, C453/056, 263-269.

Horiguchi, H., Watanabe, S., and Tsujimoto, Y., (2000). A Linear Stability Analysis of Cavitation in a Finite Blade Count Impeller. *ASME Journal of Fluids Engineering*, Vol.122, No.4, 798-805.

Kamijo, K., Yoshida, M., and Tsujimoto, Y., (1993). Hydraulic and Mechanical Performance of LE-7 LOX Pump Inducer. *AIAA Journal of Propulsion and Power*, Vol.9, No.6, 819-826.

Shimura,T., (1995). The Effects of Geometry in the Dynamic Response of the Cavitating LE-7 LOX Pump. *AIAA Journal of Propulsion and Power*, Vol.11, No.2, 330-336.

Sloteman, D.P., Cooper, P., and Dussord, J.L., (1984). Control of Backflow at the Inlets of Centrifugal Pumps and Inducers. *Proceedings of the 1st International Pump Symposium*, Texas A&M, TX, May, 9-22.

Tsujimoto, Y., Kamijio, K. and Yoshida, Y., (1993). A Theoretical Analysis of Rotating Cavitation in Inducers. *Journal of Fluids Engineering*, Vol.115, No.1, 135-141.

Tsujimoto,Y., Yoshida,Y., Acosta,A.J., Azuma,S., and Lafitte, S., (2001). Effects of Leading Edge Sweep on Unsteady Cavitation in Inducers (1st Report, Improvement of Cavity Length Due to Forward and Backward Sweep). *Transactions of JSME*, Ser.b, Vol 67, No.656, 903-910.

Yoshida,Y., Tsujimoto,Y., Acosta,A.J., Azuma,S., and Lafitte, S., (2001). Effects of Leading Edge Sweep on Unsteady Cavitation in Inducers (2nd Report, Problems of Forward and Backward Sweep). *Transactions of JSME*, Ser.b, Vol 67, No. 658, 1367-1375.

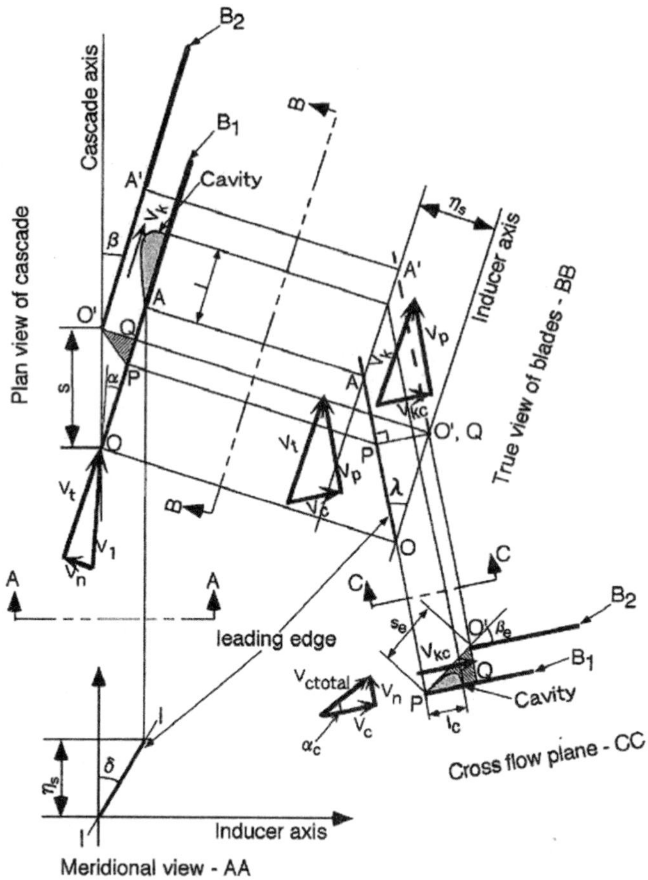

Figure 1.1 Geometry of swept cascade

Inducer	0	B50	B90	F30	F50
Sweep	Without sweep	Backward sweep	Backward sweep	Forward sweep	Forward sweep
Sweep angle at tip, λ	90 deg.	35.4 deg.	17.3 deg.	49.2 deg	30.9 deg.
Solidity at tip	2.97	2.44	1.91	2.86	3.17
Diameter, D	149.8 mm	←	←	←	←
Number of blades	4	←	←	←	←
Inlet tip blade angle, β_{t1}	7.5 deg.	←	←	←	←
Outlet tip blade angle, β_{t2}	9.0 deg.	←	←	←	←
Tip clearance, s	0.5 mm	←	←	←	←

Table 1.1 Dimensions of inducers

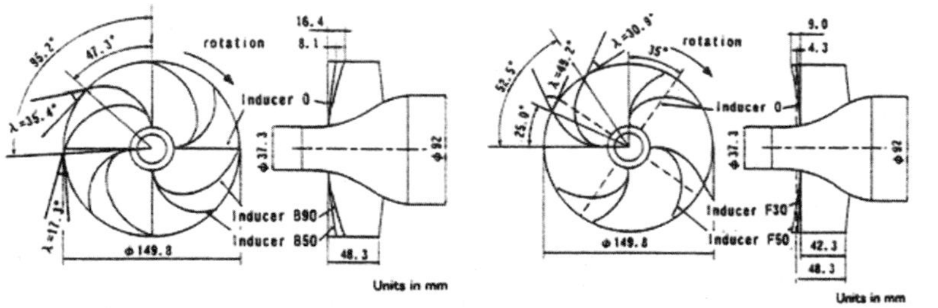

(a) Inducer B 50, B 90, and 0 (b) Inducer F 30, F 50, and 0

Figure 1.2 Geometry of test inducers

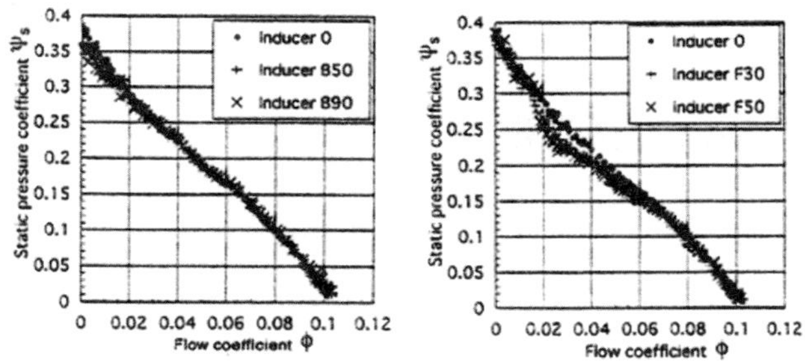

(a) Inducer B 50, B 90, and 0 (b) Inducer F 30, F 50, and 0

Figure 1.3 Non-cavitating performance of inducers

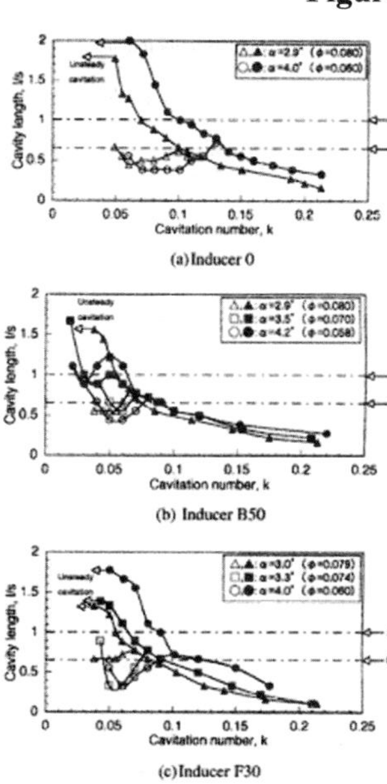

Figure 1.4 Cavity length against k

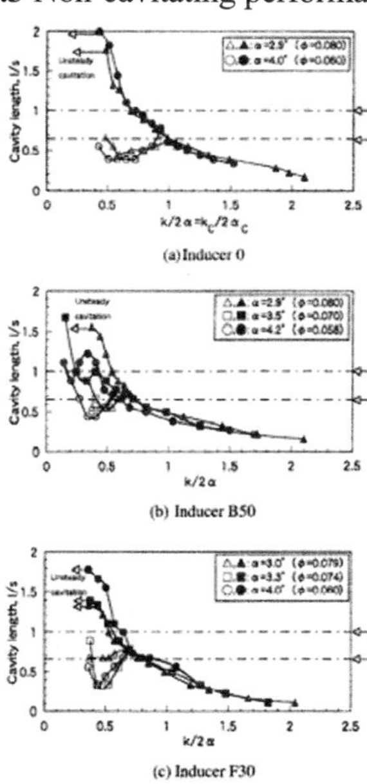

Figure 1.5 Cavity length against $k/2\alpha$

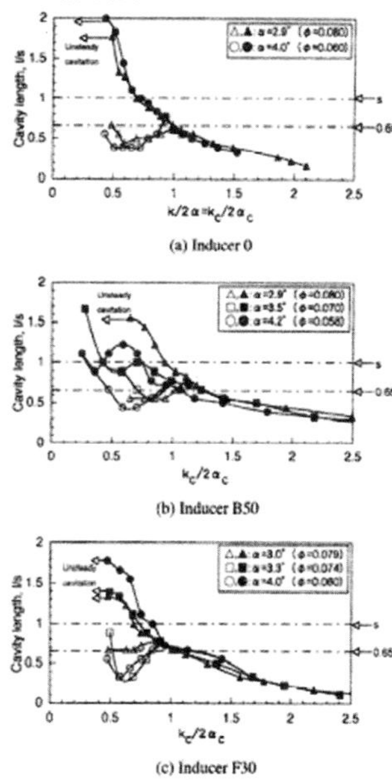

Figure 1.6 Cavity length against $k_c/2\alpha_c$

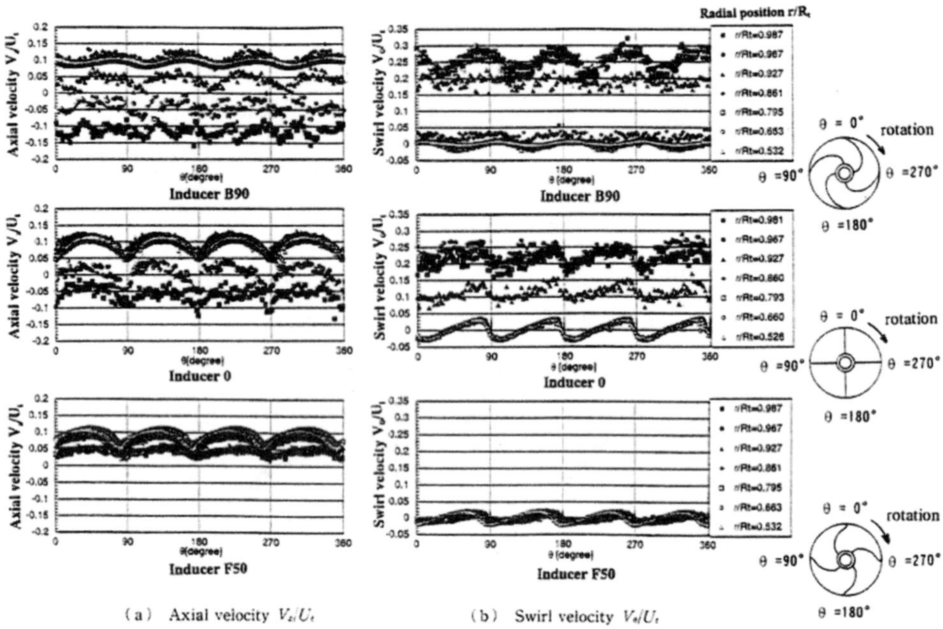

Figure 1.7 Velocity distribution at the inlet of the inducers

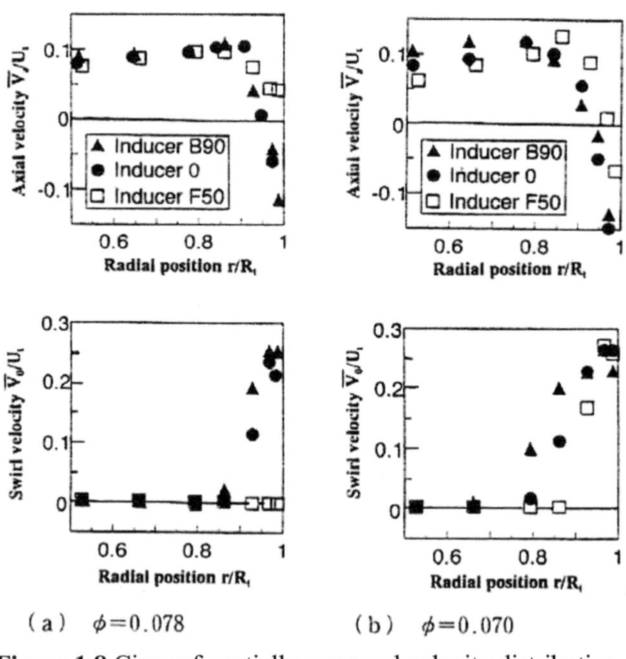

Figure 1.8 Circumferentially averaged velocity distribution

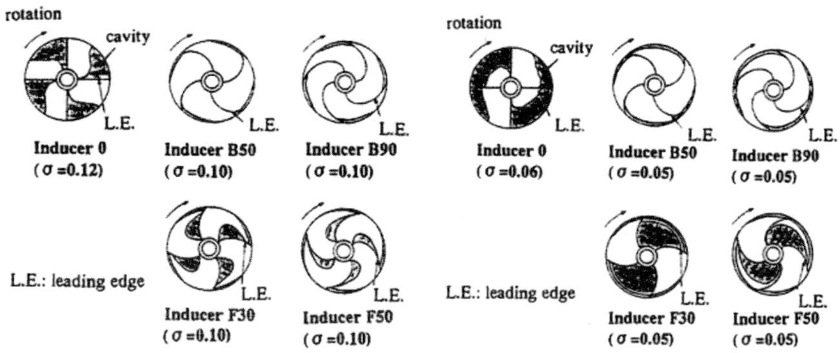

(a) Equal length cavitation (b) Alternate blade cavitation

Figure 1.9 Effect of leading edge shape on equal and alternate blade cavitation

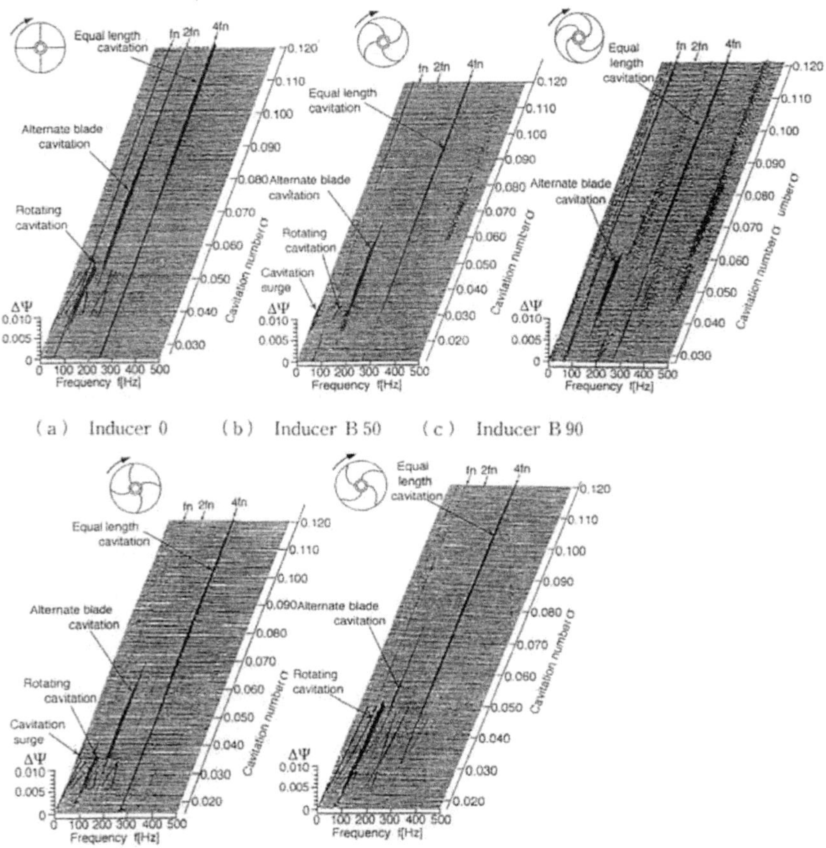

(a) Inducer 0 (b) Inducer B 50 (c) Inducer B 90

(d) Inducer F 30 (e) Inducer F 50

Figure 1.10 Spectra of inlet pressure fluctuations at $\phi = 0.080$

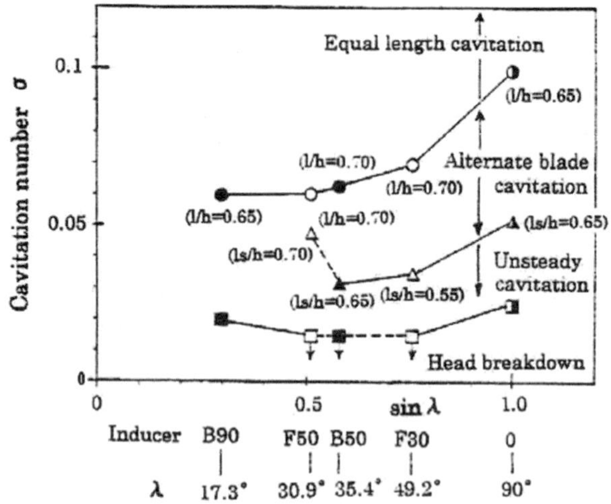

Figure 1.11 Effects of leading edge sweep on the boundary of various types of cavitation, $\phi = 0.08$.

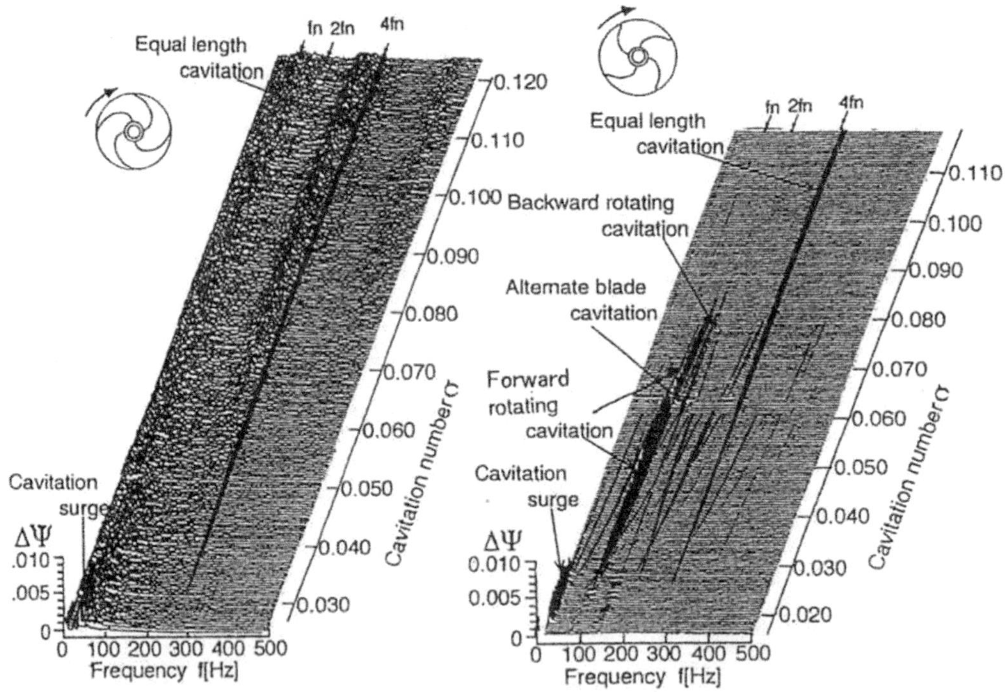

(a) Inducer B 90 (b) Inducer F 50

Figure 1.12 Spectra of inlet pressure fluctuations at reduced flow rate at $\phi = 0.070$

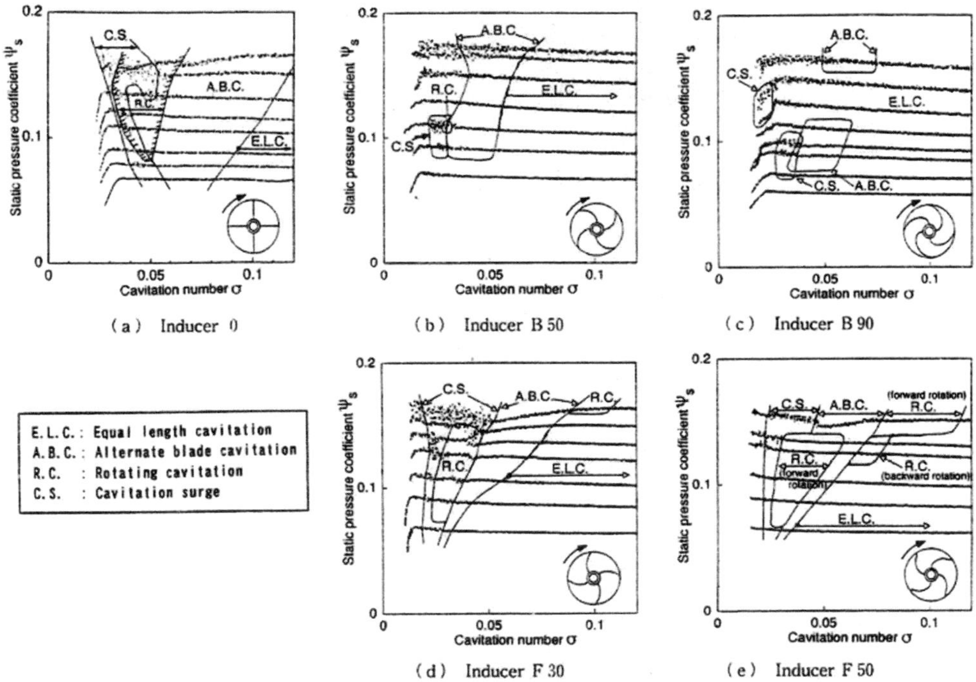

Figure 1.13 Regions of various types of unsteady cavitation in the suction performance plane

Rotational speed N, rpm	20,000
Required NPSH, m	30.0
Suction specific speed S, m, m^3/s, s^{-1}	2.10
Cavitation number σ	0.017
Number of blades	3
Inlet flow coefficient,[a] ϕ_1	0.083
Outlet flow coefficient[a] ϕ_2	0.104
Inducer head coefficient[a] Ψ	0.097
Tip diameter D_t, mm	149.8
Inlet tip blade angle β_{t1}, deg	7.50
Outlet tip blade angle β_{t2}, deg	9.50

[a]Values for 1.07 times the quantity of nominal flow.

Table 2.1 Design parameters of LE-7 LOX main pump inducer

	Inside diameter of upstream housing D_1, mm	Inside diameter of first half of liner D_2, mm	Inside diameter of latter half of liner D_3, mm
Housing A	150.8	150.8	151.4
Housing B	151.3	151.3	151.3
Housing C	153.0	151.3	151.3
Housing D	149.8	150.8	150.8
Housing E	149.8	151.8	151.8

Table 2.2 Dimensions of inducer housing

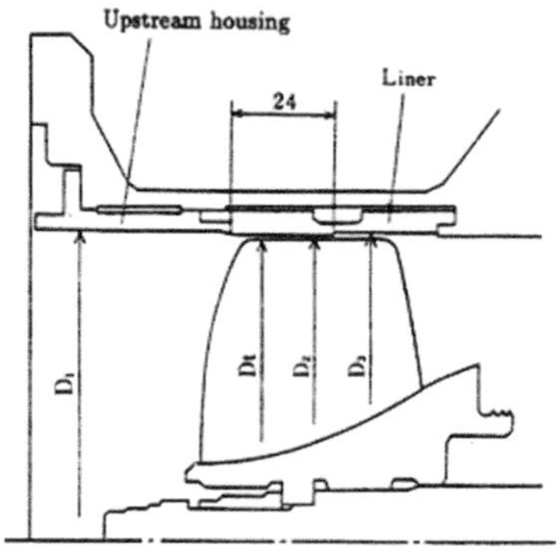

Figure 2.1 Geometry of inducer housing

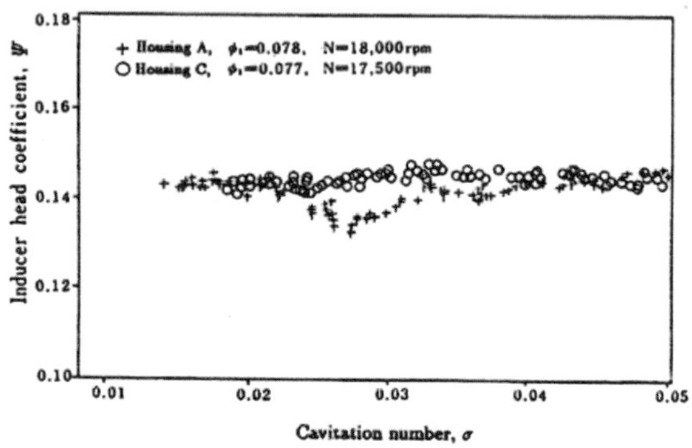

Figure 2.2 Suction performance of LE-7 LOX pump inducer

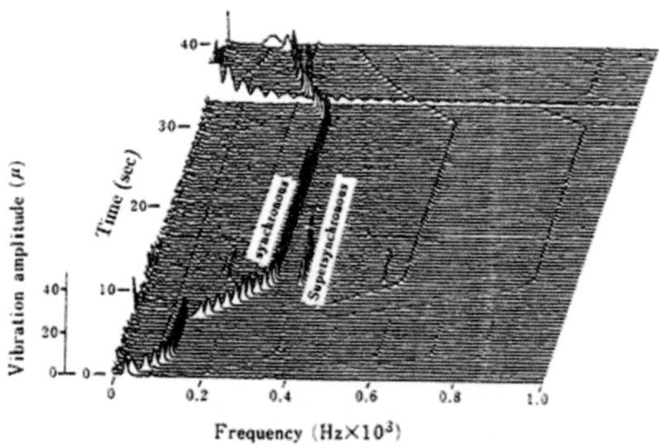

Figure 2.3 Spectrum of shaft vibration with housing A

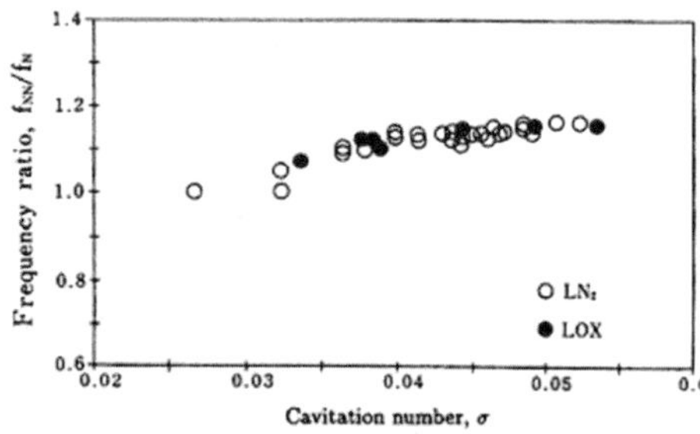

Figure 2.4 Frequency ratio corresponding to the propagation velocity ratio

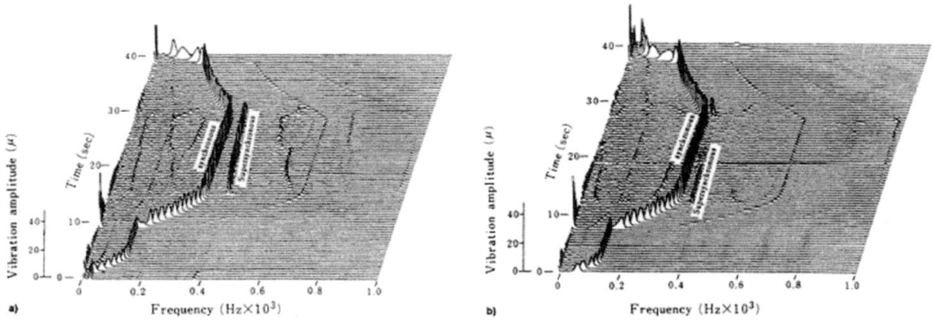

Figure 2.5 Effect of tip clearance a) housing D (clearance 0.5mm), b) housing E (clearance 1.0mm)

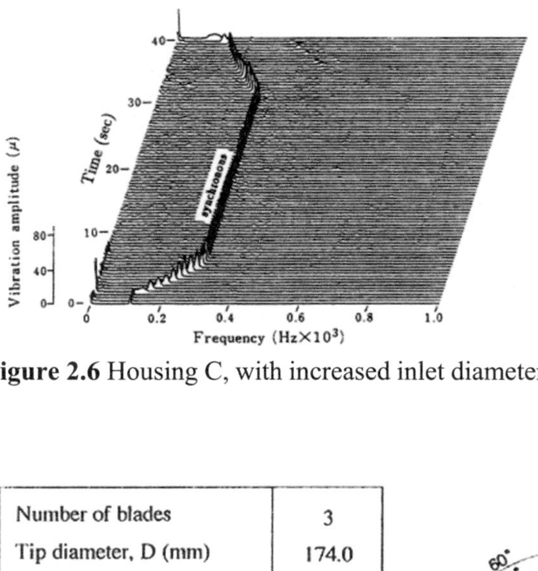

Figure 2.6 Housing C, with increased inlet diameter

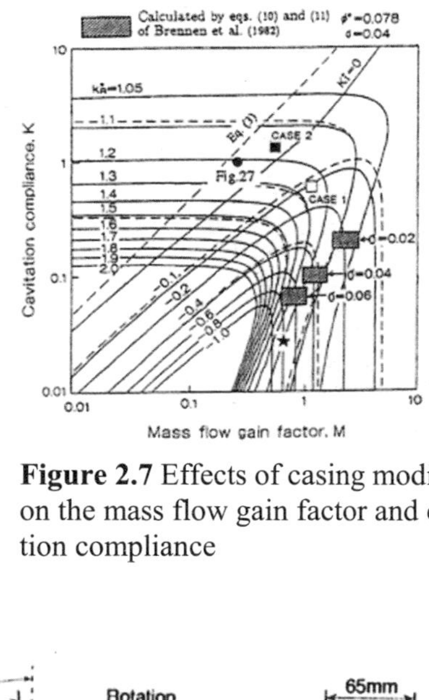

Figure 2.7 Effects of casing modification on the mass flow gain factor and cavitation compliance

Number of blades	3
Tip diameter, D (mm)	174.0
Inlet tip blade abgle, β_{t1}	6.4 °
Outlet tip blade angle, β_{t2}	11.1 °
Hub / tip ratio at inlet	0.287
Hub / tip ratio at outlet	0.460
Solidity at tip	2.1
Radial tip clearance (mm)	0.35
Design flow coefficient, ϕ_d	0.067

Table 2.3 Principal dimensions of test inducer

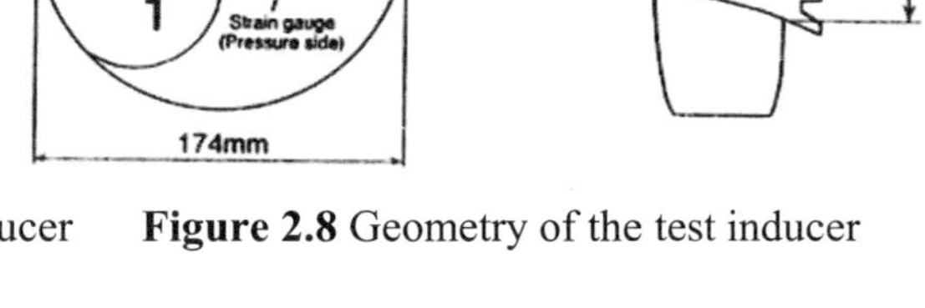

Figure 2.8 Geometry of the test inducer

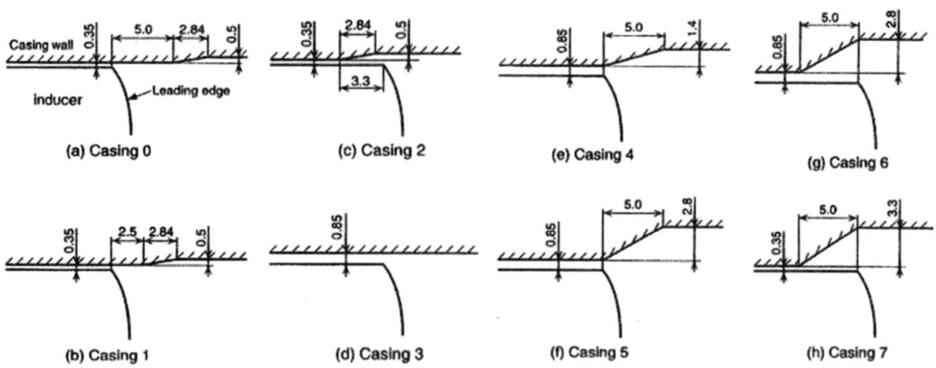

Figure 2.9 Geometries of the casings tested

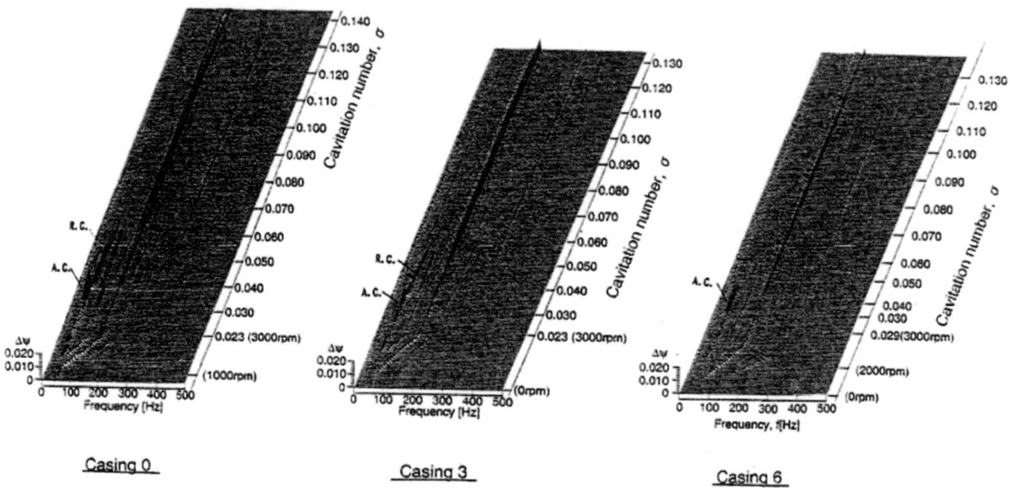

Figure 2.10 Spectra of inlet pressure fluctuations, at design flow coefficient $\phi = 0.067$, 3000rpm

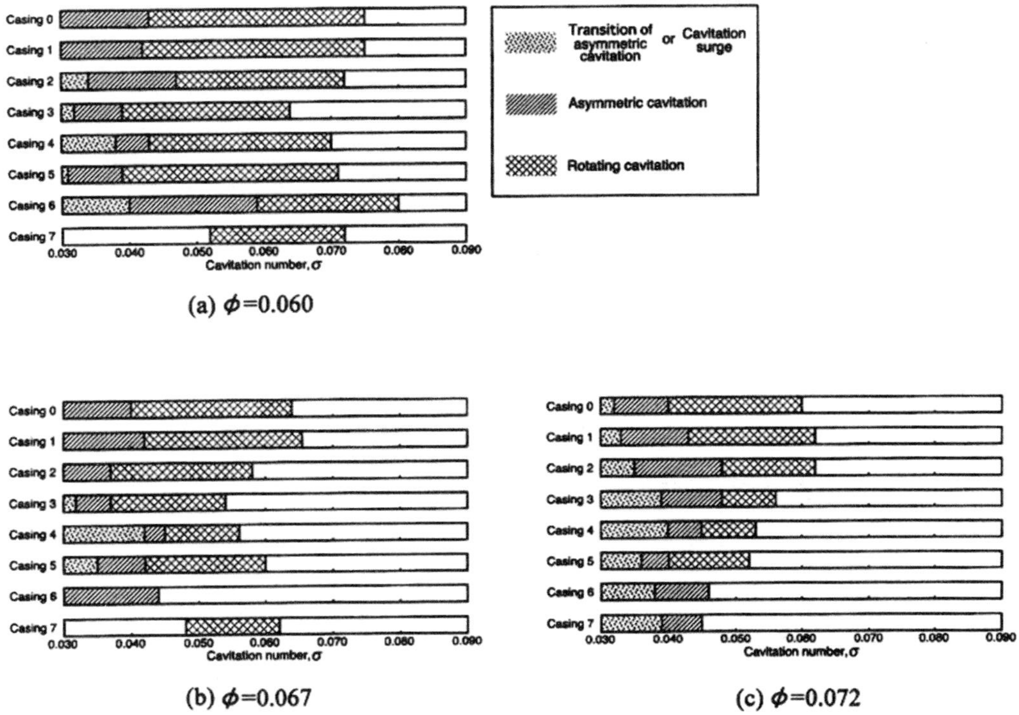

Figure 2.11 Occurrence regions of rotating cavitation, asymmetric cavitation, and cavitation surge

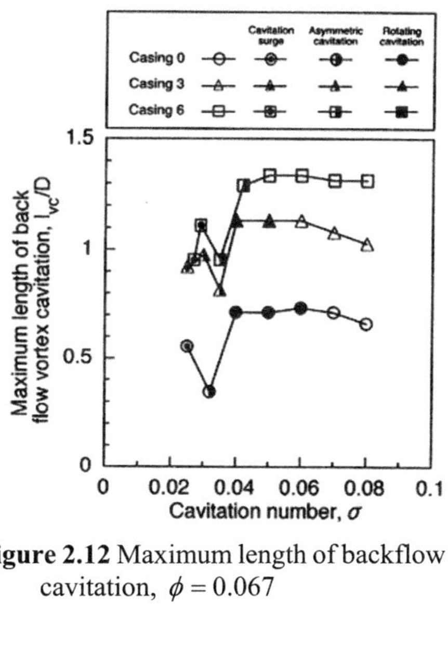

Figure 2.12 Maximum length of backflow vortex cavitation, $\phi = 0.067$

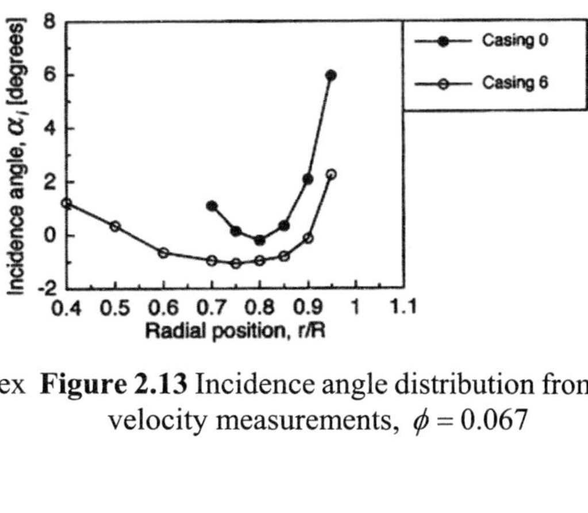

Figure 2.13 Incidence angle distribution from velocity measurements, $\phi = 0.067$

Tip Leakage and Backflow Vortex Cavitation

Yoshinobu Tsujimoto[1]

[1] Engineering Science, Osaka University, Osaka, Japan

Abstract. In addition to blade surface cavitation, tip leakage and backflow cavitations occur in real inducers. Although most of cavitation instabilities can be explained by considering only blade surface cavitation, these types of cavitation might play an important role especially for the suppression of instabilities. Those cavitations are three-dimensional in its nature and this makes it difficult to analyze them theoretically even without cavitation. Under this situation not much is known about these cavitation types. In this document, research efforts into unsteady characteristics of those cavitation types are described, although they are by no means complete as yet.

1 Tip Leakage Cavitation

There are two types of tip leakage cavitation : One is the tip leakage vortex cavitation which appears in the vortex formed by roll up of the shear layer between the leakage jet and the ambient flow, and the other is the shear layer cavitation appearing in the shear layer of the leakage jet. The former becomes stronger than the latter for the cases with larger tip clearance. The latter would be more difficult to treat since it would be affected by the population of cavitation nuclei and the level of turbulence in the shear layer. So we will focus only on the tip leakage vortex cavitation simply because it looks more tractable.

1.1 Slender Body Approximation

Rains (1954) first proposed to apply the slender body approximation to the tip leakage vortex, in which three-dimensional tip leakage flow was simulated by a two-dimensional unsteady cross flow. This method was further applied by Chen et al. (1991) and examined by comparison with experiments under non-cavitating condition. The basic idea is that the clearance velocity field can be (approximately) decomposed into independent through flow and cross flow, since chordwise pressure gradients are much smaller than normal pressure gradients in the clearance region. As in the slender body approximation in external aerodynamics, this description implies that the three-dimensional, steady, clearance flow can be viewed as a two-dimensional, unsteady flow.

Figure 1.1 explains the idea of slender body approximation applied to the tip leakage flow (Chen et al. (1991)). Consider cross flow planes *A, B, C,* and *D* at different chordwise locations *a, b, c,* and *d,* respectively, as shown in the top part of Fig. 1.1. Location *a* is at the leading edge and *d* is at the trailing edge. At station *a,* the tip clearance flow is initiated so that the flow in cross flow plane *A* might be as shown in the lower part of the figure. As one moves through the blade passage, the vortex sheet shed into the clearance rolls up so that subsequent cross sections might be as illustrated in planes *B, C,* and *D.*

The analogy proposed is that the flow pattern in different crossflow planes is similar to a two-dimensional unsteady flow. The velocities in the crossflow planes of the top part of Fig.1.1 are thus represented by the unsteady flow at the four different times shown in the lower part of the figure. If this analogy holds, it implies that evolution of the cross-plane flow structure (including tip clearance vortex strength and position) at different streamwise locations is similar to that at different times, when viewed from a moving reference frame. The transformation between time, t, and streamwise location, s, is $t = s/U$, where U is the velocity of the moving frame.

For simplicity, the above discussions are made for the case with a steady three-dimensional flow. This method can be easily extended to three-dimensional unsteady flows, by simply calculating the cross flow under the boundary conditions to which the crossflow is subjected at each instant of time, while the cross flow plane flows down from the leading edge to the location of interest.

1.2 Two-Dimensional Cross Flow

It is assumed that the leakage flow velocity is given by $U_j = \sqrt{2\Delta p/\rho}$ where Δp is the pressure difference across the blade. The cross flow is represented by the source distribution $q_j = 2U_j$ along the clearance, discrete vortices $\Gamma_k = U_j^2 \Delta t/2$ shed from the blade tip during the time interval Δt and representing the shear layer on the jet boundary. The effect of cavitation is represented by a source q_B, placed at the center of the vortex blob. The strength of the source q_B is determined as follows. We assume a cylindrical cavity with radius R. Then the strength of the source can be obtained from the relation $q_B = 2\pi R dR/dt$. The cavity radius R is determined by integrating the momentum equation corresponding to the cylindrical version of the Rayleigh-Plesset equation for the bubble dynamics. The boundary conditions on the casing and the blade surface are satisfied by placing mirror images of the singularities, as shown in Fig.1.2. Positions of the vortices Γ_k and the source q_B are renewed using the local velocity at each time step.

The validity of this model was confirmed by comparison with a two-dimensional unsteady model experiment (Watanabe et al. (2001)). It was found that location of the tip vortex can be predicted nicely even if we neglect the existence of the cavity, but without the displacement effect of the cavity represented by q_B the size of the cavity becomes too large. Figure 1.3 shows the distribution of the vortices with and without the effect of the source q_B. With the source q_B, the vortices are dispersed caused by the growth of the cavity. The pressure decrease at the center of the vortex and hence the cavity size is becomes larger if we neglect the effect of the cavity growth.

For three-dimensional steady flow, it was shown (Higashi et al. (2002)) that the present model can predict the effects of angle of attack, cavitation number, foil shape, and tip clearance on the trajectory of the tip vortex and the size of the cavity, at least qualitatively.

1.3 Three-Dimensional Unsteady Tip Leakage Cavitation

In order to study the unsteady effects on tip leakage cavitation, we consider a pitching hydrofoil (Murayama et al. (2003)). The flat plate hydrofoil with the chordlength $C = 90mm$, the span

$67mm$, and the tip clearance $\tau = 3\ mm$ was tested in a test section of $100mm \times 70mm$. The hydrofoil is arranged to pitch around the midchord with the angle of attack:

$$\alpha = \alpha_m + \Delta\alpha\ sin(\omega t)$$

where $\alpha_m = 4$ deg and $\Delta\alpha = 2$ deg. The normalized frequency is defined as $k = \omega C/U$ where the mean velocity is set to be $U = 5.0\ m/s$.

Figure 1.4 shows the pictures of tip leakage cavitation for $k = 0$, 0.45 and 0.90 at each incidence angles. The time proceeds from top to bottom and covers a complete pitching cycle. At $k = 0$, the cavity radius becomes maximum at the maximum incidence $\alpha = 6$ deg but the cavity development delays as we increase the frequency k. To examine this more explicitly, the fluctuation of cavity radius at the mid-chord $Z/C = 0.5$ and at the trailing edge $Z/C = 1.0$ is plotted in Fig. 1.5, as well as the length of the blade surface cavitation at mid-span. It is clearly shown that the cavity development delays and the maximum cavity radius becomes smaller as we increase the frequency. Note that the length of the blade surface cavitation exhibits a similar tendency.

Numerical calculations are made as follows. The unsteady pressure difference $\tilde{p}$ is estimated from the results of non-cavitating unsteady flow analysis (Fung (1969)):

$$\tilde{p} = \rho U^2 \Delta\alpha \left[-\left\{ (1+\frac{ik}{2})C(k) - \frac{ik}{2} \right\} tan\frac{\theta}{2} - 2ik\, sin\,\theta + \frac{1}{4}k^2 sin2\theta \right] e^{i\omega t}$$

where $C(k)$ is the Theodorsen function and $\xi = \cos\theta$ represents the chordwise location and $\xi = -1$ and $\xi = 1$ correspond to the leading and trailing edges, respectively. By adding the steady component we can evaluate the instantaneous pressure difference across the blade. The instantaneous flow field in a cross flow plane is obtained by assuming the pressure difference which the cross flow plane experiences while it travels from the leading edge to the location of interest.

The numerical results are shown in Fig.1.6, corresponding to the results shown in Fig.1.4. Figure 1.7 compares the total cavity volume from the leading edge to $2C$ downstream of the trailing edge. In the same way as for the local cavity radius, we observe the delay of cavity development and the decrease of maximum cavity volume, as we increase the frequency. This tendency is well simulated by the numerical calculations although the maximum cavity volume is smaller. Here we consider an inducer with the solidity C/h at tip and the number of blades N. If we assume that the uniform velocity U corresponds to the tip velocity of the inducer, we can obtain the relation $f/f_n = (N/2\pi(C/h))k$ between the non-dimensional frequency k and the ratio of the frequency f of the disturbance and the rotational frequency f_n of the impeller. For the case of an inducer with the solidity $C/h = 2.0$ and the number of blades $N = 3$, $k = 0.45$ and 0.90 corresponds to $f/f_n = 0.107$ and 0.214, respectively. Figure 1.7 shows that the phase of the tip leakage cavity volume fluctuation delays about 90^0 even at these relatively small frequencies.

Next, we consider about the reason for the delay of the cavity development. The cavity size would depend on the strength of the tip leakage vortex caused by the roll up of the vortices shed from the tip of the blade. Figure 1.8 shows the growth of the total amount of the vortices shed from the blade tip for the cases when the cross flow plane reaches the trailing edge when the incidence angle becomes $\alpha = 6$, 4(−), 2, and 4(+) deg. The plus (+) and the minus (-) signs mean the increasing and decreasing phases of angle of attack, respectively. We find that the total amount of shed vortices delays behind the incidence angle fluctuation as we proceed $\alpha = 6 \to 4(-) \to 2 \to 4(+)$ deg. This can be the cause of the delay of the cavity growth.

2 Backflow Vortices

2.1 Experimental Observation of Backflow Vortices

Figure 2.1 shows the backflow vortices visualized by cavitation (Yokota et al. (1999)). Radial location of the vortex filaments was measured by a laser sensor and is shown in Fig. 2.2. They extend axially upstream and radially inward as the flow coefficient ϕ decreases. The design flow coefficient of this inducer is 0.078. The large scatter is not caused by experimental errors but highlights the unsteady character of the phenomenon. Figure 2.3 shows the radial location of the vortices at the inlet ($z/D = 0.18$, with z measured upstream from the leading edge at root). They exist near the location where the circumferential velocity becomes nearly one half of the maximum circumferential velocity.

Figure 2.4 shows the propagation velocity of the vortices normalized by the angular velocity of the impeller. The maximum circumferential flow velocity is also shown. The propagating velocity is nearly one half of the maximum flow velocity. The results shown in Figs. 2.3 and 2.4 suggest that the vortex system is caused by the roll-up of the shear layer between the swirling backflow and straight main flow. This was proved by model experiments using axisymmetric swirling backflow and straight main flow without an impeller (Mitsuda et al. (2003)). Two vortices are shown in Fig. 2.5 observed in the model experiment in which the straight main flow is supplied from the top and the swirling backflow is supplied from the bottom through a clearance between concentric pipes.

Figure 2.6 shows the number of vortices. The number is as large as 10 at the design flow coefficient $\phi_d = 0.078$ but decreases with a decrease in flow coefficient. In order to find the mechanisms determining the number of vortices, a stability analysis was made for two-dimensional vortices placed in a circle representing the inlet pipe wall, as shown in Fig.2.7. The normal velocity on the circle is cancelled by the mirror images of the inner vortices. The analysis is in parallel with that for von Karman vortex street (Lamb (1975)) and the result is shown in Fig.2.8 with experimental data. The analysis shows that there is a maximum number of vortices which can exist stably and the maximum number is a function of the radial location of the vortices. A larger number of vortices is allowed to exist stably if the vortices are located closer to the outer boundary. The experimental data are located close to the stability boundary although they are all in the unstable region. This is perhaps because a radial location of the vortex filament at $z/D = 0.18$ has been chosen for the plot, while the vortices get closer to the pipe wall as we proceed upstream. The backflow vortex structure observed in this experiment was successfully simulated by a LES simulation (Yamanishi et al., 2007).

2.2 Simulation of Backflow

A commercial code CFX-TASCflow based on the $k-\varepsilon$ turbulence model with wall function option was applied to identify the mechanism of backflow and its response to flow rate fluctuations. It is a finite volume method but is based on a finite element approach for representing the geometry and with second order accuracy both in space and time. With an unsteady version of this RANS code, the backflow vortex structure appears at an early stage of the calculations, but it eventually dies away. So, it cannot be used to study the vortex structure but is useful for studying the origin of the backflow since it avoids the complexity caused by the vortex structure and successfully predicts the non-cavitating performance and the axial extent of the backflow (Qiao et al. (2004)). About 180,000, 240,000, and 30,000 computational cells were used for the inlet pipe, inducer, and outlet pipe, respectively. For steady calculations, the total pressure with uniform axial velocity is specified at the inlet boundary and the mass flow rate is given at the outlet. For unsteady calculations, the velocity distribution was given at the inlet and a free flow condition is applied at the outlet.

Origin of backflow. Four cases have been calculated: with and without blade leading edge sweep, and tip clearance. Both sweep and tip clearance promote backflow but sweep has a larger effect. With sweep, the axial length of backflow region is slightly increased by adding a tip clearance but significant effects of tip clearance were found for the case without sweep. Figure 2.9 shows the axial velocity in the backflow region, the tangential velocity and their products corresponding to the angular momentum flux in the backflow, in a cross section including the leading edge at the tip, for the case with sweep and tip clearance. The angular momentum flux is limited to a region very close to the casing wall and a region near the leading edge at the tip. The former region is found also for the cases without tip clearance and hence is mainly caused by the re-distribution of the swirling backflow from the latter region by the circumferential flow velocity and centrifugal force on it. The latter region is caused by an outward flow from the pressure surface of the blade, driven by the pressure difference across the blade. In addition, it was found that the flow velocity in the tip clearance can be reasonably predicted by Bernoulli's equation and the pressure difference across the blade. Thus, the mechanism for the supply of angular momentum of backflow is basically an inviscid flow process.

Response of backflow to flow rate fluctuations. In order to study the response of backflow to flow rate fluctuations, numerical calculations using the RANS code were made (Qiao et al., 2007) for the flow coefficient fluctuations

$$\phi = 0.078 + 0.01\sin(2\pi ft)$$

where 0.078 is the design flow coefficient and three cases with $f/f_n = 0.0625$, 0.125 and 0.25 (f_n is the rotational frequency of the inducer) are examined. Qualitatively the same results are obtained for these cases and the results are shown for $f/f_n = 0.125$. The unsteady calculation was started at $t = 0$ using the results of the steady flow calculation with $\phi = 0.078$ as the initial condition. The time step for the unsteady calculation was about 3.18×10^{-3} times the period of impeller rotation.

Figure 2.10 shows the axial length fluctuation of the backflow. Quasi-steady values from experiments and the results of steady flow calculations are also plotted. They agree reasonably well with each other. For the unsteady case, the amplitude is significantly decreased and the phase is delayed about 90 degrees as compared with the quasi-steady results. This is observed also for the normalized angular momentum in the upstream defined as

$$AM = \int_V \rho r^2 v_\theta dr d\theta dz / (\rho U_T D_T^4)$$

and shown in Fig.2.11, where D_T is the tip diameter and V represents the whole region at the inlet with the circumferential velocity v_θ. If we define the angular momentum supply by the backflow AMB and the angular momentum outflow on the normal flow AMN by

$$AMB = -\int_{v_z<0} \rho r^2 v_\theta v_z dr d\theta / (\rho U_T^2 D_T^3)$$

$$AMN = \int_{v_z>0} \rho r^2 v_\theta v_z dr d\theta / (\rho U_T^2 D_T^3)$$

and neglect the effects of shear stress on the boundary of the control volume, we can represent the conservation of angular momentum by

$$d(AM)/dt^* = AMB - AMN \tag{3}$$

where $t^* = t/(D_T/U_T)$ is a non-dimensional time.

Figure 2.11 also shows the plots of AMB and AMN with their quasi-steady values. We observe that

(1) The angular momentum supply by the backflow AMB from unsteady calculation is almost identical to quasi-steady calculations. This is caused by the fact that the backflow is driven by the pressure difference across the blade and the pressure difference is quasi-steady for frequencies treated here.

(2) For quasi-steady cases, AMB and AMN agree with each other except for the cases when the flow rate is small with a larger backflow region. This suggests that the size of steady backflow is determined from the balance of AMB and AMN. The unbalance observed at smaller flow rate is caused by the skin friction exerted by the pipe wall.

(3) The angular momentum transfer by the normal flow AMN for the unsteady case is nearly constant as compared with the quasi-steady results but is larger at the phases when the angular momentum in the upstream AM or the flow coefficient ϕ is larger.

Figure 2.12 examines the dynamic angular momentum balance of Eq. (3). We observe that the dynamic balance is satisfied.

The results in Figs. 2.11 and 2.12 show that the size of the backflow is determined by different mechanisms for steady and unsteady cases. For the steady case, the size of the backflow is determined from the balance of the supply of the angular momentum AMB and the outflow of the angular momentum on the normal flow AMN. For the unsteady case, the difference $AMB - AMN$ contributes to the growth of the angular momentum in the backflow region, $d(AM)/dt^*$.

To examine the phase relationships more quantitatively, we separate the quantities into steady and unsteady components such as:

$$AM = \overline{M} + \widetilde{M} e^{j\omega^* t^*}$$

$$AMB = \overline{B} + \widetilde{B} e^{j\omega^* t^*}$$

$$AMN = \overline{N} + \widetilde{N} e^{j\omega^* t^*}$$

$$\phi = \overline{\phi} + \widetilde{\phi} e^{j\omega^* t^*}$$

where $\omega^* = \omega(D_T/U_T) = 2(f/f_n)$ is a non-dimensional frequency.

Since the backflow angular momentum transfer is larger when the flow rate is smaller, we put

$$\widetilde{B} = -a\widetilde{\phi}$$

Further, the outflow of angular momentum on the normal flow would be larger if the angular momentum in the upstream and the flow rate are larger. So we put

$$\widetilde{N} = b\widetilde{M} + c\widetilde{\phi}$$

By putting these expressions into the angular momentum conservation equation (3), we obtain

$$\tilde{M} = \{-(a+c)/(b+2jf/f_n)\}\tilde{\phi} \qquad (4)$$

This clearly shows that backflow responds to the flow rate fluctuation as a first order lag element. Figure 2.13 shows the response function $\tilde{M}/\tilde{\phi}$ with the values of parameters a, b and c determined from the numerical calculations with $f/f_n = 0.25$. The results of the numerical calculations with $f/f_n = 0.0$, 0.0625, 0.125 and 0.25 are also shown in the figure. Those numerical values are close to the curve of Eq.(4), suggesting that the approximations used are adequate. For the case with $f/f_n \to \infty$, the phase of the backflow momentum advances 90 degrees ahead of the flow rate fluctuation and is delayed 90 degrees behind the quasi-steady backflow fluctuation. This is also shown in Figs. 2.10 and 2.11.

2.3 Simulation of Backflow Vortex Structure

For the simulation of backflow vortex structure, calculations based on LES (Kato et al. (1999)) were made for a simplified geometry as shown in Fig.2.14. It is a finite element method with second order accuracy both in time and space, and the standard Smagorinsky model with the Van-Driest wall function with the Smagorinsky constant $C_S = 0.15$ was used for the subgrid model. The total number of element was 1,100,000 and about 800,000 elements were used near the backflow region. The flow rate and the angular momentum supply of the backflow, *AMB*, were calculated by the RANS code with full inducer geometry, and the backflow with the same flow rate and angular momentum was introduced from the clearance between the outer and inner pipes. The velocities are specified at the main flow inlet and it is assumed that the shear stress equals to zero at the outlet. The time step of calculation corresponds to 10^{-4} rotation of the impeller. The backflow structures obtained are shown as iso-pressure surfaces in Fig.2.15. The size of the backflow region is compared with experiments in Fig.2.16. We observe reasonable agreements.

The pressure at the center of the vortices becomes lower than the inlet pressure caused by the following two reasons. One is the increase of the main flow velocity caused by the blockage effects of backflow region and the other is the centrifugal force on the vortical flow. The pressure decrease due to the blockage effect of the backflow is evaluated to be $C_{pb} = (p - p_{in})/(\rho U_t^2/2) = -0.0094$ for $\phi = 0.07$ and $C_{pb} = -0.010$ for $\phi = 0.078$.

For the evaluation of the pressure decrease due to the centrifugal force, backflow vortices are approximated by Rankine vortices with the core radius $a = \alpha \cdot 2\pi R/N$ where α is a constant, R the radial location where tangential velocity is maximum and N the number of the vortices. The total circulation of the backflow, Γ_t is estimated from $\Gamma_t = 2\pi R V_\theta$ where V_θ is the maximum tangential velocity of the backflow which is represented by $V_\theta = \beta U_t$. It is assumed that γ times the total circulation is concentrated in backflow vortices. Then, the circulation of each backflow vortex is given by $\Gamma_i = \gamma\Gamma_t/N$. Based on these assumptions, the pressure decrease due to the centrifugal force on the vortical flow is represented by $C_{pv} = -2(\beta\gamma/2\pi\alpha)^2$. The values of α and Γ_i are determined from the velocity distribution around each vortex. The results for $\phi = 0.078$ are shown in Fig. 2.17.

This result can be summarized as follows. The maximum tangential velocity of the backflow is about 25% of the inducer tip velocity ($\beta = 0.25$). About 28% of the total circulation of the backflow is concentrated in the backflow vortices ($\gamma = 0.28$). The core radius of the backflow vortices is about 12% of the circumferential distance between the vortices ($\alpha = 0.12$). As a result, the pressure in the vortices becomes lower than the ambient pressure by about 3% of the dynamic pressure of the impeller tip speed ($C_{pv} = -0.03$).

The incipient cavitation number can be evaluated by $\sigma = -C_{pb} - C_{pv}$. In experiments, backflow cavitation starts to appear at the cavitation number of order 0.1 but a definite structure can be found for $\sigma < 0.05$. Although the value obtained from the present calculations, $\sigma = 0.04$, is somewhat smaller but may be reasonable if we consider the unsteady nature of the vortices.

The volume of backflow vortex cavitation is evaluated from the volume of the region with $C_p < -\sigma$ and shown in Fig.2.18. The values of mass flow gain factor M_B and cavitation compliance K_B are determined from

$$M_B = -(\pi/3)(\partial(V/(\pi D^3/4))/\partial\phi$$

and

$$K_B = (\pi/3)(\partial(V/(\pi D^3/4))/\partial\sigma$$

In Fig. 2.19, the values of K_B and M_B thus obtained are compared with those for blade surface cavitation determined from quasi-steady calculations and of total cavitation determined from experiments (Brennen and Acosta (1976)). We find that the backflow vortex cavitation becomes important at a lower cavitation number.

2.4 Response of Backflow Vortex Structure to Flow Rate Fluctuations

The response of backflow vortex has been calculated by applying the LES model to the simplified geometry by supplying instantaneous angular momentum *AMB* determined from the unsteady calculations with the RANS code. The geometry of backflow cavitation corresponding to $\sigma = 0.01$ is shown in Fig. 2.20. The numbers in the figure show the phase of the flow rate fluctuation. For the evaluation of pressure, the effects of the inertia of the unsteady flow and of the displacement due to the backflow have been neglected by taking the averaged pressure over the cross section as the reference pressure. We observe that the cavity volume fluctuation delays behind the flow rate fluctuation in the same way as for the total angular momentum *AM* in the backflow. It was confirmed that the angular momentum relations as shown in Figs.2.10-2.13 also exists for the present case. The cavity volume fluctuation estimated for $\sigma = 0.01$ is shown in Fig.2.21. We observe that the phase delays behind the quasi-steady response by about 90 degrees for both frequencies. We also observe that the cavity volume fluctuation is larger for the case of higher frequency. This is caused by larger number of vortices for higher frequency, under which sufficient time is not allowed for the vortices to merge in response to the shift of radial location of the vortices. The response of blade surface cavitation to flow rate and inlet pressure fluctuations has been calculated by Otsuka et al. (1996). The phase delay at $f/f_n = 0.25$ is of the order of 10 degrees which is much smaller than the backflow vortex cavitation evaluated. This shows the importance of the backflow development in the response of backflow vortex cavitation.

2.5 Cavitation Surge Caused by Backflow Vortex Cavitation

The phase lag of backflow has been found experimentally by Yamamoto (1992) in a series of studies on cavitation surge of a centrifugal pump in which the backflow cavitation plays the critical role. Figure 2.22 shows the phase lag α (α corresponds to $\alpha = -Arg(\widetilde{M}/\widetilde{\phi})$ in Fig.2.13) of backflow behind the flow rate fluctuation obtained from the measurement of velocity fluctuation at the center of the inlet pipe near the pump inlet. In this figure l_m is the distance of the velocity measurement location from the pump inlet and D_1 is the diameter of the inlet pipe. We focus on the data with $l_m / D_1 = 0.59$. The phase delays behind the quasi-steady value ($\alpha = -180$ degrees) by about 90 degrees as the frequency is increased from 0 to 4Hz ($f / f_n = 4/50 = 0.08$). This agrees with the present results. Here, we discuss the effect of backflow vortex cavitation assuming that the cavity volume is in phase with the magnitude of the backflow, typically shown by AM.

We consider a case of cavity oscillation in a pipe in response to flow rate fluctuation. The velocity fluctuation in the pipe is represented by $u = \overline{U} + \widetilde{U}e^{j\omega t}$ and the cavity volume oscillation by $V_c = \overline{v}_c + \widetilde{v}_c e^{j(\omega t - \alpha)}$. If we consider that the cavity is located at a distance L from the pipe inlet where the pressure is constant, the pressure fluctuation at the cavity can be represented by $p = -\rho L \, du/dt$. Then, the displacement work done by the cavity per period can be evaluated by $E = \int_{cycle} p \cdot dV_c = -\rho \pi L \omega \widetilde{U} \widetilde{v}_c \cos\alpha$. This shows that positive work is done by the cavity when $-3\pi/2 < \alpha < -\pi/2$ and the cavity fluctuation destabilizes the system. With $\alpha = -\pi$, the cavity volume becomes smaller with higher inlet velocity and this corresponds to the case with positive mass flow gain factor. With $-\pi/2 < \alpha < \pi/2$, E becomes negative and the cavity stabilizes the system. The results in Fig.2.22 show that backflow cavitation stabilizes higher frequency oscillations and destabilizes lower frequency oscillations.

Figure 2.23 shows the resonant frequency f_0 of the system determined by excitation tests and the critical frequency f_{90} at which α becomes $-\pi/2$. The resonant frequency decreases as the cavitation number K is decreased and unstable oscillation is observed in the region of cavitation number where f_0 would be less than f_{90}. So, in this case of cavitation surge in a centrifugal pump at low flow rate, the response of backflow cavitation plays a crucial role.

In the present model, Eq. (4) shows that $\alpha = -Arg(\widetilde{M}/\widetilde{\phi})$ never becomes larger than $-\pi/2$ and backflow cavitation is always destabilizing. However, the result in Fig.2.13 shows that α is very close to $-\pi/2$ and the destabilizing effect represented by E should be very small. Since the volume of the backflow cavitation will definitely decrease as the inlet pressure is increased, backflow cavitation should have a positive cavitation compliance. As a result of a small destabilization corresponding to positive mass flow gain factor and a certain stabilizing effect of cavitation compliance, the backflow would have a stabilizing effect on higher frequency oscillations. This is clearer from the criterion

$$M\cos(\alpha + \pi) > 2(1+\sigma)\phi K$$

for the onset of cavitation surge and rotating cavitation, obtained from a one dimensional stability analysis (Tsujimoto et al. (2001)) considering the phase delay $\alpha + \pi$ of cavitation caused by the response of backflow.

3 Conclusion

It has been shown that both tip leakage vortex cavitation and backflow vortex cavitation exhibits a large delay behind quasisteady response, even at relatively small frequency of the disturbance. The delay of tip leakage vortex cavitation could be explained from the total amount of the circulation shed from the blade tip to the flow. The delay of backflow vortex cavitation was explained from the balance of the angular momentum supplied by the backflow and removed by the normal flow. So, in both cases, the unsteady behavior is controlled by relatively simple flow mechanisms. As discussed in detail for the backflow vortex cavitation, the delay causes the character shift from destabilizing to stabilizing, as the increase of the frequency of the disturbance. The critical frequency of the shift is of the order of 10% of the rotational frequency. However, more detailed study is needed to quantify the strength of the stabilizing/destabilizing effects of tip leakage and backflow vortex cavitations.

Bibliography

Brennen, C. E. and Acosta, A.J., (1976). The dynamic Transfer Function for a Cavitating Inducer," *ASME Journal of Fluids Engineering*, Vol.98, No.2, 182-191.

Chen, G.T., Greitzer, E.M., Tan, C.S., and Marble, F.E., (1991). Similarity Analysis of Compressor Tip Clearance Flow Structure. *ASME Journal of Turbomachinery*, Vol.113, 260-271.

Fung, Y.C., (1969), *Aeroelasticity*, Dover Pub. 401.

Higashi, H., Yoshida, Y., and Tsujimoto, Y., (2002). Tip Leakage Vortex Cavitation from the Tip Clearance of a Single Hydrofoil. *JSME International Journal*. Ser.B, Vol. 45, No.3, 662-671.

Kato, C., Shimizu, H., and Okamoto, T., (1999). Large Eddy Simulation of Unsteady Flow in a Mixed-Flow Pump. *Proceedings of the third ASME/JSME Joint Fluids Engineering Conference*, San Francisco, California, USA, FEDSM99-7802.

Lamb, H., (1975), *Hydrodynamics*, Cambridge University Press.

Mitsuda, K., Yokota, K., Tsujimoto, Y., and Kato, C., (2003). A Study of Vortex Structure in the Shear-Layer between Main Flow and Swirling Backflow. *Trans. JSME*, Ser.B, Vol.69, No.684, 1769-1775.

Murayama,M., Yoshida, Y., and Tsujimoto, Y., (2003). Unsteady Tip Leakage Vortex Cavitation on an Oscillating Hydrofoil. *Transactions of JSME*, Ser.B, Vol.69, No.678, 315-323.

Otsuka, S., Tsujimoto, Y., Kamijo, K., and Furuya, O., (1996). Frequency Dependence of Mass Flow Gain Factor and Cavitation Compliance of Cavitating Inducers. *ASME Journal of Fluids Engineering*, Vol.118, No.2, 400-408.

Qiao, X., Horiguchi, H., Kato, C., and Tsujimoto, Y., (2004). Effects of Leading Edge Sweep and Tip Clearance on Inlet Backflow of Turbopump Inducers. *Proceedings of the 3rd International Symposium on Fluid Machinery and Fluid Engineering*, Beijing, 141-151.

Qiao, X., Horiguchi, H., and Tsujimoto, Y., (2007). Response of Backflow to Flow Rate Fluctuations. *ASME Journal of Fluids Engineering*, Vol. 129, No. 2, 350-358.

Rains, D.A.,(1954). Tip Clearance Flows in Axial Compressors and Pumps. *California Institute of technology, Hydrodynamic and Mechanical Engineering Laboratories,* Report No.5.

Tsujimoto,Y., Kamijo,K., and Brennen, C., (2001). Unified Treatment of Cavitation Instabilities

of Turbomachines. *AIAA Journal of Propulsion and Power*, Vol.17, No.3, 636-643.
Watanabe, S., Seki, H., Higashi, H., Yokoya, K., and Tsujimoto, Y., (2001). Modeling of 2-D Leakage Jet Cavitation as a Basic Study of Tip Leakage Vortex Cavitation. *ASME Journal of Fluids Engineering*, Vol.23, No.1, 50-56.
Yamamoto, K., (1992). Instability in a Cavitating Centrifugal Pump (3rd Report: Mechanism of Low Cycle System Oscillation). *Trans. JSME, Ser.B*, Vol.58, No.545, 180-186.
Yamanishi, N., Fukao, S., Qiao, X., Kato, C., and Tsujimoto, Y., (2007). LES Simulation of Backflow Vortex Structure at the Inlet of an Inducer. *ASME Journal of Fluids Engineering*, Vol. 129, No. 3, 587-593.
Yokota, K., Kurahara, K., Kataoka, D., Tsujimoto, Y., and Acosta, A.J., (1999). A Study of Backflow and Vortex Structure at the Inlet of an Inducer. *JSME International Journal*, Ser.B, Vol.42-3, 451-459.

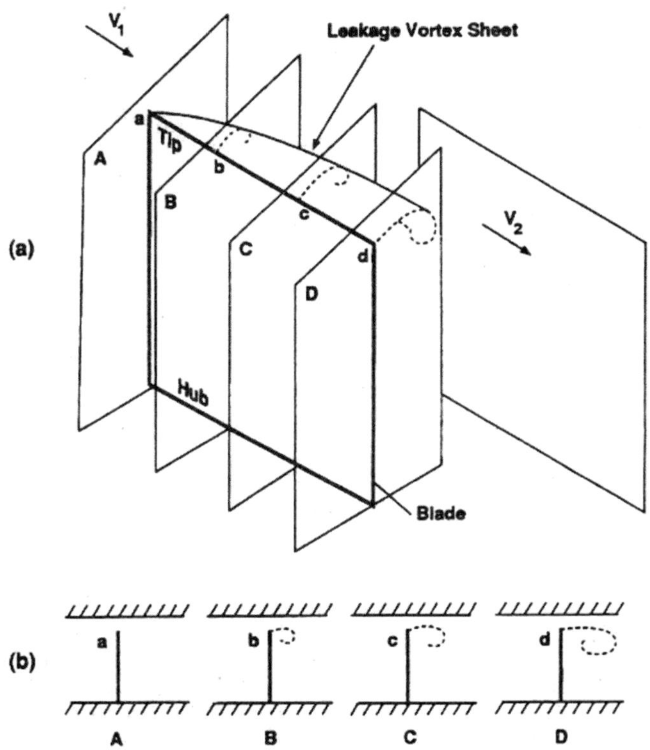

Figure 1.1 Correspondence between three-dimensional steady tip clearance flow and unsteady two-dimensional flow

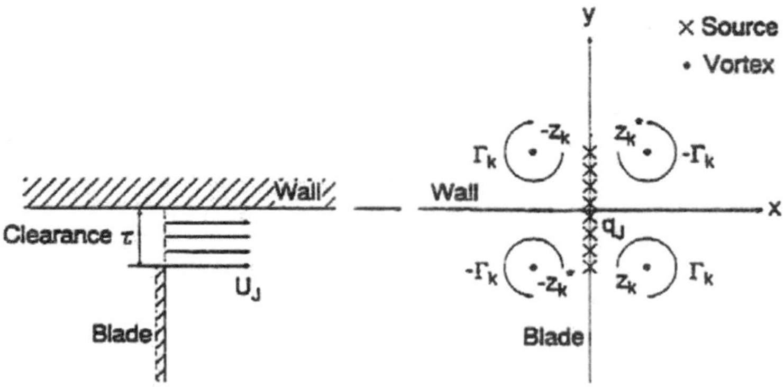

Figure 1.2 Image vortices to satisfy the boundary conditions on the casing and blade surfaces

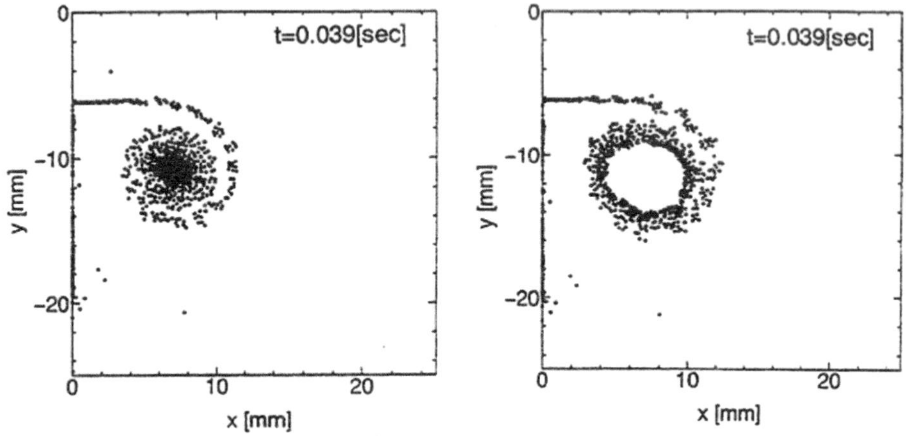

(a) Without source q_B (b) With source q_B

Figure 1.3 Distribution of vortices with and without the displacement effect of cavitation represented by the source q_B

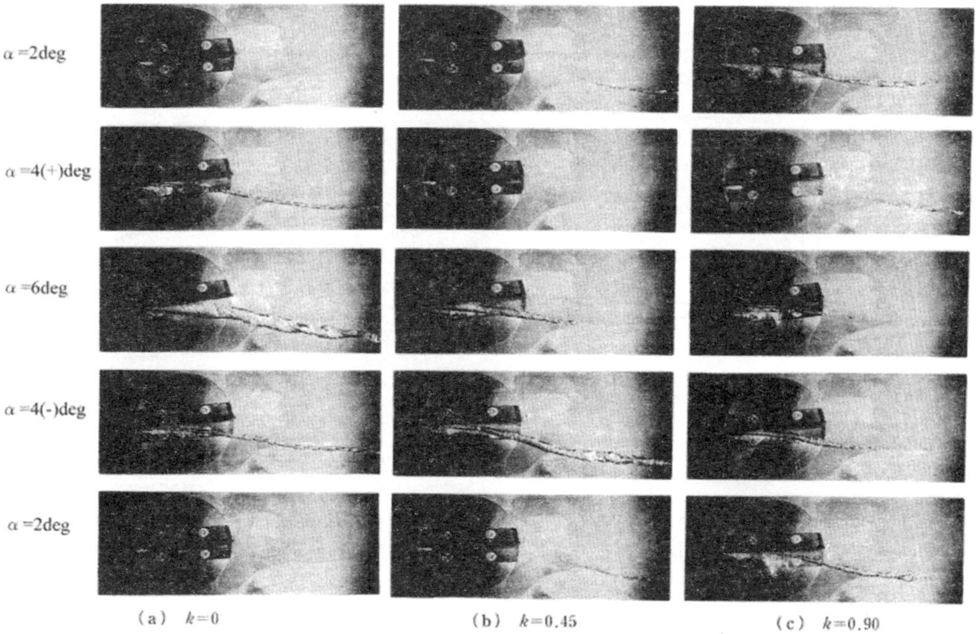

Figure 1.4 Oscillation of tip leakage vortex cavitation, $k = 0$, 0.45 and 0.90

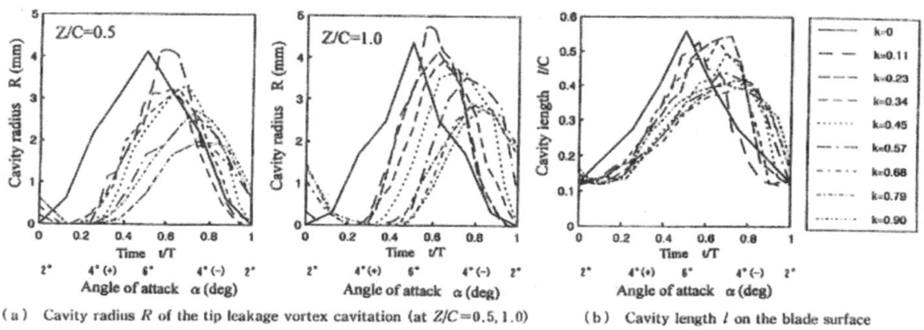

Figure 1.5 Fluctuation of cavity radius at $Z / C = 0.5$ and 1.0, and of blade surface cavity length

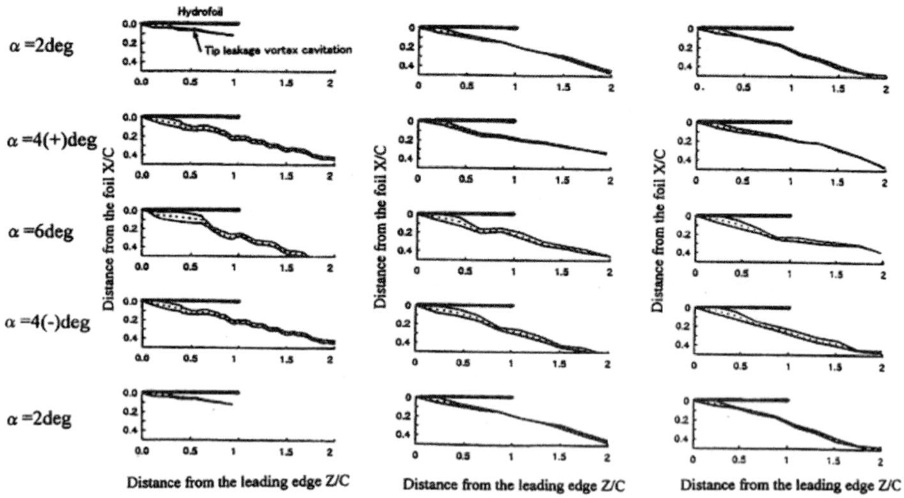

(a) $k=0$ (b) $k=0.45$ (c) $k=0.90$

Figure 1.6 Cavity oscillations predicted by the calculations

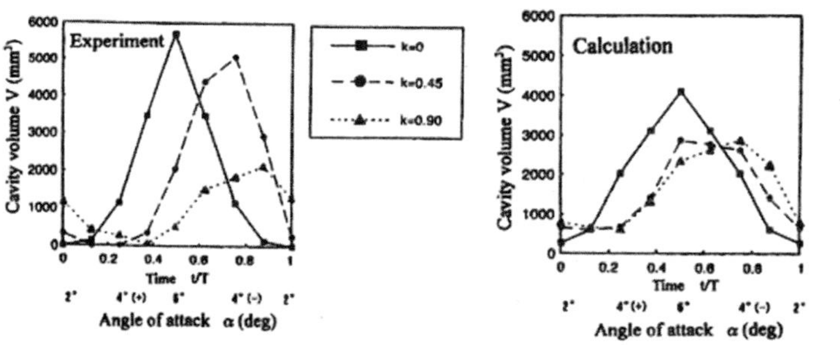

(a) Experiments (b) Calculation results

Figure 1.7 Cavity volume fluctuation, $k = 0$, 0.45, and 0.90

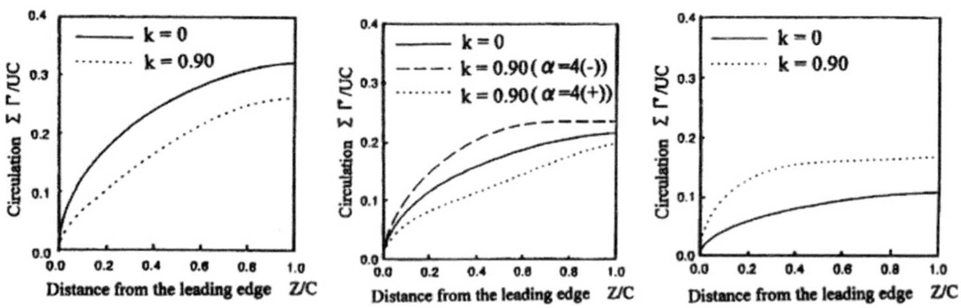

(a) α=6 deg (b) α=4 deg (c) α=2 deg

Figure 1.8 Growth of the total amount of circulation in the cross flow plane reaching the trailing edge at the instant of the angle of attack (a) $\alpha = 6\,\text{deg}$ (b) $\alpha = 4\,\text{deg}$ (c) $\alpha = 2\,\text{deg}$

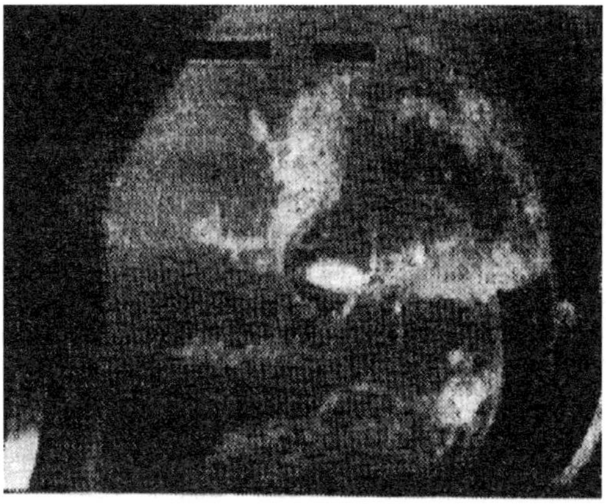

Figure 2.1 Backflow vortex cavitation

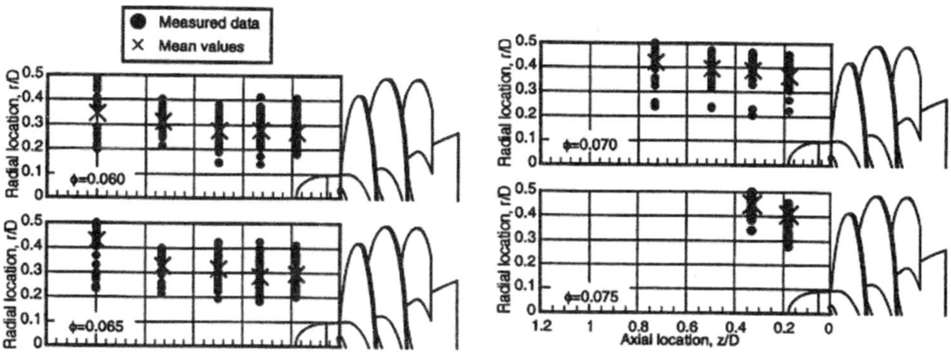

Figure 2.2 Radial location of backflow vortex filaments

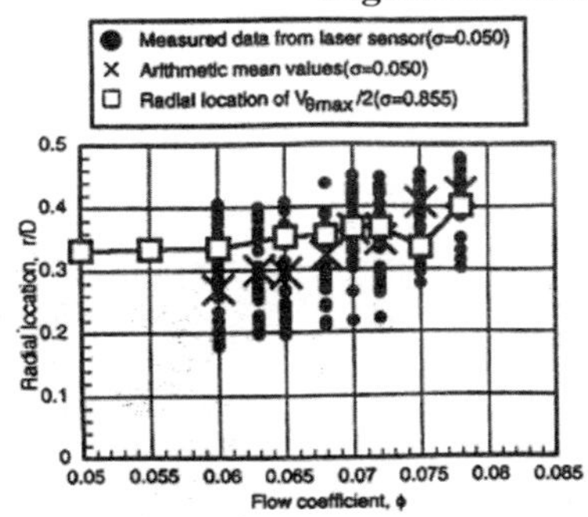

Figure 2.3 Radial location of the vortices

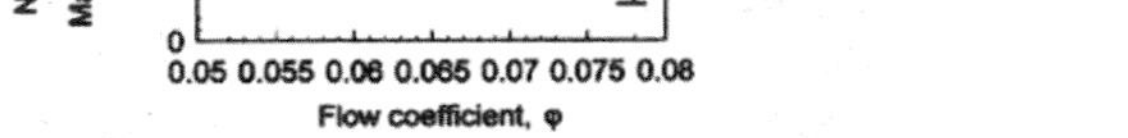

Figure 2.4 Propagation velocity ratio at the inlet

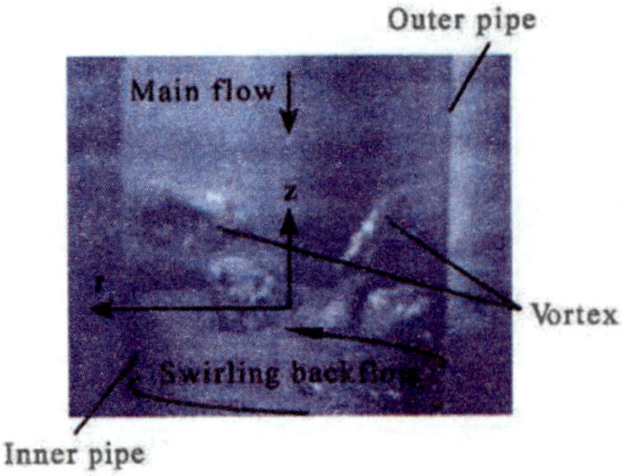

Figure 2.5 Vortex structure in the shear layer between straight main flow and swirling backflow

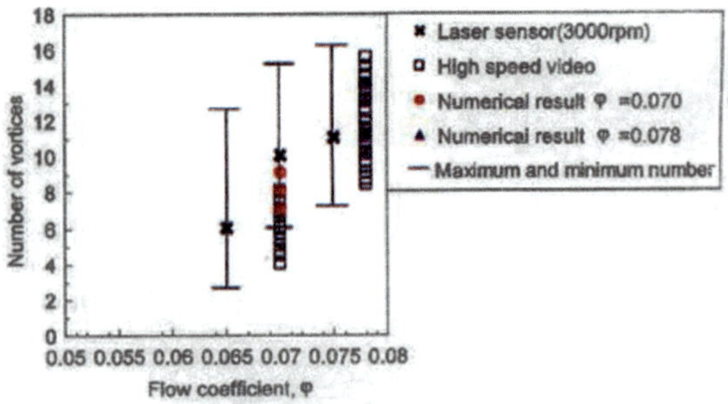

Figure 2.6 Number of vortices

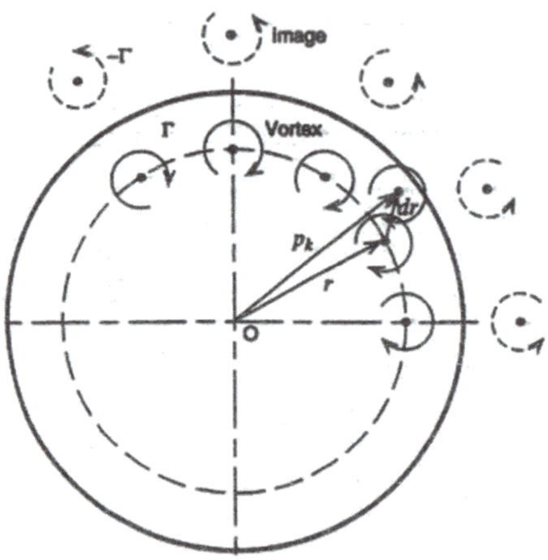

Figure 2.7 2-D stability analysis of the vortices

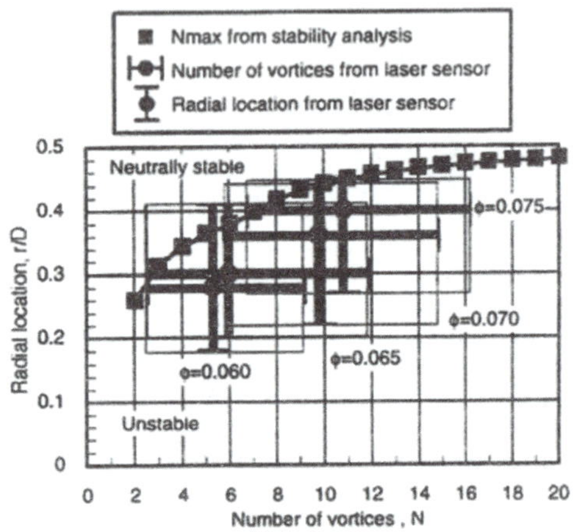

Figure 2.8 Comparison of number of vortices with stability analysis

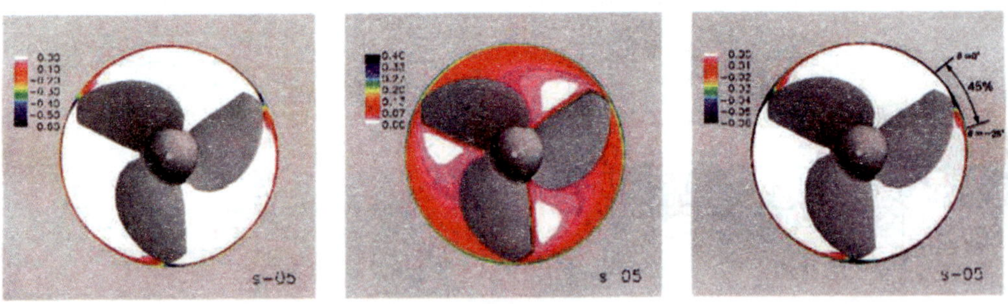

(a) Backflow velocity distribution, v_z/U_t

(b) Absolute tangential velocity distribution, $-v_\theta/U_t$

(c) Angular momentum induced by backflow, $(v_z/U_t)\cdot(-v_\theta/U_t)\cdot r/R_t$

Figure 2.9 Regions of negative axial velocity (a), absolute tangential velocity (b), and the negative angular momentum flux, (c)

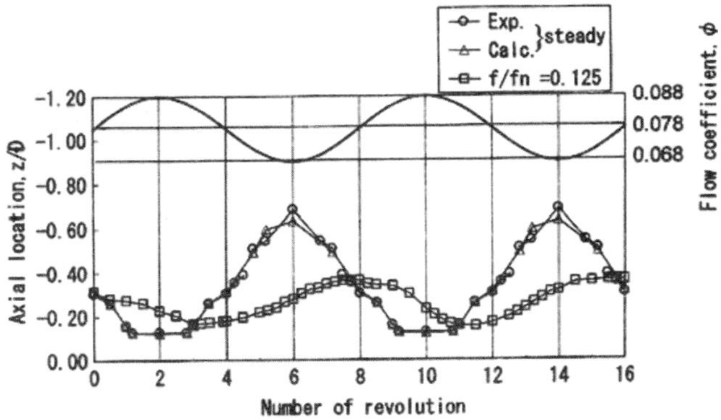

Figure 2.10 Axial location of backflow leading edge

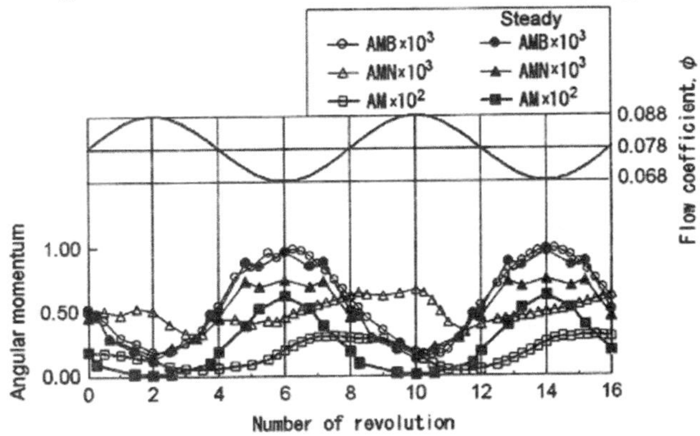

Figure 2.11 Angular momentum in the upstream, *AM* , angular momentum supply by the backflow, *AMB* , and angular momentum removal by the normal flow, *AMN*

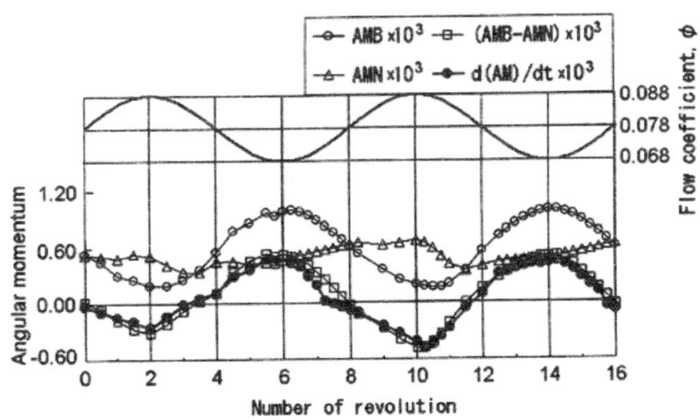

Figure 2.12 Dynamic balance of angular momentum

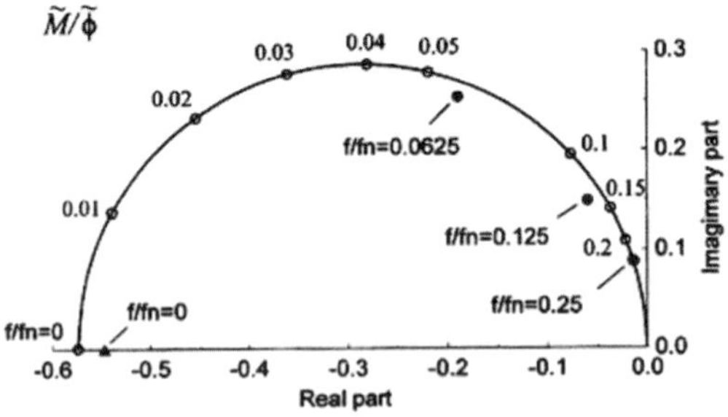

Figure 2.13 Response function $\widetilde{M}/\widetilde{\phi}$

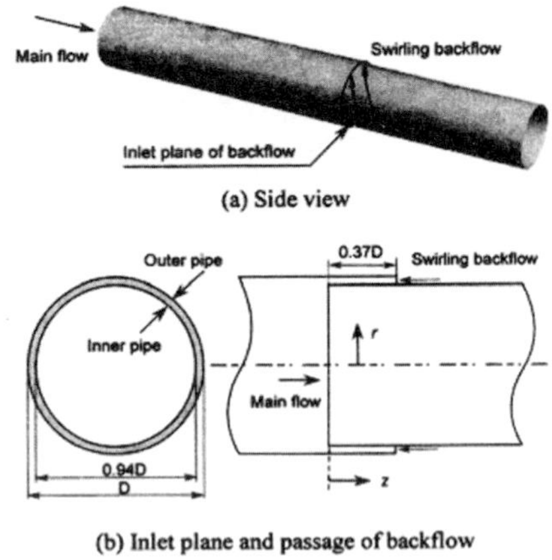

Figure 2.14 Geometry of flow region for the simulation of backflow vortex structure

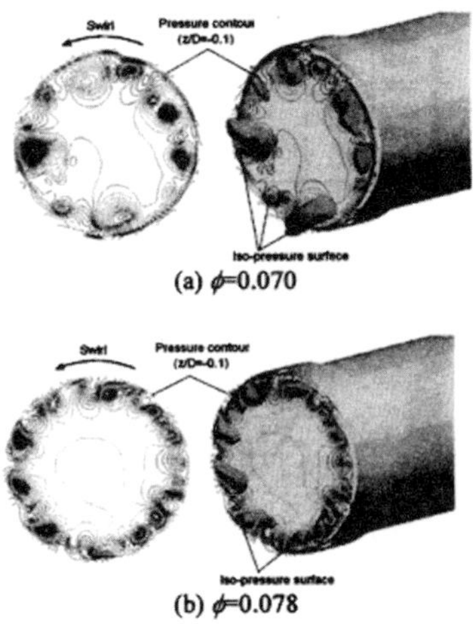

Figure 2.15 Backflow vortex structure

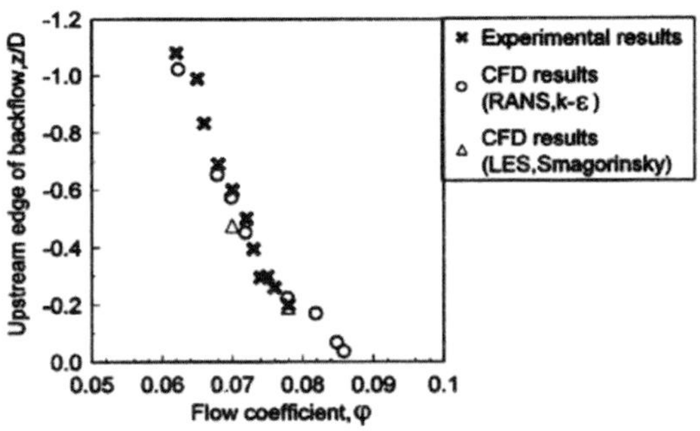

Figure 2.16 Comparison of the location of upstream edge of backflow

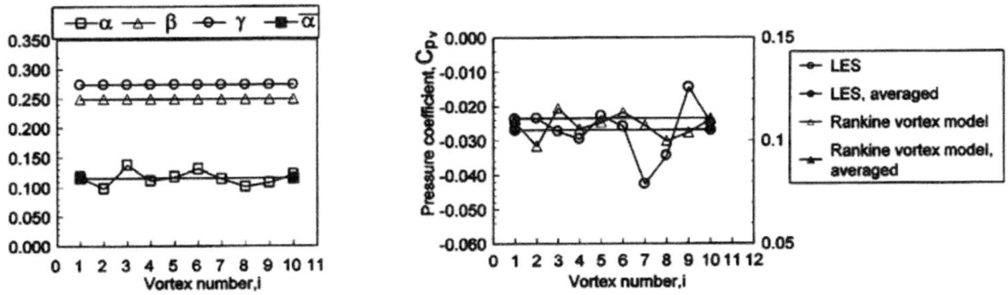

(a) Constants used in pressure calculation (b) Pressure coefficient

Figure 2.17 Values of constants for the evaluation of the pressure dip in the vortices

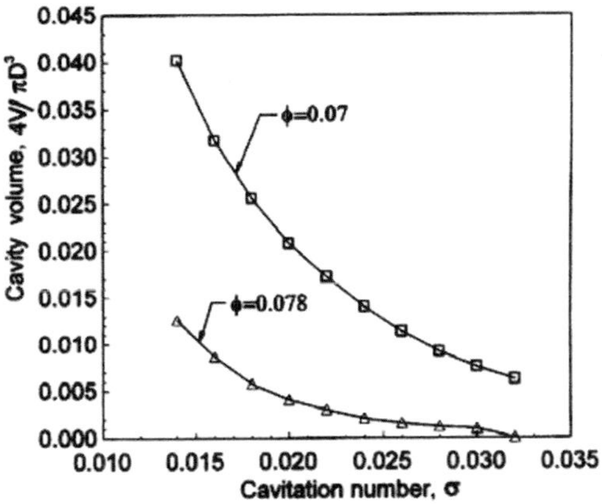

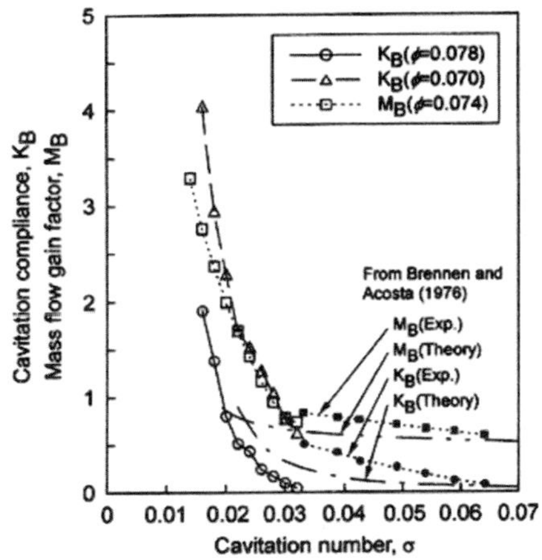

Figure 2.18 Volume of backflow vortex cavity

Figure 2.19 Mass flow gain factor and cavitation compliance of backflow vortex cavitation

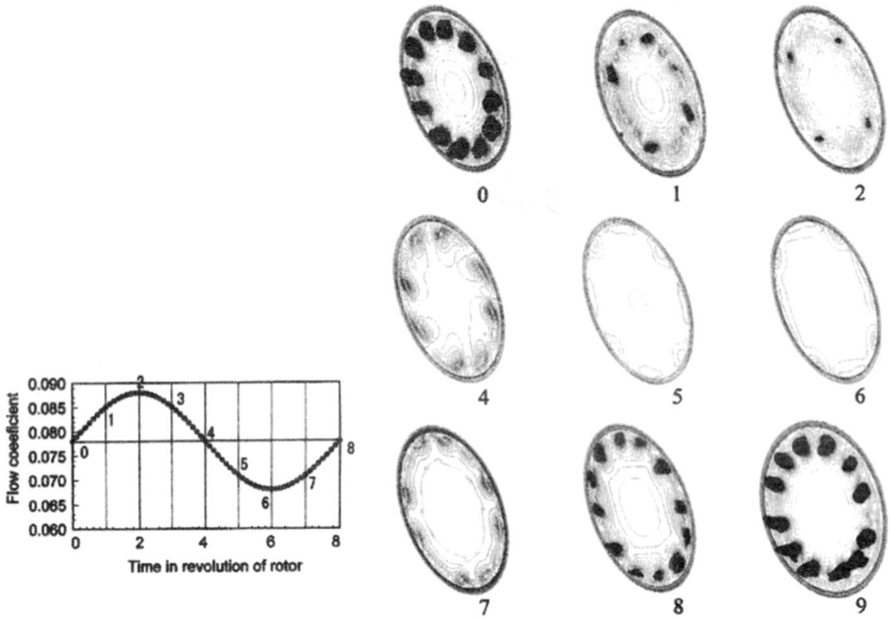

Figure 2.20 Response of backflow vortex structure to flow rate fluctuation

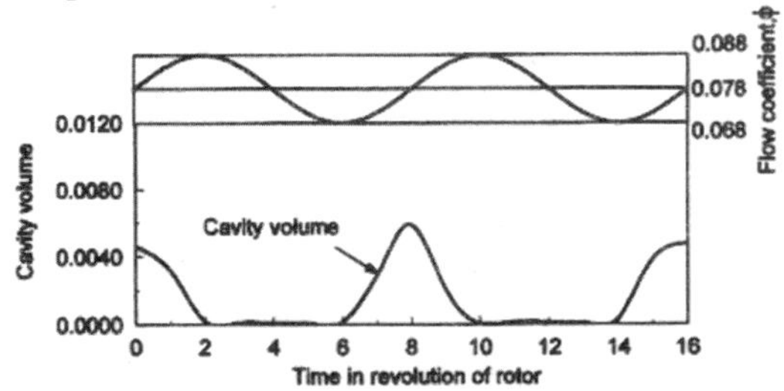

(a) f/fn=0.125

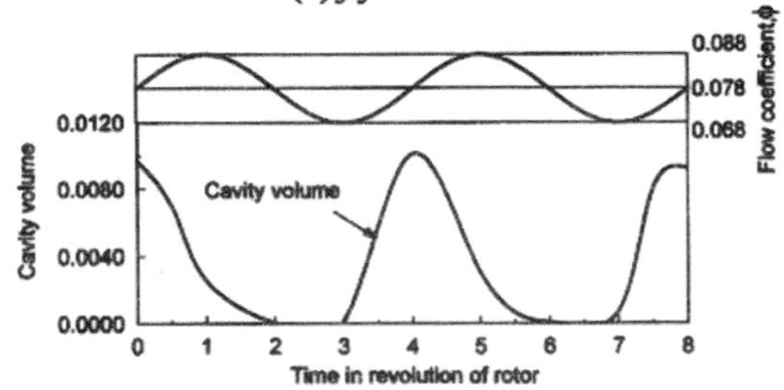

(b) f/fn=0.25

Figure 2.21 Cavity volume fluctuation in response to flow rate fluctuation

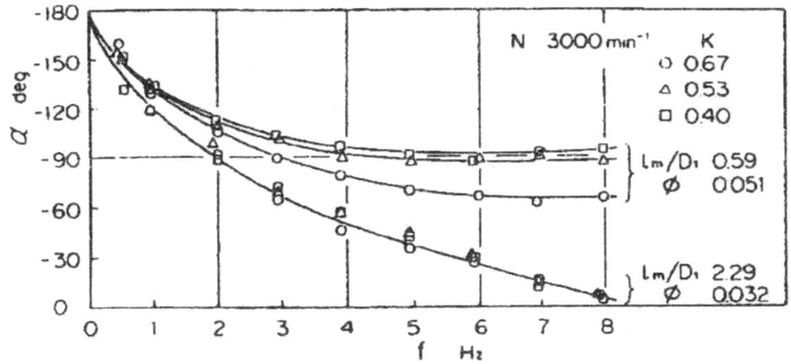

Figure 2.22 Delay of backflow evaluated from inlet velocity measurements

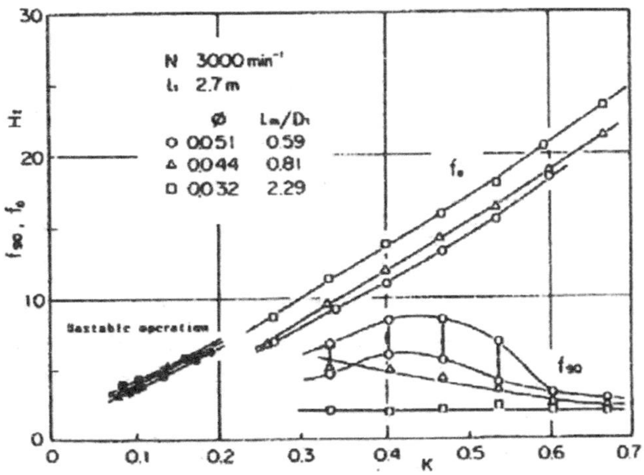

Figure 2.23 Resonant frequency and critical frequency

The Different Role of Cavitation on Rotordynamic Whirl Forces in Axial Inducers and Centrifugal Impellers

Luca d'Agostino[1]

[1] Dipartimento di Ingegneria Aerospaziale, Università di Pisa, Pisa, Italy

Abstract. The linearized dynamics of the flow in cavitating axial helical inducers and centrifugal turbopomp impellers is investigated with the purpose of illustrating the impact of the dynamic response of cavitation on the rotordynamic forces exerted by the fluid on the rotors of whirling turbopumps. The flow in the impellers is modeled as a fully-guided, incompressible and inviscid liquid. Cavitation is included through the boundary conditions on the suction sides of the blades, where it is assumed to occur uniformly in a small layer of given thickness and complex acoustic admittance, whose value depends on the void fraction of the vapor phase and the phase-shift damping coefficient used to account for the energy dissipation. Constant boundary conditions for the total pressure are imposed at the inlet and outlet sections of the impeller blade channels. The unsteady governing equations are written in rotating "body fitted" orthogonal coordinates, linearized for small-amplitude whirl perturbations of the mean steady flow, and solved by modal decomposition. In helical turbopump inducers the whirl excitation and the boundary conditions generate internal flow resonances in the blade channels, leading to a complex dependence of the lateral rotordynamic fluid forces on the whirl speed, the dynamic properties of the cavitation region and the flow coefficient of the machine. Multiple subsynchronous and supersynchronous resonances are predicted. At higher levels of cavitation the amplitudes of these resonances decrease and their frequencies approach the rotational speed (synchronous conditions). On the other hand, application of the same approach indicates that no such resonances occur in whirling and cavitating centrifugal impellers and that the rotordynamic fluid forces are almost insensitive to cavitation, consistently with the available experimental evidence. Comparison with the scant data from the literature indicates that the present theory correctly captures the observed features and parametric trends of rotordynamic forces on whirling and cavitating turbopump impellers. Hence there are reasons to believe that it can usefully contribute to shed some light on the main physical phenomena involved and provide practical indications on their dependence on the relevant flow conditions and parameters.

1 Introduction

Local flow phenomena, like tip leakage, capable of interfering with the blade loading are known to be the dominant source of rotordynamic whirl forces in compressible flow machines (Thomas, 1958; Alford, 1958; Martinez-Sanchez et al., 1995; Martinez-Sanchez and Song, 1997a, 1997b).

Previous research efforts in turbopumps mainly focused on the origin and analysis of rotordynamic impeller forces under noncavitating conditions (Chamieh et al., 1985; Jery, 1985; Jery et al., 1987; Shoji and Ohashi, 1987; Ohashi & Shoji 1987; Adkins and Brennen, 1988; Arndt et al., 1989, 1990; Tsujimoto et al., 1997; Uy and Brennen, 1999; Baskharone, 1999; Hiwata and Tsujimoto, 2002). However, it is widely recognized that cavitation in turbopumps can promote the onset of dangerous self-sustained whirl instabilities (Rosenmann, 1965) and substantially modify the behavior of fluid-induced rotordynamic forces on helical inducers (Arndt and Franz, 1986; Brennen, 1994; Bhattacharyya, 1994), where the large-scale dynamic response of the entire flow to the impeller whirl motion seems to play a significant role (d'Auria, d'Agostino and Brennen, 1995; d'Agostino and d'Auria, 1997; d'Agostino, d'Auria and Brennen, 1998; d'Agostino and Venturini, 2002, 2003). Because of their greater complexity, rotordynamic fluid forces in whirling and cavitating turbopump impellers have so far received comparatively less attention in the open literature and a satisfactory understanding of their behavior is still lacking.

The available experimental evidence indicates that cavitation affects the added mass of the rotor and significantly reduces the magnitude of the rotordynamic fluid forces on helical inducers. It is worth noting that the consequent increase of the critical speeds is especially dangerous in supercritical machines, commonly used in liquid propellant rocket feed systems. A second major effect of cavitation in helical inducers is the introduction of a complex oscillatory dependence of the rotordynamic fluid forces on the whirl frequency. This finding seems to indicate the possible occurrence of resonance phenomena in the compressible cavitating flow inside the inducer blade channels under the excitation imposed by the eccentric motion of the rotor. Earlier theoretical analyses aimed at investigating this hypothesis have addressed the case of infinitely-long whirling helical inducers with uniformly distributed traveling bubble cavitation (d'Auria et al., 1995; d'Agostino and d'Auria, 1997; d'Agostino, d'Auria and Brennen 1998). The results confirmed the presence of internal flow resonances and indicate that bubble dynamic effects do not play a major role, except, perhaps, at extremely high whirl speeds. They also suggest that the assumptions of uniformly-distributed bubbly cavitation and infinitely long inducers may contribute to explain the discrepancies between theoretical predictions and experimental data. On the other hand, no resonant phenomena seem to occur in radial impellers, where the limited available evidence indicates that cavitation only has a marginal effect on the rotordynamic whirl forces (Franz et al., 1989).

Following up on this work, we now investigate the dynamics of the unsteady flow in whirling helical inducers and radial impellers with attached blade cavitation, in order to gain some better understanding of the fundamental reasons for the different behavior of rotordynamic fluid forces in this two kinds of turbomachines. Upon introduction of suitable simplifying approximations, the flow is linearized for small-amplitude whirl motions of the rotor and solved by modal expansion. In spite of the simplifications introduced in order to obtain an efficient closed form solution, comparison with the available experimental data indicates that the proposed analyses correctly predict the main observed features and differences of the rotordynamic fluid forces in whirling and cavitating inducers and radial impellers, thus providing useful practical indications and fundamental understanding of their dependence on the relevant flow conditions and parameters.

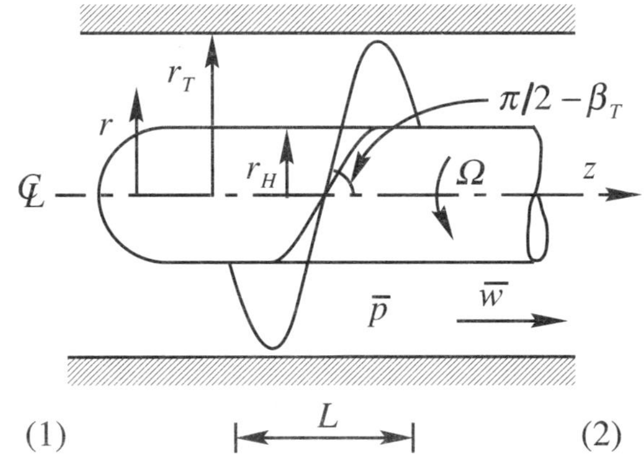

Figure 1. Schematic of the flow configuration and inducer geometry.

2 Linearized Dynamics of the Cavitating Flow in a Whirling Inducer

We first examine the dynamics of an incompressible, inviscid liquid of velocity $\boldsymbol{u}$, pressure p, and density ρ_L in a helical inducer rotating with velocity Ω and whirling on a circular orbit of small eccentricity ε at angular speed ω. A number of simplifications are introduced in order to reduce the problem to a form admitting an analytical solution. As illustrated in Figure 1, a simple helical inducer is considered, with N_B radial blades, zero blade thickness, axial length L, hub

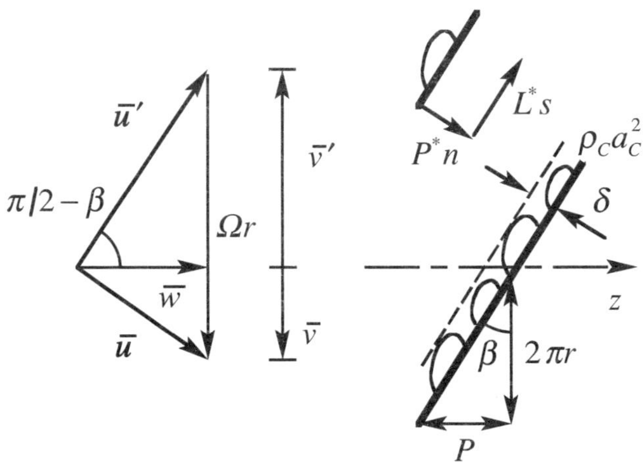

Figure 2. Schematic of the thin layer of attached cavitation pockets on the suction sides of the inducer blades.

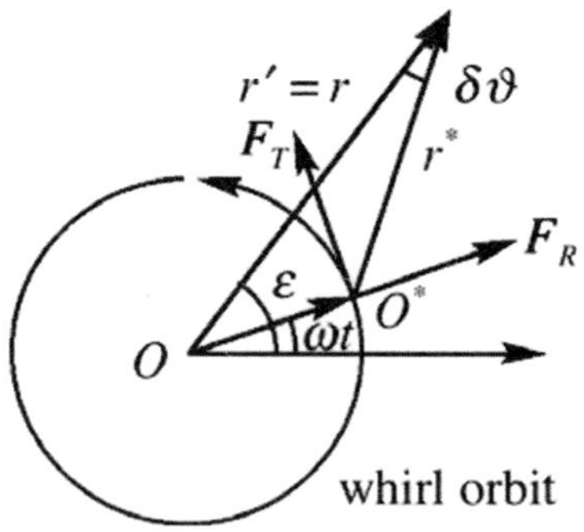

Figure 3. Schematic of whirl motion, coordinates and rotordynamic forces.

radius r_H, tip radius r_T, tip blade angle β_T, and constant pitch:

$$P = 2\pi r_T \tan\beta_T = 2\pi r^* \tan\beta$$

The flow is fully wetted everywhere except on the suction sides of the blades, where attached cavitation occurs. The mean flow velocity $\bar{\boldsymbol{u}}$ in the blade channels is specified by the flow coefficient $\phi = \bar{w}/\Omega r_T$, assuming fully-guided forced-vortex flow with zero radial velocity $\bar{u}$, uniform axial velocity $\bar{w}$, and angular velocity $\Omega_F = \bar{v}/r = \Omega\left(1-\phi\cot\beta_T\right)$. With reference to Figure 2, cavitation is thought to occur on the suction sides of the blades in the form of slowly-moving attached pockets uniformly distributed in a thin layer of given thickness $\delta << P$ and damped acoustic admittance $\rho_C a_C^2\left(1+i\zeta\right)$, where ζ is a nondimensional damping coefficient. The static pressure p_C in the cavitating layer is taken equal to the total pressure p_t of the surrounding liquid, assuming that the flow slows down without losses in the low velocity region between the cavities.

We define stationary cylindrical coordinates r,ϑ,z with center in O on the axis of the stator, rotating cylindrical coordinates r',ϑ',z' spinning at the rotor speed with center in the same point O, and rotating and whirling cylindrical coordinates r^*,ϑ^*,z^* fixed in the inducer and with center in O^* on its geometric axis, as shown in Figure 3. Then the equations of the blade surfaces are:

$$B = \vartheta^* + \frac{2\pi}{P} z^* - \vartheta_j = 0$$

where $\vartheta_j = 2\pi\left(j-1\right)/N_B$ is the angular location of the j-th blade, with $j = 1,2,...,N_B$. The flow velocities in the stationary and rotating frames are related by $\boldsymbol{u} = \boldsymbol{u}' + \boldsymbol{\Omega}\times\boldsymbol{r}$ and, to the first order in the eccentricity:

$$r^* = r - \varepsilon\cos\left(\vartheta - \omega t\right) = r' - \varepsilon\cos\left(\vartheta' - \omega' t\right)$$

$$\vartheta^* = \vartheta - \Omega t + \frac{\varepsilon}{r}\sin\left(\vartheta - \omega t\right) = \vartheta' + \frac{\varepsilon}{r}\sin\left(\vartheta' - \omega' t\right)$$

$$z^* = z' = z$$

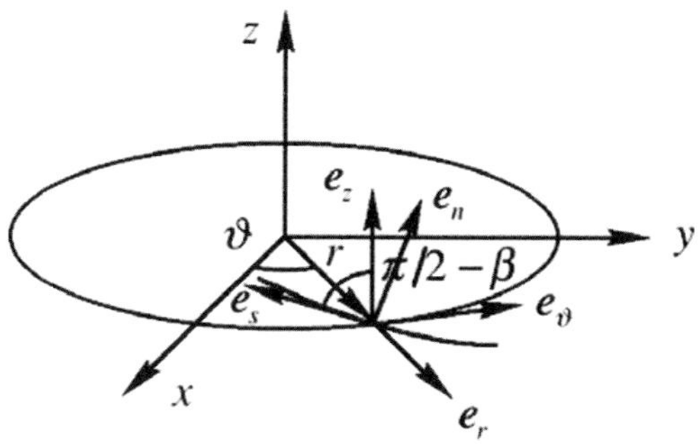

Figure 4. Schematic of the transformation from the cylindrical coordinates r, ϑ, z to the orthogonal helical coordinates r, n, s.

where $\omega' = \omega - \Omega$.

Neglecting Coriolis forces, the perturbation velocity $\tilde{\boldsymbol{u}}$ in the frame moving with the mean flow at axial velocity $\bar{w}$ and angular speed Ω_F is irrotational ($\tilde{\boldsymbol{u}} = \nabla\tilde{\varphi}$) because it originates from a resting flow with conservative centrifugal forces. Therefore, in the rotating frame the flow velocity is $\boldsymbol{u}' = \left(\boldsymbol{\Omega}_F - \boldsymbol{\Omega}\right) \times \boldsymbol{r} + \bar{\boldsymbol{w}} + \nabla\tilde{\varphi}$ and the Bernoulli's equation writes:

$$\int \frac{\partial' \boldsymbol{u}'}{\partial t} \cdot d\boldsymbol{x} + \frac{1}{2} \boldsymbol{u}' \cdot \boldsymbol{u}' - \frac{1}{2} \Omega^2 \boldsymbol{r} \cdot \boldsymbol{r} + \frac{p}{\rho_L} = \int \boldsymbol{u}' \times \left(\nabla \times \boldsymbol{u}'\right) \cdot d\boldsymbol{x}$$

where $\nabla \times \boldsymbol{u}' = 2\left(\boldsymbol{\Omega}_F - \boldsymbol{\Omega}\right)$. Evaluating this equation between the generic point and the corresponding unperturbed conditions on a vorticity line parallel to the inducer axis, the linearized governing equations for the flow perturbations (tildes) in the rotating frame are:

$$\nabla^2 \tilde{\varphi} = 0 \qquad \text{and} \qquad \frac{\partial' \tilde{\varphi}}{\partial t} + \bar{\boldsymbol{u}}' \cdot \nabla\tilde{\varphi} + \frac{\tilde{p}}{\rho_L} = 0$$

These equations must be complemented with the appropriate boundary conditions. Here, the flow velocity must satisfy the kinematic conditions $DB/Dt = 0$ on the hub, blade and casing surfaces of equations $B\left(\boldsymbol{x}, t\right) = 0$ in the relevant coordinates. In addition, the total pressure is assumed constant on the inlet and outlet sections of the inducer.

In order to simplify the boundary conditions, let introduce orthogonal helical coordinates r, n, s of unit vectors $\boldsymbol{e}_r$, $\boldsymbol{e}_n$, $\boldsymbol{e}_s$ as shown in Figure 4 with:

$$n = \left(\vartheta - \vartheta_j\right) \frac{N_B}{2\pi} + z \frac{N_B}{P}$$

$$s = z \frac{\sin^2 \beta}{P} - \left(\vartheta - \vartheta_j\right) \frac{\cos^2 \beta}{2\pi}$$

Here, for convenience, n is normalized with the channel width $P^* = \left(P/N_B\right) \cos\beta$ and s with the blade length $L^* = P/\sin\beta$ at the radius corresponding to the angle β. The rotating and body-fixed orthogonal helical coordinates r', n', s' and r^*, n^*, s^* are similarly defined in terms of r', ϑ', z' and r^*, ϑ^*, z^*. Then, the equations of the hub, blade pressure sides, blade suction sides and casing surfaces are:

$$B = r^* - r_H = 0$$

$$B = n^* = 0$$

$$B = n^* - 1 + \delta/P^* = 0$$

$$B = r - r_T = 0$$

where $\delta = \delta(t)$. From the continuity equation for the layer $\rho_C\delta \equiv \text{constant}$, the definition of $a_C^2 = dp_t/d\rho_C$ and the Bernoulli's equation $\tilde{p}_t/\rho_L = -\partial\tilde{\varphi}/\partial t$ it follows that:

$$\frac{d\delta}{dt} = \frac{\rho_L\delta}{\rho_C a_C^2}\frac{\partial^2\tilde{\varphi}}{\partial t^2}$$

With these results, expressing $\nabla^2\tilde{\varphi} = 0$ and $\partial B/\partial t + \boldsymbol{u}'\cdot\nabla B = 0$ in the rotating helical coordinates r', n', s':

$$\frac{\partial^2\tilde{\varphi}}{\partial r'^2} + \frac{1}{r'}\frac{\partial\tilde{\varphi}}{\partial r'} + \frac{N_B^2}{P^2\cos^2\beta}\frac{\partial^2\tilde{\varphi}}{\partial n'^2} + \frac{\cos^2\beta}{4\pi^2 r'^2}\frac{\partial^2\tilde{\varphi}}{\partial s'^2} = 0$$

and the linearized boundary conditions are found to be:

$$\frac{\partial\tilde{\varphi}}{\partial r'} = 0 \quad \text{on} \quad r' = r_T$$

$$\frac{\partial\tilde{\varphi}}{\partial r'} = \varepsilon(\omega - \Omega_F)\sin H \quad \text{on} \quad r' = r_H$$

$$\frac{\partial\tilde{\varphi}}{\partial n'} = \frac{\varepsilon(\omega - \Omega_F)P^2\cos^2\beta}{2\pi r' N_B}\cos H \quad \text{on} \quad n' = 0$$

$$\frac{\partial\tilde{\varphi}}{\partial n'} + K_C\frac{\partial^2\tilde{\varphi}}{\partial t^2} = \frac{\varepsilon(\omega - \Omega_F)P^2\cos^2\beta}{2\pi r' N_B}\cos H \quad \text{on} \quad n' = 1$$

$$\frac{\partial\tilde{\varphi}}{\partial t} = 0 \quad \text{on} \quad s' = s_i' = -\frac{\cos^2\beta}{2N_B},\ s_i' + N_R$$

where:

$$H = \frac{2\pi}{N_B}n'\sin^2\beta - 2\pi s' + \vartheta_j - \omega' t$$

$$K_C = \frac{\rho_L\delta P^*}{\rho_C a_C^2(1 + i\zeta)}$$

is a parameter describing the behavior of the cavitating layer, and N_R is the number of revolutions of the blade channels about the inducer axis.

With the above boundary conditions the Laplace equation for $\tilde{\varphi} = \mathrm{Re}\{\hat{\varphi}\}$ yields a well-posed boundary value problem for the complex velocity potential $\hat{\varphi}$. If the variable blade angle β is approximated by a constant value β_M at some suitable mean radius r_M, the separable solution (Lebedev, 1965) in the blade channels $0 \le n' \le 1$ is:

$$\hat{\varphi} = \hat{\varphi}_H + \hat{\varphi}_B$$

where:

$$\hat{\varphi}_H = \sum_{k=1}^{+\infty}\sum_{m=1}^{+\infty} R_{km}(r')N_k(n')S_m(s')e^{-i\omega' t}$$

$$\hat{\varphi}_B = \sum_{l=1}^{+\infty}\sum_{m=1}^{+\infty} R_{lm}(r')N_{lm}(n')S_m(s')e^{-i\omega' t}$$

are the solutions corresponding to the hub and blade excitation. In the expression of $\hat{\varphi}_H$:

$$R_{km}(r') = I_{km}\frac{K'_{q_m}(\lambda_k^* r_T)I_{q_m}(\lambda_k^* r') - I'_{q_m}(\lambda_k^* r_T)K_{q_m}(\lambda_k^* r')}{I'_{q_m}(\lambda_k^* r_H)K'_{q_m}(\lambda_k^* r_T) - K'_{q_m}(\lambda_k^* r_H)I'_{q_m}(\lambda_k^* r_T)}$$

$$N_k(n') = \cos\left(n'\sqrt{-v_k^2}\right)$$

are the coupled modal solutions corresponding to the hub excitation, where:

$$\lambda_k^* = \frac{N_B}{P\cos\beta_M}\sqrt{-v_k^2}$$

and $\sqrt{-v_k^2}$ are the (complex) principal roots of the equation:

$$\sqrt{-v^2}\,\sin\sqrt{-v^2} = -K_C\omega'^2\cos\sqrt{-v^2}$$

Similarly, in the expression of $\hat{\varphi}_B$:

$$R_{lm}(r') = Y'_{q_m}(\lambda_{lm} r_H)J_{q_m}(\lambda_{lm} r') - J'_{q_m}(\lambda_{lm} r_H)Y_{q_m}(\lambda_{lm} r')$$

$$N_{lm}(n') = \frac{I(1)\cosh(v_{lm}n') - I(0)\cosh\left[v_{lm}(n'-1)\right]}{v_{lm}\sinh v_{lm} - K_C\omega'^2\cosh v_{lm}} + $$

$$-\frac{I(0)K_C\omega'^2}{v_{lm}\sinh v_{lm} - K_C\omega'^2\cosh v_{lm}}\,\frac{\sinh\left[v_{lm}(n'-1)\right]}{v_{lm}}$$

are the coupled modal solutions corresponding to the blade excitation, where λ_{lm} are the (positive) roots of the equation:

$$J'_{q_m}(\lambda r_H)Y'_{q_m}(\lambda r_T) - Y'_{q_m}(\lambda r_H)J'_{q_m}(\lambda r_T) = 0$$

$$v_{lm} = \lambda_{lm}\frac{P\cos\beta_M}{N_B}$$

and:

$$I(n') = \frac{\varepsilon(\omega - \Omega_F)P^2\cos^2\beta_M}{\pi N_B N_R}e^{i\left[(2\pi/N_B)n'\sin^2\beta_M + \vartheta_j\right]}\times$$

$$\times\frac{\int_{r_H}^{r_T} R_{lm}(r')\,dr'}{\int_{r_H}^{r_T} R_{lm}^2(r')\,r'\,dr'}\int_{s_i'}^{s_i'+N_R} e^{-i2\pi s'}S_m(s')\,ds'$$

Finally, in the both of the expressions of $\hat{\varphi}_H$ and $\hat{\varphi}_B$:

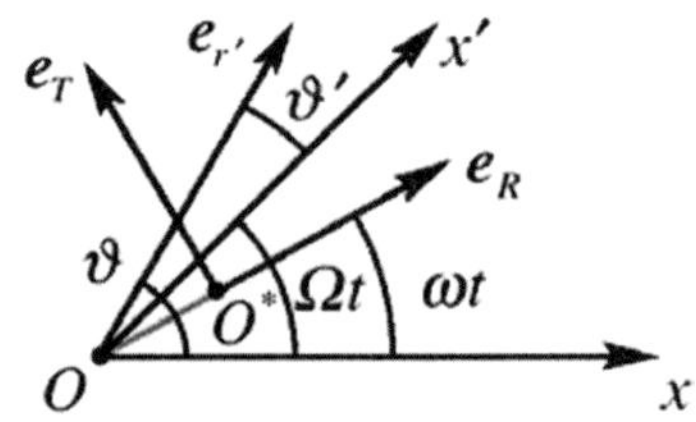

Figure 5. Schematic of the transformation form the rotating frame $O, e_{r'}, e_{\vartheta'}, e_{z'}$ to the whirling frame O^*, e_R, e_T, e_z.

$$q_m = \frac{m}{2N_R}\cos\beta_M \quad \text{and} \quad S_m(s') = \sin\frac{m\pi(s' - s'_i)}{N_R}$$

The instantaneous fluid force on the inducer is:

$$\tilde{\boldsymbol{F}} = -\oint_{S_H \cup S_B} p\left(\boldsymbol{x}'\big|_{\varepsilon\neq 0}, t\right) d\boldsymbol{S}$$

where the pressure:

$$p = \bar{p} - \rho_L\left(\frac{\partial\tilde{\varphi}}{\partial t} - \frac{\Omega_F - \Omega}{2\pi}\frac{\partial\tilde{\varphi}}{\partial s'}\right)$$

is evaluated for $\varepsilon \neq 0$ at the perturbed position of the hub and blade surfaces S_H and S_B. Thus, expanding to the first order in the eccentricity and noting that the mean pressure $\bar{p}\left(r'\big|_{\varepsilon=0}, t\right)$ makes no net contribution to the force on the inducer in its centered position, the following expression for the fluid force on the inducer is obtained:

$$\tilde{\boldsymbol{F}} \cong -\oint_{S_H \cup S_B}\left[\frac{\partial\bar{p}}{\partial r'}\bigg|_{\varepsilon=0}\left(r'\big|_{\varepsilon\neq 0} - r'\big|_{\varepsilon=0}\right) + \tilde{p}\left(\boldsymbol{x}'\big|_{\varepsilon=0}, t\right)\right] d\boldsymbol{S}$$

Here the first term is the buoyancy force on the displaced inducer due to the radial gradient of the mean pressure:

$$\frac{d\bar{p}}{dr'} = \rho_L \Omega^2 r'\left(1 - \phi^2\cot^2\beta_T\right)$$

and the second term is the force due to the pressure perturbations generated by the eccentric motion. The components of the instantaneous rotordynamic force are therefore obtained by integrating the projections of the elementary pressure forces along the axes of the whirling frame of center in O^* and unit vectors $\boldsymbol{e}_R$, $\boldsymbol{e}_T$, $\boldsymbol{e}_z$, as shown in Figure 5. Finally, further integration over a period $2\pi/\omega'$ yields the time-averaged rotordynamic force $\bar{\boldsymbol{F}}$ on the inducer.

The entire flow has therefore been determined in terms of the material properties of the two phases, the geometry of the inducer, the nature of the excitation, and the assigned quantities ϕ, δ, ρ_C, a_C and ζ.

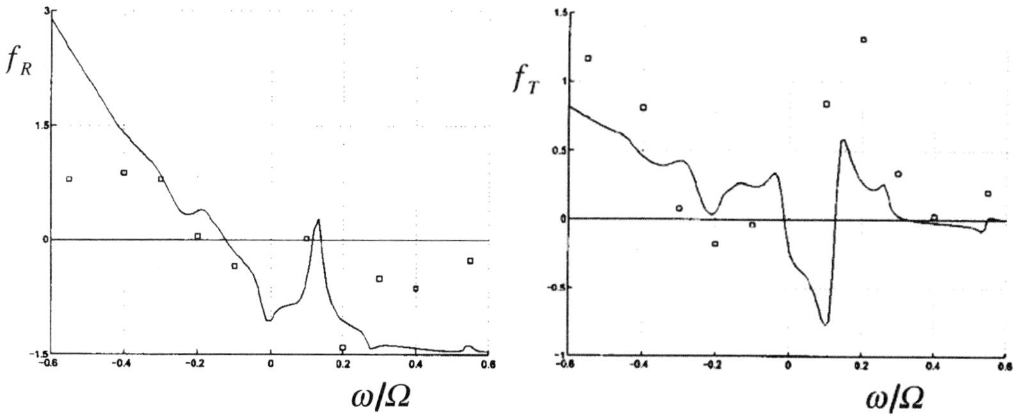

Figure 6. Nondimensional radial rotordynamic force f_R (left) and tangential rotordynamic force f_T (right) on the test inducer as functions of the ratio ω/Ω of the whirl and rotational speeds. The experimental values (circles) obtained by Bhattacharyya, 1994, under developed cavitation conditions at $\phi_1 = 0.049$ and $\sigma_1 = 0.106$ are compared with the model predictions (solid line) for $\phi = \phi_1 / \left(1 - r_H^2 / r_T^2\right) = 0.0583$, $\mathrm{Re}\left\{K_C \Omega^2\right\} = 2.5$, $\zeta = 0.045$ and $N_R =$ Œ8 .

3 Rotordynamic Forces on Whirling and Cavitating Axial Inducers

The rotordynamic fluid forces predicted by the present model will be compared with the data measured by Bhattacharyya (1994) on a three-bladed helical inducer with $r_T = 5.06$ cm, $r_H / r_T = 0.4$, $\beta_T = 9°$, $L = 2.43$ cm, $\varepsilon = 0.254$ mm. The data refer to operation in water at room temperature, rotational speed $\Omega = 3{,}000$ rpm, variable whirl speed, and several values of the inlet flow coefficient $\phi_1 = \dot{Q} / A_1 \Omega r_T$ (not corrected for hub blockage) and cavitation number $\sigma_1 = \left(p_1 - p_V\right) / \frac{1}{2} \rho_L \Omega^2 r_T^2$.

Figures 6 shows some typical experimental results for the nondimensional rotordynamic force $\tilde{f} = \bar{F} / \pi \varepsilon^2 P \rho_L \Omega^2 r_T$ on the sample inducer tested under developed cavitation at $\phi_1 = 0.049$ and $\sigma_1 = 0.106$. Notice that the radial and tangential force components do not vary quadratically with the normalized whirl speed ω/Ω, and that their behavior is characterized by multiple zero crossings. The radial force (left) is essentially positive (destabilizing) for $\omega/\Omega < -0.2$, oscillates above and below zero for $-0.2 < \omega/\Omega < 0.3$, and remains essentially negative (stabilizing) for $\omega/\Omega > 0.3$. Similar behavior was observed for other cavitation numbers (Bhattacharyya, 1994). The tangential force (right) is positive over most of the sampled frequencies, but also exhibits rapid oscillations near the origin. In all of the experiments of Bhattacharyya a peculiar feature of the tangential force is the occurrence of a sharp positive (destabilizing) peak at $\omega/\Omega \cong 0.2$, whose intensity increases at higher cavitation numbers and lower flow coefficients.

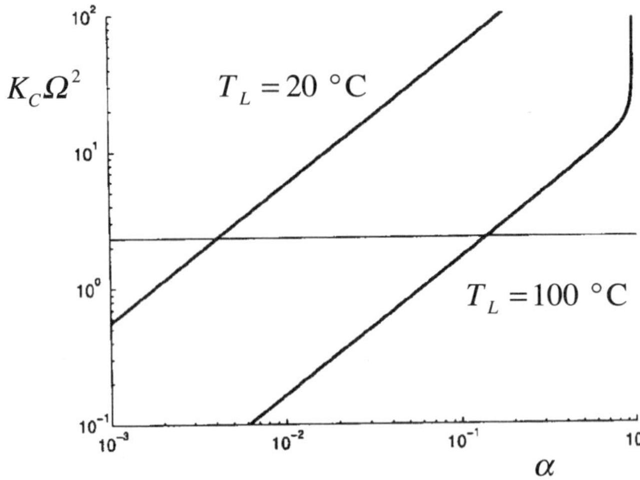

Figure 7. Nondimensional cavitation parameter $K_C\Omega^2$ as a function of the void fraction in the cavitating layer for undamped operation in water at $T_L = 20$ °C ($\delta_T/R = 0.3$, $\delta/P^* = 3\%$) and $T_L = 100$ °C ($\delta_T/R = 1$, $\delta/P^* = 10\%$).

Figures 6 also shows the rotordynamic forces predicted by the present model (continuous line) assuming $\mathrm{Re}\left\{K_C\Omega^2\right\} = 2.5$ for the real part of the cavitation parameter and $\zeta = 0.045$ for the nondimensional damping coefficient. An effective value of the nondimensional blade channel length, $N_R = 0.285$, intermediate between the geometric length of the blades and their actual overlap, has been used in the computations in order to empirically compensate for the errors introduced by the formulation in orthogonal helical coordinates. In addition, the pressure gradient of the mean flow has been evaluated for a decreased value of the swirl speed Ω_F in order to account for the gradual rotational acceleration of the flow entering the inducer. Comparison with the experimental data shows that, in spite of its approximate nature, the present theory correctly captures the observed magnitude of the rotordynamic forces and the typical features of their whirl frequency spectrum, including their stabilizing or destabilizing effects on the eccentric motion of the inducer.

The complex dependence of the lateral rotordynamic fluid forces on the whirl speed is due to the occurrence of internal resonances of the cavitating flow in the blade channels under the excitation generated by the whirl motion of the inducer. Given the functional dependence of the solution, it appears that the system has an infinite set of (generally complex) critical whirl speeds:

$$\omega'_{lm} = \omega_{lm} - \Omega = \pm\sqrt{\frac{\nu_{lm}\tanh\nu_{lm}}{K_C}}$$

symmetrically located above and below the rotational speed Ω (synchronous conditions). The critical speeds are seen to depend on the mode numbers of the flow perturbations and the parameter K_C used to characterize the occurrence of cavitation on the suction sides of the blades. The extent of cavitation increases when the value of this parameter is varied from zero, corresponding to fully-wetted flow conditions, to larger and larger values. In the special case of vanishing cavitation damping ($\zeta \to 0$), K_C tends to a real value and the boundary value problem for $\hat{\varphi}$ to become self-adjoint, with real eigenvalues λ^2 and ν^2. In the presence of damping, the series for $\hat{\varphi}$ converge rapidly even for low subcritical values of $\zeta << 1$, and only the first few modes are needed in the computations. For these modes the eigenvalues are of order unity or slightly larger. Since cavitating flows are inherently dissipative, it follows that the critical whirl speeds of practical importance tend to concentrate in two small ranges just above and below synchronous conditions as soon as the intensity of cavitation is sufficient for raising the real part of $K_C\Omega^2$ well above unity.

The relation of K_C to the extent of cavitation can be investigated with the help of a suitable flow model. Here we make use of the classical quasi-homogeneous isenthalpic cavitation model with thermal effects described by Brennen (1995) and modified by Rapposelli and d'Agostino (2001) to account for the concentration of active nuclei. The behavior of the real part of $K_C\Omega^2$ with the local void fraction α is illustrated in Figure 7 for water at room and boiling temperature. The results depend parametrically on the blade channel blockage δ/P^* and the ratio δ_T/R between the thermal boundary layer thickness surrounding a spherical cavity and the radius of the cavity itself. The parameter δ_T/R is nearly constant during the thermally controlled growth of cavitating bubbles and its value can be estimated as a function of the flow conditions, the thermophysical properties of the two phases, and the concentration of active cavitation nuclei (Rapposelli and d'Agostino, 2001). Here the choice of a higher blockage δ/P^* at the boiling temperature reflects the greater penetration of cavitation in the liquid at elevated temperatures. From Figure 8 it appears that the value $\mathrm{Re}\left\{K_C\Omega^2\right\} = 2.5$ previously used for the prediction of the rotordynamic forces would correspond to void fractions ranging from $4 \cdot 10^{-2}$ at room temperature to 10^{-1} at boiling conditions. The average separation between bubbles would then be on the order of 2 to 3 diameters, not unrealistic mean values for typical cavitation on inducer blades.

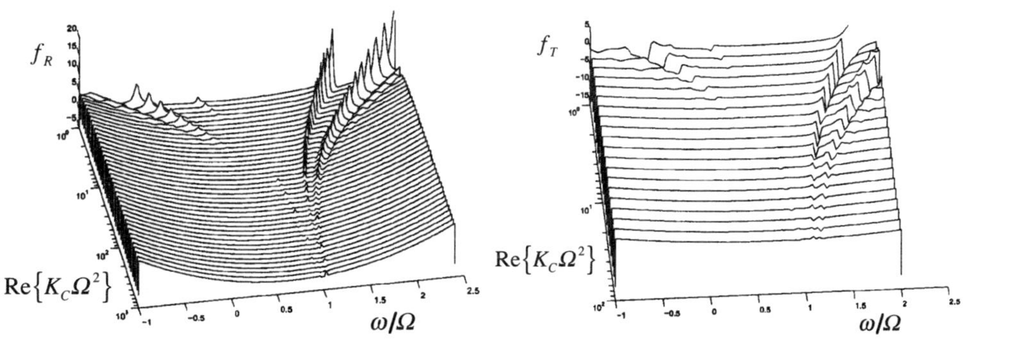

Figure 8. Waterfall plots of the nondimensional radial rotordynamic force f_R (left) and tangential rotordynamic force f_T (right) on the test inducer functions of the ratio ω/Ω of the whirl and rotational speeds and the real part of the nondimensional cavitation parameter, $K_C\Omega^2$. The flow coefficient is $\phi = 0.0583$, the nondimensional damping coefficient is $\zeta = 0.045$ and the effective length of the blade channels is $N_R = 0.285$.

The influence of the cavitation parameter on the solution is illustrated by the waterfall plots of Figure 8. The figure clearly shows that the degree of cavitation has a major impact in locating the critical speeds and determining the magnitude of the rotordynamic forces as functions of the whirl speed. Two sets of subsynchronous and supersynchronous resonances are predicted. At higher values of the cavitation parameter the amplitudes of the resonances decrease, as their frequencies approach synchronous conditions. At low values of $\mathrm{Re}\left\{K_C \Omega^2\right\} << 1$ the void fraction is likely to violate the condition:

$$\left(\alpha\delta\right)_{\min} \approx \varepsilon \sin \beta_M$$

for the survival of the cavitating layer during a complete oscillation cycle of the whirl motion. With typical choices of the relevant quantities, the minimum void fraction is estimated to be $\alpha_{\min} \cong 10^{-2} \div 10^{-1}$. Hence, using the results of Figure 7, the physically significant solutions of the present theory are restricted to minimum values of the cavitation parameter:

$$\mathbb{R}\left\{K_C \Omega^2\right\}_{\min} \approx 1 \text{ to } 10$$

corresponding to room temperature and boiling conditions, respectively. In this range Figure 8 indicates the presence of two relatively weak subsynchronous critical speeds near $\omega/\Omega \approx 0.5$, and a second couple of considerably more intense supersynchronous critical speeds in the vicinity of $\omega/\Omega \approx 1.5$. The spectral locations of these critical speeds evocatively overlap with the reported ranges of free-whirl instabilities in cavitating turbopumps.

As a final comment, comparison of the data reported in Figure 6 with the results of Figure 8 indicates that the scant experimental information currently available on the behavior of rotordynamic forces in cavitating turbopumps only covers a limited portion of the frequency spectrum, and probably not the most significant one in connection with the onset of cavitation-induced whirl instabilities.

4 Linearized Dynamics of the Cavitating Flow in a Whirling Centrifugal Impeller

With a similar approach, we next examine the dynamics of an incompressible, inviscid liquid of velocity $\boldsymbol{u}$, pressure p, and density ρ_L in a centrifugal pump impeller rotating with velocity Ω and whirling on a circular orbit of small eccentricity ε at angular speed ω. A number of idealizations are introduced in order to obtain an analytical solution. Figure 9 (left) illustrates the simple centrifugal pump considered, with N_B logarithmic-spiral blades of equation $r\,d\vartheta/r = -\tan\beta$, zero blade thickness, axial length b, hub radius r_H, tip radius r_T, blade angle β.

Also in this case the flow is fully wetted everywhere except on the suction sides of the blades, where attached cavitation occurs. The mean flow velocity $\bar{\boldsymbol{u}}$ in the blade channels (on the right in Fig. 9) is specified by the flow coefficient $\phi = \bar{u}_T/\Omega r_T$, assuming fully-guided forced-vortex flow with zero axial velocity $\bar{w}$, radial velocity $\bar{u} = r_T \bar{u}_T/r$, and tangential velocity:

$$\bar{v}^2 = \Omega^2\phi^2\left(r_T^4/r^2\right)\tan^2\beta + \Omega^2 r^2 - 2\Omega^2 r_T^2 \phi \tan\beta$$

Also in this case cavitation is thought to occur on the suction sides of the blades in a thin layer of given variable thickness δ (constant coordinate n, see below) and acoustic admittance $\rho_C a_C^2$. For simplicity, it is also assumed that the static pressure p_C in the cavitating layer is nearly equal to the total pressure p_t of the surrounding liquid.

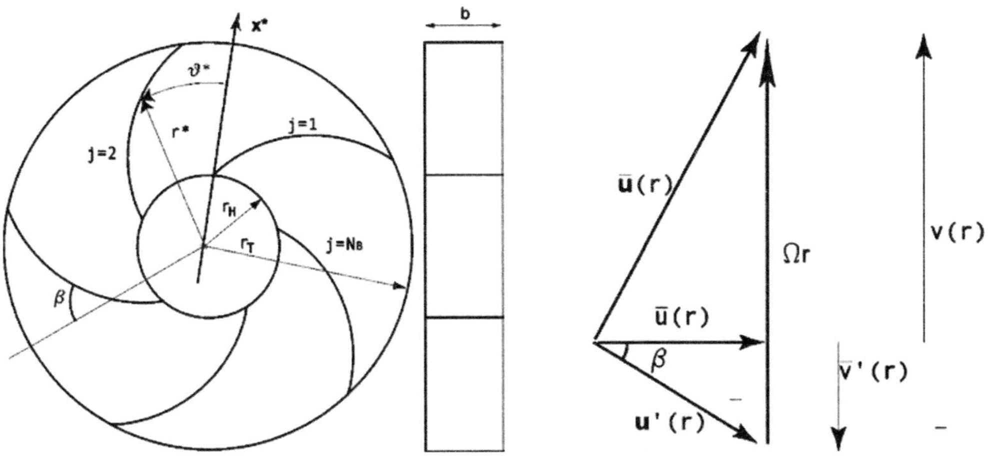

Figure 9. Schematic of the radial impeller geometry (left) and flow velocity triangle (right).

We define again stationary cylindrical coordinates r, ϑ, z with center in O on the axis of the stator, rotating cylindrical coordinates r', ϑ', z' spinning at the rotor speed with center in the same point O, and rotating and whirling cylindrical coordinates r^*, ϑ^*, z^* fixed in the inducer and with center in O^* on its geometric axis. Then the equations of the blade surfaces are:

$$B = \ln\frac{r}{r_H} + \left(\vartheta - \vartheta_j\right)\cot\beta = 0$$

where $\vartheta_j = 2\pi\left(j-1\right)/N_B$ is the angular location of the j-th blade, with $j = 1,2,...,N_B$. Once more the flow velocities in the stationary and rotating frames are related by $\boldsymbol{u} = \boldsymbol{u}' + \boldsymbol{\Omega}\times\boldsymbol{r}$ and, to the first order in the eccentricity (Fig. 3):

$$r^* = r - \varepsilon\cos\left(\vartheta - \omega t\right) = r' - \varepsilon\cos\left(\vartheta' - \omega' t\right)$$

$$\vartheta^* = \vartheta - \Omega t + \frac{\varepsilon}{r}\sin\left(\vartheta - \omega t\right) = \vartheta' + \frac{\varepsilon}{r}\sin\left(\vartheta' - \omega' t\right)$$

$$z^* = z' = z\,.$$

where $\omega' = \omega - \Omega$.

The perturbation velocity $\tilde{\boldsymbol{u}}$ generated by the blade motion is irrotational ($\tilde{\boldsymbol{u}} = \nabla\varphi$) because the flow originates from a uniform stream. Therefore, in the rotating frame the flow velocity is $\boldsymbol{u}' = \boldsymbol{u} - \boldsymbol{\Omega}\times\boldsymbol{r}$ and the Bernoulli's equation writes:

$$\int\frac{\partial'\boldsymbol{u}'}{\partial t}\cdot d\boldsymbol{x} + \frac{1}{2}\boldsymbol{u}'\cdot\boldsymbol{u}' - \frac{1}{2}\Omega^2\boldsymbol{r}\cdot\boldsymbol{r} + \frac{p}{\rho_L} = C\left(t\right)$$

where $C\left(t\right)$ is the unsteady Bernoulli's constant. Hence the linearized governing equations for the flow perturbations (tildes) at any given point in the rotating frame are:

$$\nabla^2\tilde{\varphi} = 0 \qquad \text{and} \qquad \frac{\partial'\tilde{\varphi}}{\partial t} + \bar{\boldsymbol{u}}'\cdot\nabla\tilde{\varphi} + \frac{\tilde{p}}{\rho_L} = 0$$

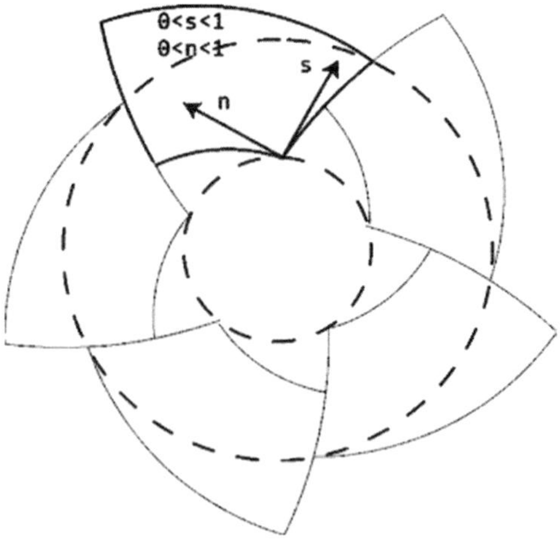

Figure 10. Schematic of the logarithmic-spiral coordinates.

Here the flow velocity must satisfy the kinematic conditions $Db/Dt = 0$ on the hub, blade and casing surfaces of equations $b(\boldsymbol{x},t) = 0$ in the relevant coordinates. In addition, the total pressure is assumed constant on the inlet and outlet sections of the inducer.

In order to simplify the derivation of the solution, let us introduce orthogonal spiral coordinates n, s (Campos and Gil, 1995; Visser, 1999) as shown in Figure 10, with:

$$n = \frac{N_B}{2\pi}\left(\vartheta - \vartheta_j\right) + \frac{N_B}{2\pi}\ln\left(\frac{r}{r_H}\right)\tan\beta$$

$$n = -\frac{\sin\beta\cos\beta}{\ln\left(r_T/r_H\right)}\left(\vartheta - \vartheta_j\right) + \frac{\ln\left(r/r_H\right)}{\ln\left(r_T/r_H\right)}\cos^2\beta$$

For convenience, n and s are normalized to map a channel into a rectangle $(0,1)\times(0,1)$. Rotating and body-fixed orthogonal spiral coordinates n', s' and n^*, s^* are similarly defined in terms of r', ϑ' and r^*, ϑ^*. The third dimension z is easily added. Then, the equations of the inlet, blade pressure side, blade suction side and outlet surfaces are:

$$B = s^* = 0\,.$$

$$B = n^* = 0\,.$$

$$B = n^* - 1 + \delta = 0\,.$$

$$B = s^* - 1 = 0\,.$$

where $\delta = \delta(t)$.

From the continuity equation for the layer $\rho_C\delta \equiv \text{constant}$, the definition of $a_C^2 = dp/d\rho_C$ and the Bernoulli's equation $\tilde{p}/\rho_L \cong -\partial\tilde{\varphi}/\partial t$ it follows that:

$$\frac{d\delta}{dt} = \frac{\rho_L\delta}{\rho_C a_C^2}\frac{\partial^2\tilde{\varphi}}{\partial t^2}$$

With these results, and expressing $\nabla^2\tilde{\varphi}=0$ and $\partial B/\partial t+\bar{\boldsymbol{u}}'\cdot\nabla B=0$ in the rotating helical coordinates $r', n',' s'$:

$$\nabla'^2\tilde{\varphi}=\frac{\cos^2\beta}{\ln^2\left(r_T/r_H\right)}\frac{\partial^2\tilde{\varphi}}{\partial s'^2}+\frac{N_B^2}{4\pi^2\cos^2\beta}\frac{\partial^2\tilde{\varphi}}{\partial n'^2}=0$$

and the linearized boundary conditions are found to be:

$$\frac{\partial\tilde{\varphi}}{\partial t}=0 \quad \text{on} \quad s'=1 \quad \text{and} \quad s'=0$$

$$\frac{\partial\tilde{\varphi}}{\partial n'}=\varepsilon\frac{2\pi}{N_B}\left[\omega' r_H\left(\frac{r_T}{r_H}\right)^{s'}\cos^2\beta\sin\beta\left(\frac{\cos\Theta'}{\sin\beta}+\frac{\sin\Theta'}{\cos\beta}\right)+\right.$$

$$\left.+\left(\frac{r_T}{r_H}\right)^{1-s'}\bar{u}_T\sin\Theta'\right] \quad \text{on} \quad n'=0$$

$$\frac{N_B}{2\pi}K_E\frac{\partial^2\tilde{\varphi}}{\partial t^2}+\frac{N_B}{2\pi r_H^2\left(r_T/r_H\right)^{2s'}\cos^2\beta}\exp\left(-\frac{4\pi}{N_B}\sin\beta\cos\beta\right)\frac{\partial\tilde{\varphi}}{\partial n'}=$$

$$-\varepsilon\omega'\frac{\sin\beta}{r_H\left(r_T/r_H\right)^{s'}}\exp\left(-\frac{2\pi}{N_B}\sin\beta\cos\beta\right)\left(\frac{\cos\Theta'}{\sin\beta}+\frac{\sin\Theta'}{\cos\beta}\right)+$$

$$+\varepsilon\frac{\bar{u}_T}{r_H^2}\left(\frac{r_T}{r_H}\right)^{1-3s'}\exp\left(-\frac{6\pi}{N_B}\sin\beta\cos\beta\right)\frac{\sin\Theta'}{\cos^2\beta} \quad \text{on} \quad n'=1$$

Here:

$$\Theta'=\vartheta'-\omega' t=\vartheta_j+\frac{2\pi}{N_B}n'\cos^2\beta-s'\ln\left(\frac{r_T}{r_H}\right)\tan\beta-\omega' t$$

and:

$$K_E=\frac{4\pi^2\rho_L\delta}{N_B^2\rho_C a_C^2}\cos^2\beta$$

is a parameter describing the dynamic behavior of the cavitating layer and the extent of cavitation.

With the above boundary conditions the Laplace equation for $\tilde{\varphi}=\operatorname{Re}\{\hat{\varphi}\}$ yields a well-posed boundary value problem for the complex velocity potential $\hat{\varphi}$. The separable solution in the blade channels is:

$$\hat{\varphi}=\sum_{m=1}^{+\infty}\left\{\left[c_1\cosh\left(n'\sqrt{v_m^2}\right)+c_2\sinh\left(n'\sqrt{v_m^2}\right)\right]\sin\left(m\pi s'\right)\right\}e^{-i\omega' t}$$

with eigenvalues:

$$\sqrt{v_m^2}=i\frac{2m\pi^2\cos^2\beta}{N_B\ln\left(r_T/r_H\right)}$$

The instantaneous fluid force on the inducer is then:

$$\tilde{\boldsymbol{F}} = \int_{\text{blades}} \tilde{p}\left(\boldsymbol{r}'\big|_{\varepsilon\neq 0}, t\right) d\boldsymbol{S}$$

where, with second order error in the perturbations, the pressure:

$$\tilde{p} = -\rho_L \frac{\partial \tilde{\varphi}}{\partial t} - \rho_L \overline{\boldsymbol{u}}' \cdot \nabla \tilde{\varphi}$$

is evaluated for at the unperturbed position of the impeller ($\varepsilon = 0$). Because no hub is present, no buoyancy force acts on the displaced inducer due to the radial gradient of the mean pressure. The components of the instantaneous rotordynamic force are therefore obtained by integrating the projections of the elementary pressure forces along the axes of the whirling frame of center in O^* and unit vectors $\boldsymbol{e}_R, \boldsymbol{e}_T$, oriented as the eccentricity and its normal in the direction of the whirl motion. Finally, further integration over a period $2\pi/\omega'$ yields the time-averaged rotordynamic force $\boldsymbol{F}$ on the inducer.

The flow has then been determined in terms of the material properties of the two phases, the geometry of the inducer, the nature of the excitation, and the assigned quantities ϕ, δ, ρ_C, a_C.

5 Rotordynamic Forces on Whirling and Cavitating Radial Impellers

In centrifugal pumps, the measured rotordynamic fluid forces in the presence of cavitation are nearly equal to those for fully-wetted flows (Jery, 1987; Franz et al., 1989). Here the rotordynamic fluid forces predicted by the present model are compared with the data measured by Jery (1987) at Caltech on a five-bladed centrifugal pump with $r_T = 81$ mm, $r_H = 40$ mm, $\beta = 23°$, $b = 16$ mm, $\varepsilon = 0.126$ mm, without cavitation. The data refer to operation in water at room temperature, flow coefficient $\phi = 0.092$, rotational speed $\Omega = 1000$ rpm, and variable whirl speed. The results obtained by Franz (1989) for a cavitating radial impeller with a volute are also very similar.

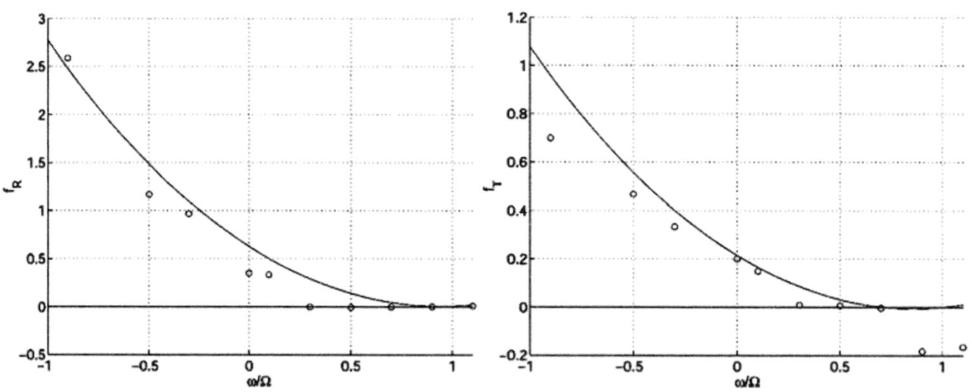

Figure 11. Comparison of the normalized radial (left) and tangential (right) rotordynamic forces, f_R and f_T, obtained from the present theory (continuous line) and the experimental results of Jery, 1987 (divided by six, circles) for a centrifugal impeller with $\Omega = 1000$ rpm, $N_B = 5$, $\phi = 0.060$, $\beta = 23°$, $r_T = 81$ mm, $b = 16$ mm, $r_H/r_T = 0.5$ and $\Omega^2 K_E = 2$.

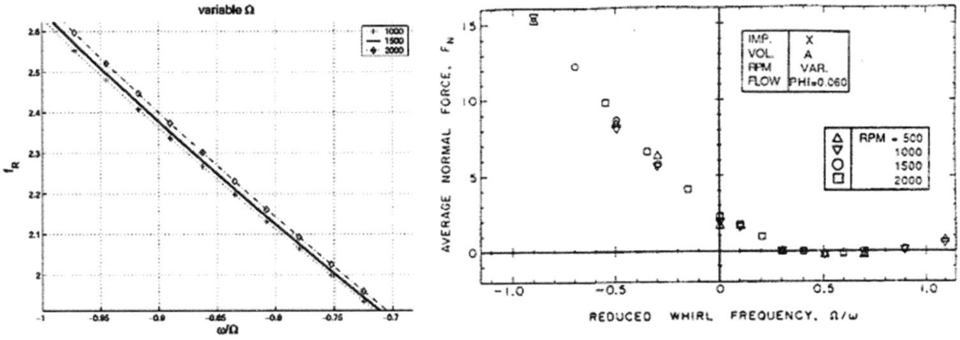

Figure 12. Comparison of the normalized radial rotordynamic force, f_R, obtained from the present theory (left) for $K_E = 2 \cdot 10^{-5}$ sec^2 and several rotational speeds Ω with the experimental results (right) of Jery, 1987, for a centrifugal impeller with $N_B = 5$, $\phi = 0.060$, $\beta = 23°$, $r_T = 81$ mm, $b = 16$ mm and $r_H / r_T = 0.5$.

The calculated rotordynamic forces shown in the figures have been nondimensionalized by $\pi \varepsilon \rho_L r_T^2 \Omega^2 b$. Comparison with the experimental results by Jery (1987) in Figure 11 shows that the calculated forces are about six times smaller than experimentally measured, but their familiar quadratic behavior with the whirl speed is well captured by the theoretical results and the vertex of the parabola is correctly located. The reasons for the observed discrepancy have not been identified with certainty, but they are likely to be mostly related to the approximate nature of the boundary conditions at the inlet and outlet sections of the impeller. In their present form these boundary conditions do not realistically account for the dynamic response of the flow in the impeller eye and diffuser (or volute). The inclusion of these effects would introduce significant additional contributions to the inertial reaction of the flow on the impeller, increasing the magnitude of the rotordynamic forces.

Clearly with present notations rotordynamic forces are destabilizing when the radial component is positive and, for the onset of asynchronous whirl, when the tangential component has the same sign as the whirl speed ω. Hence, with reference to Figure 11, the predicted radial force is generally destabilizing except near synchronous conditions ($\omega \cong \Omega$), while the tangential force would promote subsynchronous shaft motions in the range of whirl speeds $0 < \omega < 0.7$. Also notice that both components of the rotordynamic force are relatively small in the vicinity of $\omega / \Omega = 0.5$, corresponding to the familiar "whip conditions" of journal bearings (Newkirk and Taylor, 1925; Hori, 1959).

Present results for radial turbopumps are also radically different from those obtained for cavitating inducers. In this case both the experiments of Bhattacharyya et al. (1997) and our previous theoretical investigations based on the same approach used herein (d'Agostino, d'Auria and Brennen (1998), d'Agostino and Venturini (2002, 2003) showed a more complex dependence of the rotordynamic forces on the whirl speed. The spectral response of these forces as functions of the whirl frequency displayed a number of multiple peaks, which the theory indicated to be related with the occurrence of internal resonances of the cavitating flow in the blade channels under the excitation provided by the eccentric motion of the inducer. From the mathematical standpoint,

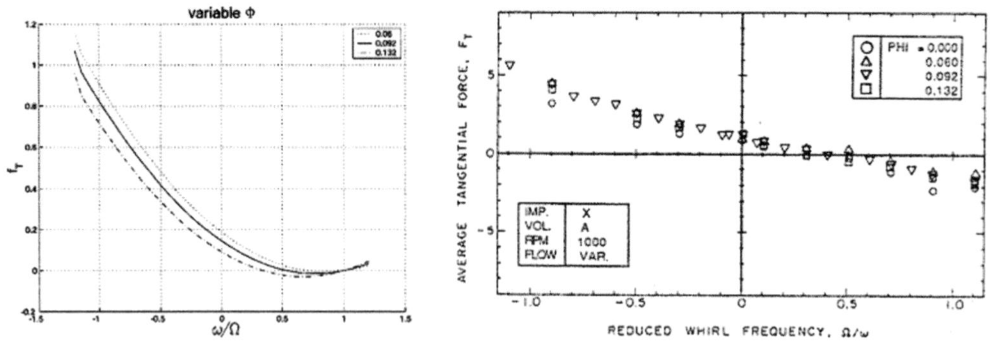

Figure 13. Comparison of the normalized tangential rotordynamic force, f_T, obtained from the present theory (left) for $\Omega^2 K_E = 2$ and several rotational speeds Ω with the experimental results (right) of Jery, 1987, for a centrifugal impeller with $N_B = 5$, $\phi = 0.060$, $\beta = 23°$, $r_T = 81$ mm, $b = 16$ mm and $r_H / r_T = 0.5$.

these resonances are the consequence of the (nearly) real nature of the flow eigenvalues, which leads to an infinite set of lowly-damped critical whirl speeds, symmetrically located above and below the rotational speed Ω (synchronous conditions). Physically, the peaks of the rotordynamic forces are due to the occurrence of standing pressure waves with frequency-dependent wavelength in the blade channels. Hence, at some specific excitation frequencies the wavelength of the resonant flow perturbations is an odd multiple of the blade channel revolution around the hub. In this case the pressure distribution acts in a strong and spatially coherent fashion on the inducer, leading to the intensification of the resulting forces.

Rotordynamic forces on radial impellers, on the other hand, do not peak at any whirl frequency. Mathematically, in this case the critical whirl speeds are (nearly) imaginary:

$$\omega'_m = \pm i\pi e^{-\frac{2\pi}{N_B}\sin\beta\cos\beta} \sqrt{\frac{2m}{K_E N_B r_H r_T \ln\left(r_T/r_H\right)} \tan\frac{2m\pi^2\cos^2\beta}{N_B \ln\left(r_T/r_H\right)}}$$

Physically, in radial impellers the presence of the blades prevents the formation of synchronous pressure waves with significant extension in the azimuthal direction, capable of reacting in a coherent fashion on the impeller.

Notice that the flow solution depends on the parameter K_E, whose relationship with the extent of cavitation has already been investigated in our earlier work (d'Agostino and Venturini, 2002) with the help of a quasi-homogeneous isenthalpic cavitation model with thermal effects (Rapposelli and d'Agostino, 2001). However, as mentioned earlier, rotordynamic forces are only weakly dependent on the extent of cavitation and the value of K_E.

The capability of the model of qualitatively capturing the main phenomena controlling the development of rotordynamic fluid forces in whirling centrifugal impellers is confirmed by Figures 12 and 13, which illustrate the sensitivity of the solution to changes of the rotational speed Ω and flow coefficient ϕ. In both cases the predicted impact of these parameters is small, consistently with typical experimental results from Jery (1987) shown in the right diagram of the figure. The forces in Figure 12 have been calculated with fixed K_E and therefore variable $E = \Omega^2 K_E$. With variable Ω and constant K_E the computed curves overlap, showing that K_E

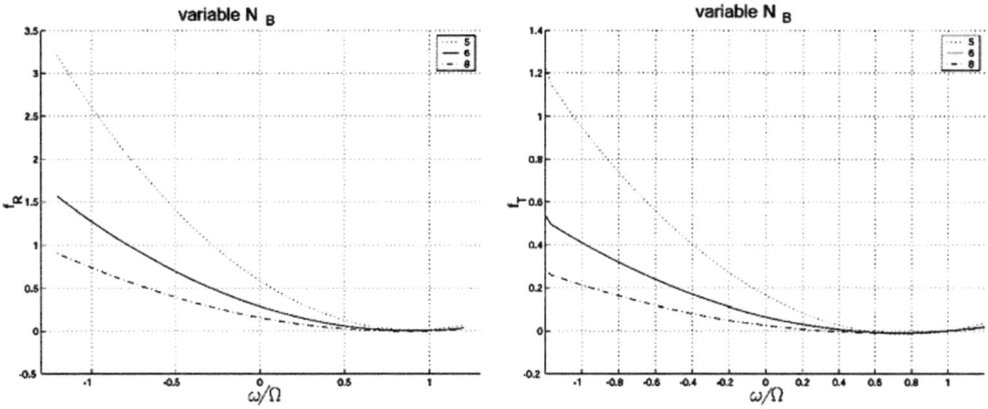

Figure 14. Normalized radial (left) and tangential (right) rotordynamic forces, f_R and f_T, predicted by the present theory as functions of the whirl ratio ω/Ω for a centrifugal impeller with variable number of blades $N_B = 5$ (dotted line), 6 (solid line) and 8 (dash-dotted line), $\Omega = 1000$ rpm, $\phi = 0.092$, $\beta = 3\ ^\circ$, $r_T = 81$ mm, $r_H = 40$ mm, $b = 16$ mm and $\Omega^2 K_E = 1$.

is a well-suited similarity parameter for cavitation effects. Besides, the curves computed for constant Ω and variable K_E (not shown here) almost overlap, confirming that in radial impellers the rotordynamic forces are practically insensitive to cavitation, in accordance with the experimental data by Franz et al. 1989.

Figure 13 shows the negligible influence of the flow coefficient ϕ on the rotordynamic forces. Also this finding agrees well with the experimental data. However, it should be emphasized that different values of Ω and ϕ correspond to very different rotordynamic forces in axial inducers, and that our approach to cavitation modeling correctly reflect this aspect (d'Agostino and Venturini, 2002; Venturini, 2003).

The present theory can also be used to investigate the dependence of rotordynamic whirl forces on the impeller geometry. Specifically, Figures 14, 15 and 16 illustrate the predicted effects of the number of blades, N_B, the blade angle, β, and the hub-to-tip radius ratio, r_H/r_T. As expected, the magnitude of rotordynamic forces decreases as the number of blades increases, but their stabilizing/destabilizing nature is not significantly affected (Figure 14). In this respect it is worth noting that the accuracy of the model increases with N_B because the spiral coordinate system more closely approximates the actual geometry of the impeller when the blade channels are narrower.

Figure 15 shows that both the radial and tangential components of the rotordynamic force decrease at lower blade angles. At higher values of $\beta = 40°$ the radial force is destabilizing only for negative whirl, and the tangential force undergoes two zero crossings, being potentially destabilizing only for supersynchronous whirl ($\omega/\Omega > 1$), where, however, the radial force is not capable of sustaining the eccentricity of the impeller.

Finally, Figure 16 shows that rotordynamic forces and their stabilizing/destabilizing properties are relatively insensitive to the hub-to-tip radius ratio, at least in the range of values meaningful for radial impellers.

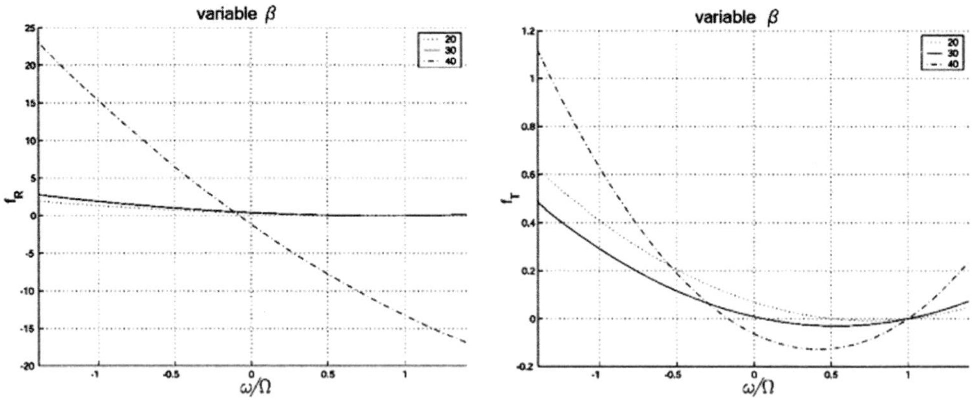

Figure 15. Normalized radial (left) and tangential (right) rotordynamic forces, f_R and f_T, predicted by the present theory as functions of the whirl ratio ω/Ω for a centrifugal impeller with variable blade angle $\beta = 20°$ (dotted line), 30° (solid line) and 40° (dash-dotted line), $N_B = 7$, $\Omega = 1000$ rpm, $\phi = 0.092$, $r_T = 81$ mm, $r_H = 40$ mm, $b = 16$ mm and $\Omega^2 K_E = 1$.

6 Limitations

We now briefly examine the restrictions imposed to the present theory by the various simplifying approximations that have been made. Specifically we shall discuss the limitations due to the assumption of thin-layer cavitation, to the neglect of Coriolis forces, to the applicability of the formulation in orthogonal helical coordinates to the analysis of cavitating impellers, and to the use of the linear perturbation approach in deriving the solution.

The assumption of thin-layer cavitation implies that the thickness of the cavitating region is significantly smaller than the blade channel width and that its properties can be approximated as constant over the entire length of the blades for the purpose of evaluating the rotordynamic forces. Although clearly none of these conditions is rigorously met in cavitating impellers, comparison with earlier results obtained by d'Agostino and his collaborators (1997, 1998) for uniformly distributed bubbly cavitation shows that the predicted values of the rotordynamic forces are remarkably independent on the precise geometry of flow cavitation.

The neglect of Coriolis forces implies that $\Omega_F << \Omega$, a condition that is approximately satisfied in moderately loaded impellers.

For the formulation in orthogonal helical coordinates to be valid the geometric length of the blades should be comparable to their actual overlap, which is rarely the case in low blade angle inducers. Formally, this is probably one of the most stringent limitations of the present analysis and can only be partially circumvented by artificially introducing an empirical "effective length" of the blade channels.

Finally, the perturbation approach simply requires that $\varepsilon << r_T$, a condition that can safely be assumed in the analysis of whirl instabilities of cavitating turbomachines.

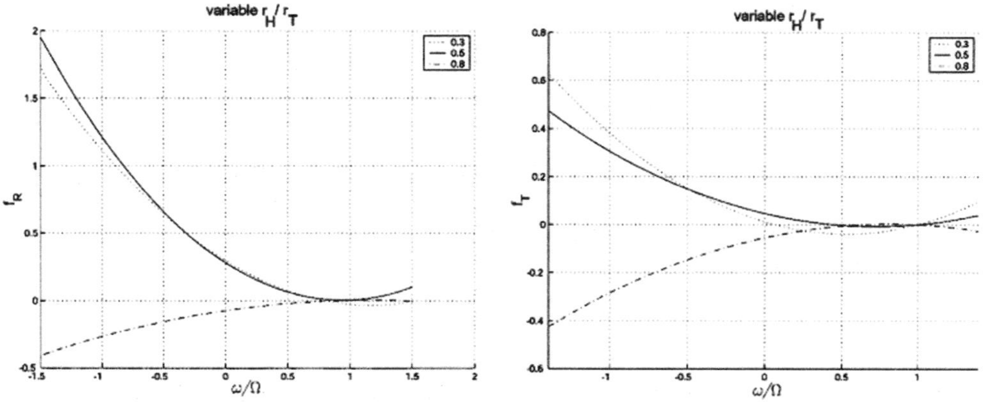

Figure 16. Normalized radial (left) and tangential (right) rotordynamic forces, f_R and f_T, predicted by the present theory as functions of the whirl ratio ω/Ω for a centrifugal impeller with variable hub-to-tip radius ratio $r_H/r_T = 0.3$ (dotted line), 0.5 (solid line) and 0.8 (dash-dotted line), $N_B = 7$, $\Omega = 1000$ rpm, $\phi = 0.092$, $\beta = 23°$, $r_T = 81$ mm, $b = 16$ mm and $\Omega^2 K_E = 1$.

7 Summary and Conclusions

This investigation reveals a number of important flow phenomena occurring in whirling and cavitating helical inducers. The results clearly indicate that blade cavitation drastically modifies the rotordynamic forces exerted on the inducer by the surrounding fluid. The dynamic response of the cavitating flow to the periodic excitation imposed by the whirl motion generates multiple subsynchronous and supersynchronous flow resonances in the blade channels, interfering with the more regular spectral behavior of the rotordynamic fluid forces, typical of noncavitating operation. The extent of cavitation has a major impact in locating the critical speeds and determining the intensity of flow-induced rotordynamic forces. At higher levels of cavitation the amplitudes of the flow resonances decrease, and their frequencies approach the rotational speed of the inducer (synchronous conditions).

On the other hand, the present theory predicts that blade cavitation does not appreciably modify the rotordynamic fluid forces on whirling and centrifugal impellers, in accordance with the experimental evidence and in striking contrast with the observed behavior of axial inducers. Comparison with the results of the analysis of cavitating inducers confirms that the contribution of cavitation to the rotordynamic whirl forces is only significant when the standing pressure waves excited in the blade channels by the impeller motion are capable of exerting synchronous and coherent actions on the rotor. For this to happen:

- the blade channels must be long enough in the azimuthal direction for the pressure wave to become at least partly coherent with the channel rotation around the axis: only in this case the resulting forces do not average out and generate appreciable fluid reactions on the rotor;
- possibly the cavitating flow in the blade channels must become resonant, in order to maximize the amplitude of the pressure fluctuations.

The first condition can never be satisfied in radial impellers due to the limited azimuthal extension of the blade channels. This geometric limitation is the essential reason for the different behavior rotordynamic whirl forces in cavitating radial and axial impellers. Besides, according to the present theory, no resonance phenomena can occur in whirling and cavitating centrifugal impellers because the natural frequencies of the flow are essentially imaginary.

In spite of its approximate nature, the present theory correctly captures the main observed features and different behavior of the rotordynamic forces in axial inducers and radial impellers. There is therefore reason to believe that it contributes some useful fundamental insight into the complex physical phenomena responsible for the onset and sustain of free-whirl instabilities in cavitating turbopumps.

8 Acknowledgements

The present work has been partially supported by the Agenzia Spaziale Italiana under a 1999 grant for fundamental research. The author would like to acknowledge the help of Dr. Marco Venturini-Autieri, and express his gratitude to Prof. Mariano Andrenucci, Director of Alta S.p.A., Ospedaletto (Pisa), Italy, and to Prof. Renzo Lazzeretti of the Dipartimento di Ingegneria Aerospaziale, Università degli Studi di Pisa, Pisa, Italy, for their constant and friendly encouragement.

9 Nomenclature

a	sound speed
A	cross-sectional flow area
b	axial length of the radial impeller
B	boundary equation
c	specific heats, constant
$\boldsymbol{e}$	unit vector
f	nondimensional force
F	force
i	imaginary unit
I	modified Bessel function of the first kind, integral
j	blade index
J	Bessel function of the first kind
k	hub excitation mode index
K	modified Bessel function of the second kind
K_C, K_E	cavitating layer parameters
L	axial length of the inducer
L^*	blade axial length
l	blade excitation mode index
m	streamwise mode index
n	blade-to-blade helical/spiral coordinate
N	blade-to-blade mode function
N_B	number of blades
N_R	number of blade revolutions

O, O^*	origin of coordinate systems
p	pressure
p_t	total or stagnation pressure
P	blade axial pitch
P^*	blade-to-blade distance
$\dot{Q}$	volume flow rate
r	radial coordinate, radius
$\boldsymbol{r}$	radial vector
R	radial mode function, cavity radius
s	streamwise helical/spiral coordinate
S	surface, streamwise mode function
t	time
T	temperature, aspect ratio
u	radial velocity component
$\boldsymbol{u}$	velocity vector
v	azimuthal velocity component
w	axial velocity component
x	abscissa
y	ordinate
Y	Bessel function of the second kind
z	axial coordinate
α	void fraction
β	blade angle
δ	cavitation layer thickness
δ_T	thermal boundary layer thickness
ε	whirl eccentricity
ζ	nondimensional damping coefficient
ϑ	azimuthal angle
λ	radial eigenvalue
μ	streamwise eigenvalue
ν	blade-to-blade eigenvalue
ρ	density
σ	cavitation number
φ	velocity potential
ϕ	flow coefficient
ω	whirl angular velocity
Ω	inducer rotational speed

Subscripts and Superscripts

B	blade
C	cavitation
F	mean flow
H	hub
L	liquid

M	mean
R	radial
T	tangential, blade tip
V	vapor
$\bar{q}$	unperturbed value of q
$\tilde{q}$	perturbation value of q
$\hat{q}$	complex representation of $\tilde{q}$
q'	derivative, value of q in the rotating frame
q^*	value of q in the inducer-fixed frame
1	inducer inlet
2	inducer outlet

Bibliography

Adkins, D.R. and Brennen, C.E., 1988, "Analysis of Hydrodynamic Radial Forces on Centrifugal Pump Impeller", *ASME Journal of Fluids Engineering*, Vol. 110, pp. 20-28.

Alford J.S., 1958, "Protecting Turbomachinery from Self-Excited Rotor Whirl", *ASME J. Eng. Power*, Vol. 87, pp. 333-334.

Arndt, N., and Franz, R., 1986, "Observation of Hydrodynamic Forces on Several Inducers Including the SSME LPOTP", *Report No. E249.3*, Division of Engineering and Applied Sciences, California Institute of Technology, Pasadena, California.

Arndt N. et al., 1989, "Rotor-Stator Interaction in a Diffuser Pump", *ASME J. Turbomachinery*, Vol. 111, pp. 213-221.

Arndt N. et al., 1990, "Experimental Investigation of Rotor-Stator Interaction in a Centrifugal Pump with Several Vaned Diffusers", *ASME J. Turbomachinery*, Vol. 112, pp. 98-108.

Baskharone, E.A., 1999, "Swirl Brake Effect on the Rotordynamic Stability of a Shrouded Impeller", *ASME Journal of Turbomachinery*, Vol. 121, pp. 127-133.

Bhattacharyya, A., 1994, "Internal Flows and Force Matrices in Axial Flow Inducers", *Ph.D. Thesis*, Division of Engineering and Applied Science, California Institute of Technology, Pasadena California.

Bhattacharyya, A., Acosta, A.J., Brennen, C.E., and Caughey, T.K., 1997, "Rotordynamic Forces in Cavitating Inducers", *ASME Journal of Fluids Engineering*, Vol. 199, No. 4, pp. 768-774, ISSN 0098-2202.

Brennen, C. E., 1994, "Hydrodynamics of Pumps", *Concepts ETI, Inc.,* P.O. Box 643, Norwich, Vermont 05055, USA, ISBN 0-933283-07-5, and *Oxford University Press*, Walton St., Oxford OX2 6DP, England, ISBN 0-19-856442-2.

Brennen, C., E., 1995, "Cavitation and Bubble Dynamics", *Oxford University Press.*

Campos L.M.B.C. and Gil P.J.S., 1995, "On Spiral Coordinates with Application to Wave Propagation", *J. Fluid Mechanics*, Vol. 301, pp. 153-173.

Chamieh D.S. et al., 1985, "Experimental Measurements of Hydrodynamic Radial Forces and Stiffness Matrices for a Centrifugal Pump Impeller", *ASME Journal of Fluids Engineering*, Vol. 107, pp. 307-315.

d'Agostino L. and d'Auria F., 1997, "Three-Dimensional Analysis of Rotordynamic Forces on Whirling and Cavitating Inducers", *ASME Paper FEDSM97-3335, 1997 ASME Fluids Engineering Division Summer Meeting*, Vancouver, British Columbia, Canada, June 22-26.

d'Agostino L., d'Auria F. and Brennen C.E., 1998, "A Three-Dimensional Analysis of Rotordynamic Forces on Whirling and Cavitating Helical Inducers", *ASME Journal of Fluids Engineering*, Vol. 120, pp. 698-704.

d'Agostino L. and Venturini-Autieri M.R., 2002, "Three-Dimensional Analysis of Rotordynamic Fluid Forces on Whirling and Cavitating Finite-Length Inducers", *9th Int. Symp. on Transport Phenomena and Dynamics of Rotating Machinery (ISROMAC-9),* Honolulu, HI, USA, February 10-14.

d'Agostino L. and Venturini-Autieri M.R., 2003, "Rotordynamic Fluid Forces on Whirling and Cavitating Radial Impellers", *CAV 2003, 5th International Symposium on Cavitation*, November 1-4, Osaka, Japan.

d'Auria, F., d'Agostino, L., Brennen, C.E., 1995, "Bubble Dynamics Effects on the Rotordynamics Forces in Cavitating Inducers", *1995 ASME Cavitation and Multiphase Flow Forum*, Hilton Head Island, South Carolina, USA, FED Vol. 201, pp. 47-54.

Franz R. et al., 1989, "The Rotordynamic Forces on a Centrifugal Pump Impeller in the Presence of Cavitation", *ASME-F.E.D. 81*, 205-212.

Hiwata A. and Tsujimoto Y., 2002, "Theoretical Analysis of Fluid Forces on and Open-Type Centrifugal Impeller in Whirling Motion", *ASME J. Fluids Eng.*, Vol. 124, pp. 342-347.

Hori Y., 1959, "The Theory of Oil Whip", *ASME J. Appl. Mech.*, Vol. 26, pp. 189-198.

Jery, B., Acosta A. J., and Caughey, T. K.,, 1985, "Forces on Centrifugal Pump Impellers", *Proc. 2nd Int. Pump Symp.*, Houston, Texas, USA, April 29-May 2.

Jery, B. 1987, "Experimental Study of Unsteady Hydrodynamic Force Matrices on Whirling Centrifugal Pump Impellers", *Report No. E200.22*, Division of Engineering and Applied Sciences, California Institute of Technology, Pasadena, California, USA.

Lebedev, N.N., 1965, "Special Functions and Their Applications", *Prentice Hall.*

Martinez-Sanchez M., Jaroux B., Song S.J. and Yoo S., 1995, "Measurement of Turbine Blade-Tip Rotordynamic Excitation Forces", *ASME J. Turbomachinery*, Vol. 117, pp. 384-393.

Martinez-Sanchez M. and Song S.J., 1997a, "Rotordynamic Forces Due to Turbine Tip Leakage-Part I: Blade Scale Effects", *ASME J. Turbomachinery*, Vol. 119, pp. 695-703.

Martinez-Sanchez M. and Song S.J., 1997b, "Rotordynamic Forces Due to Turbine Tip Leakage-Part II: Radius Scale Effects and Experimental Verification", *ASME J. Turbomachinery*, Vol.119, pp. 704-717.

Newkirk B.L. and Taylor H.D., 1925, "Shaft Whipping due to Oil Action in Journal Bearing", *General Electric Review*, August 1925, pp. 559-568.

Ohasi H. and Shoji H., 1987, "Lateral Fluid Forces on Whirling Centrifugal Impeller (2nd Report: Experiment in Vaneless Diffuser)", *ASME Journal of Fluids Engineering*, Vol. 109, pp. 100-106.

Rapposelli, E., and d'Agostino, L., 2001, "A Modified Isenthalpic Model of Cavitation in Plane Journal Bearings", *CAV2001, International Symposim on Cavitation*, Pasadena, California USA, June 20-23.

Rosenmann, W., 1965, "Experimental Investigations of Hydrodynamically Induced Shaft Forces with a Three Bladed Inducer", *Proc. ASME Symp. on Cavitation in Fluid Machinery*, pp. 172-195.

Shoji, H. and Ohashi, H., 1987, "Lateral Fluid Forces on Whirling Centrifugal Impeller (1st Report: Theory)", *ASME Journal of Fluids Engineering*, Vol. 109, pp. 94-99.

Thomas J.H., 1958, "Instabile Eigenschwingungen von Turbinenlaufern Angefacht durch die Spaltstroemung in Stopfubuchsen und Bechauchflug (Unstable Natural Vibrations of Turbine Rotors Induced by Clearance Flows in Glands and Bladings)", *Bull. De. L.A.I.M.,* Vol. 71, pp. 1039-1063.

Tsujimoto, Y, Yoshida, Y., Ohashi, H., Teramoto, N. and Ishizaki, S., 1997, "Fluid Force Moment on a Centrifugal Impeller Shroud in Precessing Motion", *ASME Journal of Fluids Engineering*, Vol. 119, pp. 366-371.

Uy, R.V., and Brennen, C.E., 1999, "Experimental Measurements of Rotordynamic Forces Caused by Front Shroud Pump Leakage", *ASME Journal of Fluids Engineering*, Vol. 121, pp. 633-637.

Venturini-Autieri M.R., 2003, "Analisi delle Forze Rotodinamiche in Turbomacchine Cavitanti", *Tesi di Laurea, Dipartimento di Ingegneria Aerospaziale*, Università di Pisa, Pisa, Italy.

Visser F.C., 1999, "On the Asymptotic Solution of the Poisson Equation Describing the Two-Dimensional Incompressible Inviscid Flow in a Rotating Centrifugal Impeller", *Zeitschrift für Angewandte Mathematik und Mechanik*, Vol. 79 (5), pp. 353-356.

A hyperbolic non equilibrium model for cavitating flows

Richard Saurel*‡ and Fabien Petitpas‡

‡ Polytech'Marseille, Aix-Marseille University and SMASH Project UMR CNRS 6595 - IUSTI - INRIA, 5 rue E. Fermi, 13453 Marseille Cedex 13, France.

* Also, University Institute of France, same address

Abstract A hyperbolic two-phase flow model involving five partial differential equations is built for liquid-gas interface modelling. The model is able to deal with interfaces of simple contact where normal velocity and pressure are continuous as well as transition fronts where heat and mass transfer occur, involving pressure and velocity jumps. These fronts correspond to extra waves into the system. The model involves two temperatures and entropies but a single pressure and a single velocity. The closure is achieved by two equations of state that reproduce the phase diagram when equilibrium is reached. Relaxation toward equilibrium is achieved by temperature and chemical potential relaxation terms whose kinetics is considered infinitely fast only at specific locations, typically at evaporation fronts. Doing so, metastable states are involved for locations far from these fronts. Specific numerical hyperbolic and relaxation solver are built to solve the non-conservative system. Computational tests are done in 1D and 2D and are compared to experimental observations.

1 Introduction

When a liquid initially in thermodynamic equilibrium undergoes strong rarefaction waves, it may reach a metastable state where the temperature is higher than the saturated one at the final pressure of the expanded state. Then the superheated liquid releases its metastable energy (stored as internal energy) very quickly, even explosively, producing either pure vapor or liquid-vapor mixture at thermodynamic equilibrium, flowing at high velocity. This phenomenon is often called cavitation. Such situations appear frequently in nature and in many industrial applications, as liquid flows around hypervelocity projectiles, airfoils, or inside nozzles such as fuel injector systems. Cavitation in these systems always produces strong disturbances. In most applications, cavitation appears as a multidimensional process due to geometrical effects. This multidimensional character complicates both experiments and theoretical approaches. By using one-dimensional expansion tubes many authors (Thompson et al. (1987); Frost et al. (1991); Kurschat et al. (1992); Simões-Moreira and Shepherd (1999); Reinke and Yadigaroglu (2001)) succeeded in isolating the main phenomenon we propose to summarize hereafter. These experiments consisted in connecting a vertical tube filled with a liquid in thermodynamic equilibrium at atmospheric pressure (or higher) to a very low-pressure chamber (Figure 1).

As soon as the membrane between the liquid and the vacuum chamber is ruptured, rarefaction waves propagate through the liquid producing a superheated liquid (Figure 2).

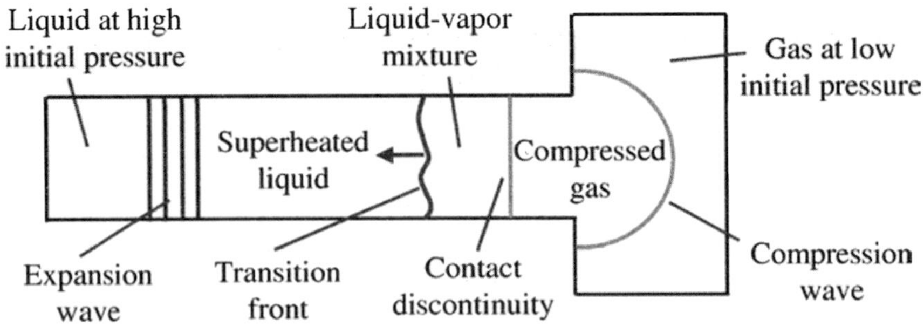

Figure 1. Simplified experimental setup of the expansion tube in Simões-Moreira and Shepherd (1999) and associated waves.

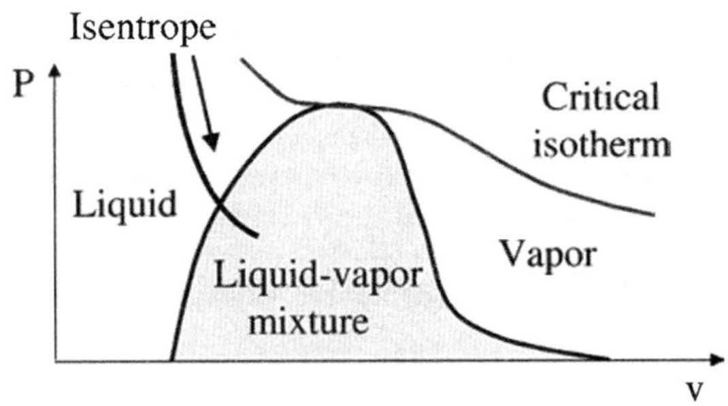

Figure 2. Liquid thermodynamic path associated to the expansion wave producing a superheated liquid

Then a subsonic phase-transition front propagates through the superheated liquid producing a high velocity liquid-vapor mixture in thermodynamic equilibrium towards the low-pressure chamber. The front velocity is approximately 1 m/s while the ejected mixture velocity is of the order of 100 m/s.

These experimental observations indicate that both liquid and vapor compressibilities must be considered. They also report the presence of an acoustic wave (expansion wave) preceding the evaporation front, itself preceding a contact discontinuity and a compression wave as represented in the Figure 1. The corresponding waves pattern is shown in the (x, t) diagram of Figure 3.

Expansion effects in industrial systems are often due to geometrical effects, as for example in nozzles where cavitation appears at locations where the pressure is very low and the liquid superheated (Figure 4).

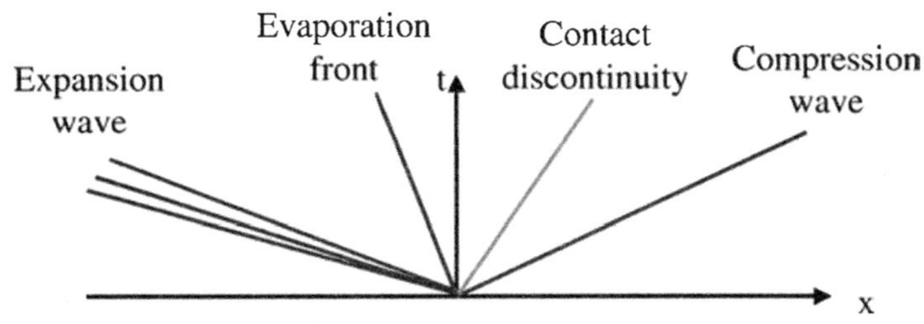

Figure 3. A typical waves pattern of cavitating systems

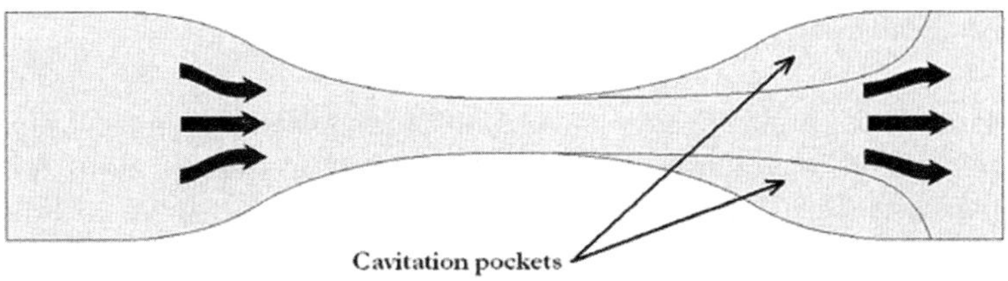

Figure 4. Cavitating flow in a nozzle

The aim of the present paper is to develop a model able to deal with metastable states and evaporation front dynamics, as well as the other waves shown in Figure 3. The model must also be able to deal with interfaces separating a liquid and non condensable gases. Indeed, mass transfer occurs only under specific thermodynamic conditions that do not necessarily correspond to the local thermodynamic state at the interfaces. Thus the model must be able to deal with interfaces of simple contact as well as evaporating interfaces.

Before entering in the details of the novel approach let us examine existing models to deal with evaporation fronts, interfaces and cavitation pockets.

2 Review of existing models

Essentially two classes of models are available in the literature. The first class corresponds to pressure and temperature equilibrium models, the second class being related to temperature non-equilibrium models.

Pressure and temperature equilibrium models

Four models at least belong to this category, with increasing level of complexity.

- The mixture Euler equations with a cubic equation of state (EOS). A prototype of such EOS is the Van der Waals one. With such models a loss of hyperbolic occurs in the two-phase region. In other words, the square sound speed becomes negative and waves propagation has no physical sense.

- The mixture Euler equations can also be used with a tabular EOS or a combination of pure phase EOSs with assumption of pressure, temperature and chemical potentials equilibrium in the two-phase region (see for example Saurel et al. (1999) and Liou and Edwards (1999)). Such model removes the preceding loss of hyperbolicity. However, the model now involves discontinuous sound speed when the state ranges from single phase to two-phase. This discontinuous behavior is not observed experimentally. Also, the mixture evolving under thermodynamic equilibrium, metastable states are omitted. Last, the model is unable to deal with material interfaces separating a liquid and non-condensable gases. It thus has a restricted area of application.

- The mixture Euler equations are sometimes augmented by a mass fraction equation with a relaxation term (Faucher et al. (2000)), as frequently done for mixtures of reacting gases. This four equations model is unable to solve interfaces between a liquid and a non-condensable gas (water/air for example). This is due to its isothermal closure not compatible with interface conditions (equal pressures and normal velocities but not temperatures). Moreover, determination of the chemical relaxation term is an issue.

- Models derived from the second-gradient theory, described in Cahn and Hilliard (1958) and Jamet et al. (2001). In this approach, the interface is described as a diffuse zone and hyperbolicity is preserved. Contrarily to preceding models, a length scale appears and this capillary zone has to be resolved. This result in severe restrictions: the capillary zone length is typically of the order of one nano-meter. To do practical simulations, at the scale on an individual bubble for example, the interface has to be thickened. Such procedure has important consequences on the model ability to deal with metastable states. Also, interfaces with large density gradients are very difficult to solve, as well as interfaces separating immiscible fluids (water/air for example).

Temperature non-equilibrium models

- The most conventional temperature non-equilibrium model is the 6 equations model obtained for example with averaging methods (Ishii (1975) or Delhaye and Boure (1982)). It involves a balance equation of mass, momentum and energy per fluid. Contrarily to preceding models, the mixture evolves with two velocities. It is present in the literature with two options:
  - i) One of the phases is considered incompressible, the mixture pressure being those of the gas phase. Such option is not compatible with the presence of rarefaction waves in the liquid phase, as mentioned in the introduction.
  - ii) Both phases are compressible and the closure is achieved with the pressure equilibrium assumption (Butler et al. (1982)).

The 6 equation model, with both options has a restricted domain of hyperbolicity. This results in a restricted validity to problems where transient propagation is not essential. Moreover, it is unable to solve interface problems (water/air for example).

- The 7 equations model (Baer and Nunziato (1986)) and its variants is unconditionally hyperbolic and is able to deal with a wide range of applications. It is composed of the same 6 equations as previously, augmented by an evolution equation for the volume fraction of one of the constituents. Its ability to solve interface problems as well as fluid mixtures with several velocities was demonstrated in Saurel and Abgrall (1999), Saurel and Métayer (2001) and Saurel et al. (2003). This model has been extended to the propagation of evaporation fronts in cavitating systems Métayer et al. (2005). The four waves present in the Figure 3 are correctly involved and the model is able to deal with metastable states. It has been used for practical computations and validated against experiments of supercavitation. However it is quite complex to code and does not provide any information about the evaporation front's structure, as the front is treated as a discontinuity.

- The last model involves a single pressure and velocity. It is composed of two mass balance equations, one mixture momentum equation, one mixture energy equation and a volume fraction equation. This 5 equations model (Kapila et al. (2001), Perigaud and Saurel (2005)) is unconditionally hyperbolic. Its ability to solve interface problems with compressible fluids was demonstrated in Perigaud and Saurel (2005). Also, it was extended in the same reference to capillary and viscous effects. Contrarily to Cahn and Hilliard (1958) and Jamet et al. (2001) model, the interface has no length scale. This has important consequences on numerical resolution as well as its ability to deal with large density ratios. Another important difference is also present regarding thermodynamic closure. Two pure phase EOSs are used instead of a mixture EOS, like Van der Waals one. The model remains out of thermal equilibrium and the presence of temperature and chemical potentials differences will be used hereafter to derive a relaxation model able to deal with evaporation fronts.

This last model is the starting point of the present analysis. Its derivation in the presence of heat and mass transfer is detailed hereafter. Its thermodynamic closure, based on two equations of state is detailed. The building of kinetic terms that make the system relax toward equilibrium is done on the basis of the entropy inequality. With this kinetic closure, and special treatment of stiff relaxation, the model is shown to be able to compute evaporation fronts by solving their internal structure. Metastable states are involved, as well as shock, rarefaction and interface dynamics. Dynamic appearance of cavitation pockets is studied in the following limit situations:

- simple mechanical pocket growth,
- mechanical pocket growth with heat and mass exchanges.

3 Building of the flow model in absence of mass transfer

The present approach is based on a five equations model with single pressure and single velocity but two temperatures and entropies. Mass transfer effects will be introduced in this model by temperature and chemical potential relaxation effects. Before doing such extension, the flow model in absence of mass transfer is derived. It is obtained in the asymptotic limit of a non equilibrium two-phase, two-pressure and two-velocity hyperbolic multiphase flow model. The following asymptotic reduction follows the lines of Kapila et al. (2001).

3.1 The 'parent' model

The starting point of the present analysis is the model of Saurel et al. (2003), Chinnayya et al. (2004) where a symmetric formulation of the Baer and Nunziato (1986) model is proposed:

$$\left\{\begin{array}{l}
\dfrac{\partial \alpha_1}{\partial t} + \vec{u}_I \cdot \vec{\nabla}\alpha_1 = \mu(p_1 - p_2), \\[2ex]
\dfrac{\partial \alpha_1\rho_1}{\partial t} + div(\alpha_1\rho_1\vec{u}_1) = 0, \\[2ex]
\dfrac{\partial \alpha_1\rho_1\vec{u}_1}{\partial t} + div(\alpha_1\rho_k\vec{u}_k \otimes \vec{u}_k) + \vec{\nabla}(\alpha_1 p_1) = p_I\vec{\nabla}\alpha_1 + \lambda(\vec{u}_2 - \vec{u}_1) \\[2ex]
\dfrac{\partial \alpha_1\rho_1 E_1}{\partial t} + div(\alpha_1(\rho_1 E_1 + p_1)\vec{u}_1) \\
\qquad\qquad = p_I\vec{u}_I \cdot \vec{\nabla}\alpha_1 + \lambda\vec{u}_I \cdot (\vec{u}_2 - \vec{u}_1) - p_I\mu(p_1 - p_2) + Q_1 \\[2ex]
\dfrac{\partial \alpha_2\rho_2}{\partial t} + div(\alpha_2\rho_2\vec{u}_2) = 0, \\[2ex]
\dfrac{\partial \alpha_2\rho_2\vec{u}_2}{\partial t} + div(\alpha_2\rho_k\vec{u}_k \otimes \vec{u}_k) + \vec{\nabla}(\alpha_2 p_2) = p_I\vec{\nabla}\alpha_2 - \lambda(\vec{u}_2 - \vec{u}_1) \\[2ex]
\dfrac{\partial \alpha_2\rho_2 E_2}{\partial t} + div(\alpha_2(\rho_2 E_2 + p_2)\vec{u}_2) \\
\qquad\qquad = p_I\vec{u}_I \cdot \vec{\nabla}\alpha_2 - \lambda\vec{u}_I \cdot (\vec{u}_2 - \vec{u}_1) + p_I\mu(p_1 - p_2) - Q_1
\end{array}\right. \tag{3.1}$$

We denote respectively by $\alpha_k, \rho_k, \vec{u}_k, p_k, E_k$ and e_k the volume fraction, the density, the velocity vector, the pressure, the total energy and the internal energy of the phase k. The total energy is defined by:

$$E_k = e_k + \frac{1}{2}\vec{u}_k \cdot \vec{u}_k$$

The heat transfer term Q_1 is simply modeled by $Q_1 = H(T_2 - T_1)$ where H denotes a temperature relaxation parameter, involving surface exchange and heat transfer coefficients.

This system guarantees conservation for the mixture and is frame invariant. Interaction terms appear in the right hand side and express the effects which drive the system back to mechanical equilibrium by the way of the relaxation coefficients:

$$\mu = \frac{S_I/V}{Z_1 + Z_2} \text{ and } \lambda = Z_1 Z_2 \mu$$

where Z represents the acoustic impedance ($Z = \rho c$) and S_I/V the specific exchange surface.

$$\begin{aligned} p_I &= \tfrac{Z_1 p_2 + Z_2 p_1}{Z_1 + Z_2} + sign(\frac{\partial \alpha_1}{\partial x}) \frac{(u_2 - u_1) Z_1 Z_2}{Z_1 + Z_2}, \\ u_I &= \tfrac{Z_1 u_1 + Z_2 u_2}{Z_1 + Z_2} + sign(\frac{\partial \alpha_1}{\partial x}) \frac{p_2 - p_1}{Z_1 + Z_2}, \end{aligned} \tag{3.2}$$

This model involves average interface variables (3.2) that allow a symmetric closure of the equations and are in agreement with the entropy inequality. Another important feature is present in these formulas. When the volume fraction is discontinuous, as for example at material interfaces, interface conditions (equality of normal velocities and pressures) are automatically fulfilled (Abgrall and Saurel (2003)). This system is unconditionally hyperbolic with the characteristic waves speeds: $u_k, u_k + c_k, u_k - c_k$ for each phase k and the interface velocity u_I.
For the present application, this system involves unnecessary effects (two velocities and two pressures) and a reduced model is preferred. In the reduced model, the introduction of mass transfer effects will be done rigorously while such an extension is not obvious with system (3.1). Also, it is always important to determine the simplest model involving the pertinent physics. Reduction of system (3.1) is done in the following subsection.

3.2 Reduced model

The two-phase flow model (3.1) can be written under primitive variables formulation:

$$\frac{\partial U}{\partial t} = F(U) + \Phi(U)$$

With:

$$U = \begin{pmatrix} \alpha_1 \\ \alpha_1 \rho_1 \\ \vec{u}_1 \\ p_1 \\ \alpha_2 \rho_2 \\ \vec{u}_2 \\ p_2 \end{pmatrix}, F(U) = \begin{pmatrix} -\vec{u}_1 \cdot \vec{\nabla} \alpha_1 \\ -div(\alpha_1 \rho_1 \vec{u}_1) \\ -\vec{u}_1 \cdot \vec{\nabla} \vec{u}_1 - \frac{1}{\rho_1} \vec{\nabla} p_1 + \frac{p_I - p_1}{\alpha_1 \rho_1} \vec{\nabla} \alpha_1 \\ -\rho_1 c_1^2 div(\vec{u}_1) + \frac{\Gamma_1}{\alpha_1} \left[p_I - \rho_1^2 \left(\frac{\partial e_1}{\partial \rho_1} \right)_{p_1} \right] (\vec{u}_I - \vec{u}_1) \cdot \vec{\nabla} \alpha_1 \\ -div(\alpha_2 \rho_2 \vec{u}_2) \\ -\vec{u}_2 \cdot \vec{\nabla} \vec{u}_2 - \frac{1}{\rho_2} \vec{\nabla} p_2 + \frac{p_I - p_2}{\alpha_2 \rho_2} \vec{\nabla} \alpha_2 \\ -\rho_2 c_2^2 div(\vec{u}_2) + \frac{\Gamma_2}{\alpha_2} \left[p_I - \rho_2^2 \left(\frac{\partial e_2}{\partial \rho_2} \right)_{p_2} \right] (\vec{u}_I - \vec{u}_2) \cdot \vec{\nabla} \alpha_2 \end{pmatrix}$$

And

$$\Phi(U) = \begin{pmatrix} \mu(p_1 - p_2) \\ 0 \\ \frac{\lambda}{\alpha_1\rho_1}(\vec{u}_2 - \vec{u}_1) \\ -\mu\frac{\Gamma_1}{\alpha_1}\left[p_I - \rho_1^2\left(\frac{\partial e_1}{\partial \rho_1}\right)_{p_1}\right](p_1 - p_2) + \lambda\frac{\Gamma_1}{\alpha_1}(\vec{u}_I - \vec{u}_1)(\vec{u}_2 - \vec{u}_1) + \frac{\Gamma_1}{\alpha_1}Q_1 \\ 0 \\ -\frac{\lambda}{\alpha_2\rho_2}(\vec{u}_2 - \vec{u}_1) \\ \mu\frac{\Gamma_2}{\alpha_2}\left[p_I - \rho_2^2\left(\frac{\partial e_2}{\partial \rho_2}\right)_{p_2}\right](p_1 - p_2) - \lambda\frac{\Gamma_2}{\alpha_2}(\vec{u}_I - \vec{u}_2)(\vec{u}_2 - \vec{u}_1) - \frac{\Gamma_2}{\alpha_2}Q_1 \end{pmatrix}$$

Where c_k represents the speed of sound of phase k:

$$\forall k = 1,2 \qquad c_k = \frac{\frac{p_k}{\rho_k^2} - \left(\frac{\partial e_k}{\partial \rho_k}\right)_{p_k}}{\left(\frac{\partial e_k}{\partial p_k}\right)_{\rho_k}},$$

and Γ_k represents the Gruneisen coefficients of phase k :

$$\forall k = 1,2 \qquad \Gamma_k = \frac{1}{\rho_k\left(\frac{\partial e_k}{\partial p_k}\right)_{\rho_k}}$$

The reduced system is obtained in the limit of stiff mechanical relaxation:

$$\mu = \frac{1}{\epsilon}\ ,\ \lambda = \frac{Z_1 Z_2}{\epsilon} \quad \text{where} \quad \epsilon \to 0^+$$

Note that heat transfer effects are not considered in the same limit, as we are seeking a temperature non equilibrium model.
With these notations, the preceding system becomes:

$$\frac{\partial U}{\partial t} = F(U) + \frac{1}{\epsilon}\Psi(U) + Q(U) \tag{3.3}$$

With:

$$\Psi(U) = \begin{pmatrix} (p_1 - p_2) \\ 0 \\ \frac{Z_1 Z_2}{\alpha_1 \rho_1}(\vec{u}_2 - \vec{u}_1) \\ -\frac{\Gamma_1}{\alpha_1} K_1 (p_1 - p_2) + \frac{Z_1 Z_2 \Gamma_1}{\alpha_1}(\vec{u}_I - \vec{u}_1)(\vec{u}_2 - \vec{u}_1) \\ 0 \\ -\frac{Z_1 Z_2}{\alpha_2 \rho_2}(\vec{u}_2 - \vec{u}_1) \\ \frac{\Gamma_2}{\alpha_2} K_2 (p_1 - p_2) - \frac{Z_1 Z_2 \Gamma_2}{\alpha_2}(\vec{u}_I - \vec{u}_2)(\vec{u}_2 - \vec{u}_1) \end{pmatrix}, Q(U) = \begin{pmatrix} 0 \\ 0 \\ 0 \\ \frac{\Gamma_1}{\alpha_1} Q_1 \\ 0 \\ 0 \\ -\frac{\Gamma_2}{\alpha_2} Q_1 \end{pmatrix}$$

F and Ψ are regular enough functions close to the mechanical equilibrium state, U_0, to allow a Taylor expansion. Assuming that, close to this equilibrium, state U can be expressed as:

$$U^\epsilon = U_0 + \epsilon U_1 + O(\epsilon^2)$$

System (3.3) becomes:

$$\frac{\partial U_0}{\partial t} + \epsilon \frac{\partial U_1}{\partial t} = F(U_0) + \epsilon \frac{\partial F(U)}{\partial U}(U_0)U_1 + \frac{1}{\epsilon}\Psi(U_0) + \frac{\partial \Psi(U)}{\partial U}(U_0)U_1$$
$$+Q(U_0) + \epsilon \frac{\partial Q(U)}{\partial U}(U_0)U_1$$

When ϵ tend to 0^+, we have:

$$\forall \epsilon > 0, \Psi(U_0) = 0 \tag{3.4}$$

And, the system under reduced form becomes:

$$\frac{\partial U_0}{\partial t} = F(U_0) + \frac{\partial \Psi(U)}{\partial U}(U_0)U_1 + Q(U_0) \tag{3.5}$$

Condition (3.4) implies that in the asymptotic state, the system has to verify:

$$\vec{u}_1^0 = \vec{u}_2^0 = \vec{u}^0$$

and

$$p_1^0 = p_2^0 = p^0$$

These constraints used in (3.5) gives the reduced model including heat transfer effects. After some calculations, a convenient form is obtained:

$$\begin{cases} \dfrac{\partial \alpha_1}{\partial t} + \vec{u} \cdot \vec{\nabla} \alpha_1 = K div(\vec{u}) + \dfrac{\alpha_1 \alpha_2}{\alpha_2 \rho_1 c_1^2 + \alpha_1 \rho_2 c_2^2} \left(\dfrac{\Gamma_1}{\alpha_1} + \dfrac{\Gamma_2}{\alpha_2} \right) Q_1 \\ \dfrac{\partial \alpha_1 \rho_1}{\partial t} + div(\alpha_1 \rho_1 \vec{u}) = 0 \\ \dfrac{\partial \alpha_2 \rho_2}{\partial t} + div(\alpha_2 \rho_2 \vec{u}) = 0 \\ \dfrac{\partial \rho \vec{u}}{\partial t} + div(\rho \vec{u} \otimes \vec{u}) + \vec{\nabla} p = 0 \\ \dfrac{\partial \rho E}{\partial t} + div\left(\vec{u}(\rho E + p)\right) = 0 \end{cases} \tag{3.6}$$

where $K = \alpha_1\alpha_2(\rho_2 c_2^2 - \rho_1 c_1^2)/(\alpha_2\rho_1 c_1^2 + \alpha_1\rho_2 c_2^2)$ and the subscript "0" has been omitted for the sake of clarity.
The mixture variables are defined by: $\rho = \alpha_1\rho_1 + \alpha_2\rho_2$ and $E = Y_1 E_1 + Y_2 E_2$, where $Y_k = \frac{\alpha_k \rho_k}{\rho}$ denotes the mass fraction of the phase k.

4 Thermodynamic closure

In order to circumvent the difficulty of models having negative square speed of sound in the two-phase region, the present model uses two equations of state (EOS). Each fluid possesses its own EOS of "stiffened gas" (SG) type. This EOS is the simplest prototype that contains the main physical properties of pure fluids: attractive and repulsive molecular effects.
Doing so, each fluid has its own thermodynamics and in particular its own entropy. In the present modeling of mass transfer, detailed in the next section, relaxation towards equilibrium is achieved by a kinetic process, contrarily to Van der Waals modeling where mass transfer is a thermodynamic path. It is the reason why the present modeling preserves hyperbolicity during mass transfer.
However, when equilibrium is reached, conventional properties of the phase diagram have to be recovered (latent heat of vaporization, saturation temperature) that depend on pressure or temperature. In other words, the two pure fluids EOS must be connected by some constraints. These constraints are used for the determination of the various constants involved in these EOSs.

For each phase the thermodynamic state is determined by the SG EOS that reads:

$$\begin{cases} e(p,v) = \frac{p+\gamma p_\infty}{(\gamma-1)} v + q \\ v(p,T) = \frac{(\gamma-1)C_v T}{p+p_\infty} \\ h(T) = \gamma C_v T + q \\ g(p,T) = (\gamma C_v - q')T - C_v T ln \frac{T^\gamma}{(p+p_\infty)^(\gamma-1)} + q \end{cases} \tag{4.1}$$

where e, $v = 1/\rho$, p, T, h and g are respectively the internal energy, the specific volume, the pressure, the temperature, the enthalpy and the free enthalpy of the considered phase. The constants, characteristic of each fluid are: γ, p_∞, C_v, q and q'.
A method to determine these parameters in gas-liquid systems is given in Métayer et al. (2004). The coupling of gas and liquid parameters is particularly examined. This method is summarized hereafter.

4.1 Method to determine SG EOS parameters

Parameters of the SG EOS are determined from experimental curves for each fluid. In the case we are interesting in (liquid in presence of its vapor), we need the saturation curves. The experimental data needed are : $p = p_{sat}(T)$, $h_{l,exp}(T)$, $h_{g,exp}(T)$, $v_{l,exp}(T)$, $v_{g,exp}(T)$ and the latent heat of vaporization $L_{v,exp}(T) = h_{g,exp}(T) - h_{l,exp}(T)$.

1) From the expression of enthalpy in (4.1), we have:

$$\forall k = l, g \qquad \frac{dh_k}{dT} = \gamma_k C_{v,k} = C_{p,k}$$

Which permits to determine the average heat capacity coefficients $C_{p,k}$ by a linear approximation between two reference states 0 and 1 as:

$$\forall k = l, g \qquad C_{p,k} = \frac{h_{k,exp}(T_1) - h_{k,exp}(T_0)}{T_1 - T_0}$$

This stage also allows the calculation of the reference energies:

$$\forall k = l, g \qquad q_k = h_{k,exp}(T_0) - C_{p,k} T_0$$

2) Then, by using the experimental curve $p = p_{sat}(T)$, the specific volume is expressed as:

$$\forall k = l, g \qquad v_k(T) = \frac{(C_{p,k} - C_{v,k})T}{p_{sat}(T) + p_{\infty,k}} \tag{4.2}$$

Under logarithmic differenciation, we have:

$$\forall k = l, g \qquad dln\left(v_k(T)\right) = dln(T) - dln\left(p_{sat}(T) + p_{\infty,k}\right)$$

Integration of this equation between the two reference states 0 and 1 yields:

$$\forall k = l, g \qquad ln\left(v_k(T_1)\right) - ln\left(v_k(T_0)\right)$$
$$= ln(T_1) - ln(T_0) - ln\left(p_{sat}(T_1) + p_{\infty,k}\right) + ln\left(p_{sat}(T_0) + p_{\infty,k}\right)$$

These expression allows the calculation of the coefficients $p_{\infty,k}$:

$$\forall k = l, g \qquad p_{\infty,k} = \frac{v_k(T_0)T_1 p_{sat}(T_0) - v_k(T_1)T_0 p_{sat}(T_1)}{v_k(T_1)T_0 - v_k(T_0)T_1}$$

3) Then equation (4.2) applied to the reference state 0, allows the approximation of $C_{v,k}$:

$$\forall k = l, g \qquad C_{v,k} = C_{p,k} - \frac{v_k(T_0)}{T_0}\left(p_{sat}(T_0) + p_{\infty,k}\right)$$

4) The approximation for γ_k follows :

$$\forall k = l, g \qquad \gamma_k = \frac{C_{p,k}}{C_{v,k}}$$

5) At thermodynamic equilibrium the two chemical potentials have to be equal ($g_g = g_l$ with the definition of g given by (4.1)). It implies:

$$ln\left(p + p_{\infty,v}\right) = A + \frac{B}{T} + Cln(T) + Dln\left(p + p_{\infty,l}\right) \tag{4.3}$$

where A, B, C and D depends on the SG EOS parameters:

$$A = \frac{C_{p,l} - C_{p,g} + q'_g - q'_l}{C_{p,g} - C_{v,g}}, B = \frac{q_l - q_g}{C_{p,g} - C_{v,g}}, C = \frac{C_{p,g} - C_{p,l}}{C_{p,g} - C_{v,g}}, \text{ and } D = \frac{C_{p,l} - C_{v,l}}{C_{p,g} - C_{v,g}}$$

Relation (4.3) is non linear but permits the computation of the theoretical curve $p = p_{sat}(T)$. Such computation needs the two entropy constants q'. By convention we take $q'_l = 0J.kg^{-1}.K^{-1}$ and choose q'_g in order to have the best fit between theoretical and experimental curves.

This algorithm for the determination of SG EOS parameters is very accurate provided that experimental curves saturation curves are quasi-linear. It means that the two reference states have to be sufficiently close. Near the critical point, restrictions appear. But far from this point, wide ranges of temperatures and pressures can be covered as shown hereafter.

4.2 Results for water and dodecane

As an illustration, results concerning liquid and vapor water and liquid and vapor dodecane are presented.
For water, the chosen temperature range is $298 - 473K$. The corresponding experimental data (Oldenbourg (1989)) are :

$p_{sat}(T_0) = 3166Pa$, $h_{l,exp}(T_0) = 104.7\ 10^3 J/Kg$, $h_{g,exp}(T_0) = 2473.4\ 10^3 J/Kg$, $v_{g,exp}(T_0) = 42.4 m^3/Kg$,
$p_{sat}(T_1) = 15.5\ 10^5 Pa$, $h_{l,exp}(T_1) = 851.6\ 10^3 J/Kg$, $h_{g,exp}(T_1) = 2733.7\ 10^3 J/Kg$, $v_{g,exp}(T_1) = 0.124 m^3/Kg$.
The results of the preceding algorithm are summarized in Table 1 and the comparison between experimental and theoretical curves are shown in Figure 5.

	$p_{\infty,k}(Pa)$	$C_p(J/kg.K)$	$C_v(J/kg.K)$	γ	$q(J/kg)$	$q'(J/kg.K)$
liquid	10^9	4267	1816	2,35	-1167.10^3	0
vapor	0	1487	1040	1,43	2030.10^3	-23.10^3

Table 1. SG EOS parameters for liquid and vapor water

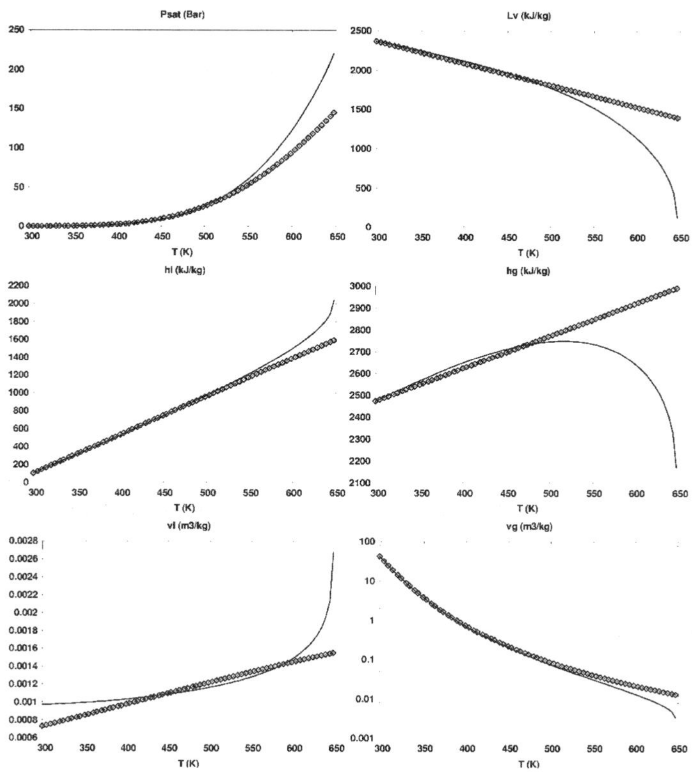

Figure 5. Saturation curves for watter: experimental curves (lines) and theoretical curves (symbols). The comparison has to be done in the range of temperature 298-473 K.

The same algorithm was used for dodecane, in the temperature range $298 - 473K$. The corresponding SG EOS parameters are summarized in Table 2 . The same accuracy when compared with experimental data was observed.

	$p_{\infty,k}(Pa)$	$C_p(J/kg.K)$	$C_v(J/kg.K)$	γ	$q(J/kg)$	$q'(J/kg.K)$
liquid	4.10^8	2534	1077	2,35	-755.10^3	0
vapor	0	2005	1956	1,025	-237.10^3	-24.10^3

Table 2. SG EOS parameters for liquid dodecane and its vapor

4.3 Mixture SG EOS

With the help of the phases EOS the mixture EOS is readily obtained. From the mixture internal energy definition:

$$\rho e = \alpha_1 \rho_1 e_1 + \alpha_2 \rho_2 e_2$$

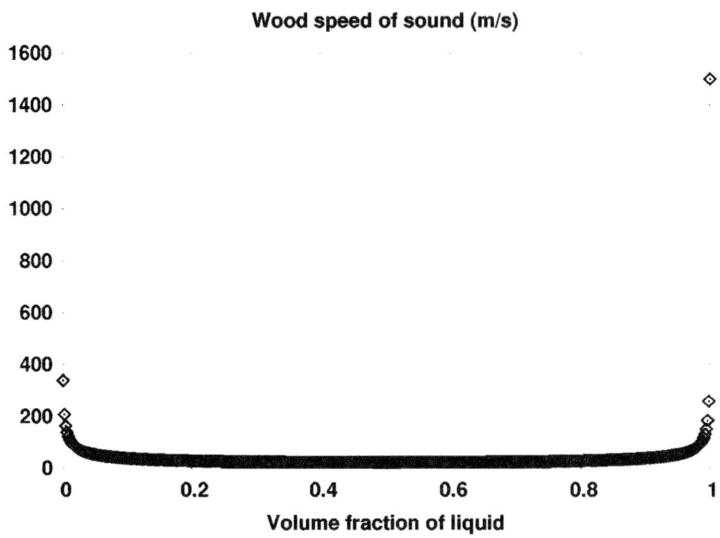

Figure 6. Wood's speed of sound versus liquid volume fraction for a liquid-vapor water mixture.

By using SG EOS (4.1), each product $\rho_k e_k$ can be written as:

$$\rho_k e_k = \frac{p_k + \gamma_k p_{\infty,k}}{\gamma_k - 1} + \rho_k q_k$$

Under pressure equilibrium we obtain the closure relation for system (3.6):

$$p(\rho, e, \alpha_1, \alpha_2, Y_1, Y_2) = \frac{\rho(e - Y_1 q_1 - Y_2 q_2) - \left(\frac{\alpha_1 \gamma_1 p_{\infty,1}}{\gamma_1 - 1} + \frac{\alpha_2 \gamma_2 p_{\infty,2}}{\gamma_2 - 1}\right)}{\frac{\alpha_1}{\gamma_1 - 1} + \frac{\alpha_2}{\gamma_2 - 1}} \tag{4.4}$$

With this mixture EOS, the flow model (3.6) admits the Wood's speed of sound:

$$\frac{1}{\rho c_w^2} = \frac{\alpha_1}{\rho_1 c_1^2} + \frac{\alpha_2}{\rho_2 c_2^2} \tag{4.5}$$

that present a non monotonic behavior versus volume fraction, as shown in Figure 6. Thus system (3.6) is strictly hyperbolic with the characteristic waves speeds: $u + c_w$, $u - c_w$ and u.

5 Model with mass transfer

System (3.6) describes a compressible two-phase flow mixture in mechanical equilibrium but out of thermal equilibrium. The next step is now to introduce mass transfer effects. The addition of mass transfer modifies the mass equation of each fluid:

$$\frac{\partial \alpha_1 \rho_1}{\partial t} + div(\alpha_1 \rho_1 \vec{u}) = \dot{m}_1 = \rho\, \dot{Y}_1$$

$$\frac{\partial \alpha_2 \rho_2}{\partial t} + div(\alpha_2 \rho_2 \vec{u}) = -\, \dot{m}_1 = -\rho\, \dot{Y}_1$$

where $\dot{m}_1$ is the mass flux from fluid 2 to fluid 1. Its expression has to be determined. The presence of mass transfer implies changes in the volume fraction. We assume that the volume fraction equation becomes:

$$\frac{\partial \alpha_1}{\partial t} + \vec{u} \cdot \vec{\nabla} \alpha_1 = K div(\vec{u}) + \frac{\alpha_1 \alpha_2}{\alpha_2 \rho_1 c_1^2 + \alpha_1 \rho_2 c_2^2} \left(\frac{\Gamma_1}{\alpha_1} + \frac{\Gamma_2}{\alpha_2} \right) Q_1 + \frac{\dot{m}_1}{\rho_I} \tag{5.1}$$

where the interfacial density ρ_I has also to be determined.
Determination of the expressions for mass transfer $\dot{m}_1$ and interface density ρ_I is based on the analysis of the mixture entropy production. To this end, the first step is to determine the entropy equations for each fluid.

5.1 Determination of the entropy equations of the phases

The entropy equations are determined as solutions of an algebraic system built on the basis of:

- energy conservation of the mixture,

- pressure equilibrium between phases.

Let us first examine the constraint given by energy conservation to the entropy equations. By using the energy and momentum equations of system (3.6) a simpler form of the energy equation is obtained:

$$\frac{de}{dt} + p\frac{dv}{dt} = 0 \tag{5.2}$$

where the mixture internal energy is defined by $e = Y_1 e_1 + Y_2 e_2$ and the mixture specific volume is given by $v = Y_1 v_1 + Y_2 v_2$. Thus equation (5.2) becomes:

$$Y_1 \left(\frac{de_1}{dt} + p \frac{dv_1}{dt} \right) + Y_2 \left(\frac{de_2}{dt} + p \frac{dv_2}{dt} \right) + (h_1 - h_2)\, \dot{Y_1} = 0$$

where $h_k = e_k + p v_k$ is the enthalpy of the phase k.
By using the Gibbs identity for each phase k, we have:

$$\frac{de_k}{dt} + p \frac{dv_k}{dt} = T_k \frac{ds_k}{dt}$$

The mixture energy conservation now becomes:

$$Y_1 T_1 \frac{ds_1}{dt} + Y_2 T_2 \frac{ds_2}{dt} + (h_1 - h_2)\, \dot{Y_1} = 0 \tag{5.3}$$

This last equation involves the two functions $\frac{ds_1}{dt}$ and $\frac{ds_2}{dt}$ that we want to determine.

A second equation is provided by the mechanical equilibrium condition:

$$p_1(\rho_1, s_1) = p_2(\rho_2, s_2), \tag{5.4}$$

that reads under differential form:

$$\left(\frac{\partial p_1}{\partial \rho_1} \right)_{s_1} \frac{d\rho_1}{dt} + \left(\frac{\partial p_1}{\partial s_1} \right)_{\rho_1} \frac{ds_1}{dt} = \left(\frac{\partial p_2}{\partial \rho_2} \right)_{s_2} \frac{d\rho_2}{dt} + \left(\frac{\partial p_2}{\partial s_2} \right)_{\rho_2} \frac{ds_2}{dt}$$

The definition of the speed of sound and the Gruneisen coefficient for each phase reads:

$$\forall k = 1, 2 \qquad \left(\frac{\partial p_k}{\partial \rho_k} \right)_{s_k} = c_k^2 \qquad \text{and} \qquad \left(\frac{\partial p_k}{\partial s_k} \right)_{\rho_k} = \rho_k \Gamma_k\, T_k$$

And the mechanical equilibrium condition (5.4) becomes:

$$c_1^2 \frac{d\rho_1}{dt} + \rho_1 \Gamma_1\, T_1 \frac{ds_1}{dt} = c_2^2 \frac{d\rho_2}{dt} + \rho_2 \Gamma_2\, T_2 \frac{ds_2}{dt} \tag{5.5}$$

Equations (5.3) and (5.5) form a system of two equations with the two unknowns functions $\frac{ds_1}{dt}$ and $\frac{ds_2}{dt}$:

$$\begin{cases} Y_1 T_1 \frac{ds_1}{dt} + Y_2 T_2 \frac{ds_2}{dt} + (h_1 - h_2)\, \dot{Y_1} = 0 \\ c_1^2 \frac{d\rho_1}{dt} + \rho_1 \Gamma_1\, T_1 \frac{ds_1}{dt} = c_2^2 \frac{d\rho_2}{dt} + \rho_2 \Gamma_2\, T_2 \frac{ds_2}{dt} \end{cases}$$

Resolution of this system yields:

$$\begin{cases} \frac{\alpha_1 \rho_1 \alpha_2 \rho_2}{\rho} T_1 \left(\frac{\Gamma_1}{\alpha_1} + \frac{\Gamma_2}{\alpha_2} \right) \frac{ds_1}{dt} = Y_2 \left(c_2^2 \frac{d\rho_2}{dt} - c_1^2 \frac{d\rho_1}{dt} \right) - \rho_2 \Gamma_2 (h_1 - h_2)\, \dot{Y_1} \\ \frac{\alpha_1 \rho_1 \alpha_2 \rho_2}{\rho} T_2 \left(\frac{\Gamma_1}{\alpha_1} + \frac{\Gamma_2}{\alpha_2} \right) \frac{ds_2}{dt} = -Y_1 \left(c_2^2 \frac{d\rho_2}{dt} - c_1^2 \frac{d\rho_1}{dt} \right) - \rho_1 \Gamma_1 (h_1 - h_2)\, \dot{Y_1} \end{cases} \tag{5.6}$$

Let us replace the variation $\frac{d\rho_1}{dt}$ and $\frac{d\rho_2}{dt}$ by variations of the volume fraction and velocity divergence. The mass equations of the fluids can be written as:

$$\frac{d\rho_1}{dt} = -\frac{\rho_1}{\alpha_1}\frac{d\alpha_1}{dt} - \rho_1 div(\vec{u}) + \frac{\rho \dot{Y_1}}{\alpha_1}$$

$$\frac{d\rho_2}{dt} = \frac{\rho_2}{\alpha_2}\frac{d\alpha_1}{dt} - \rho_2 div(\vec{u}) - \frac{\rho \dot{Y_1}}{\alpha_2}$$

System (5.6) now becomes:

$$\left\{\begin{array}{l} \frac{\alpha_1\rho_1\alpha_2\rho_2}{\rho} T_1 \left(\frac{\Gamma_1}{\alpha_1} + \frac{\Gamma_2}{\alpha_2}\right)\frac{ds_1}{dt} = Y_2\left[\left(\frac{\rho_1 c_1^2}{\alpha_1} + \frac{\rho_2 c_2^2}{\alpha_2}\right)\frac{d\alpha_1}{dt} - \left(\rho_2 c_2^2 - \rho_1 c_1^2\right) div(\vec{u})\right] \\ \qquad -\rho Y_2\left(\frac{c_1^2}{\alpha_1} + \frac{c_2^2}{\alpha_2}\right)\dot{Y_1} - \rho_2\Gamma_2\left(h_1 - h_2\right)\dot{Y_1} \\ \\ \frac{\alpha_1\rho_1\alpha_2\rho_2}{\rho} T_2 \left(\frac{\Gamma_1}{\alpha_1} + \frac{\Gamma_2}{\alpha_2}\right)\frac{ds_2}{dt} = -Y_1\left[\left(\frac{\rho_1 c_1^2}{\alpha_1} + \frac{\rho_2 c_2^2}{\alpha_2}\right)\frac{d\alpha_1}{dt} - \left(\rho_2 c_2^2 - \rho_1 c_1^2\right) div(\vec{u})\right] \\ \qquad +\rho Y_1\left(\frac{c_1^2}{\alpha_1} + \frac{c_2^2}{\alpha_2}\right)\dot{Y_1} - \rho_1\Gamma_1\left(h_1 - h_2\right)\dot{Y_1} \end{array}\right. \tag{5.7}$$

The introduction of the volume fraction equation (5.1) results in entropy equations that are function only of heat exchange, mass transfer and interfacial density ρ_I:

$$\left\{\begin{array}{l} Y_1\frac{ds_1}{dt} = \frac{H(T_2 - T_1)}{\rho T_1} + \\ \qquad \frac{\dot{Y_1}}{T_1\left(\frac{\Gamma_1}{\alpha_1} + \frac{\Gamma_2}{\alpha_2}\right)}\left[\frac{\frac{\rho_1 c_1^2}{\alpha_1} + \frac{\rho_2 c_2^2}{\alpha_2}}{\rho_I} - \left(\frac{c_1^2}{\alpha_1} + \frac{c_2^2}{\alpha_2}\right) - \frac{\Gamma_2}{\alpha_2}\left(h_1 - h_2\right)\right] \\ \\ Y_2\frac{ds_2}{dt} = -\frac{H(T_2 - T_1)}{\rho T_1} + \\ \qquad \frac{\dot{Y_1}}{T_2\left(\frac{\Gamma_1}{\alpha_1} + \frac{\Gamma_2}{\alpha_2}\right)}\left[-\frac{\frac{\rho_1 c_1^2}{\alpha_1} + \frac{\rho_2 c_2^2}{\alpha_2}}{\rho_I} + \left(\frac{c_1^2}{\alpha_1} + \frac{c_2^2}{\alpha_2}\right) - \frac{\Gamma_1}{\alpha_1}\left(h_1 - h_2\right)\right] \end{array}\right. \tag{5.8}$$

5.2 Mixture entropy inequallity

The second principle of thermodynamics applied to the mixture reads:

$$\frac{\partial \rho s}{\partial t} + div(\rho s\vec{u}) \geq 0$$

where the mixture entropy is defined by $s = Y_1 s_1 + Y_2 s_2$.
Introducing the expressions (5.8) in this inequality leads to:

$$\frac{H(T_2 - T_1)^2}{\rho} + (\overline{g}_2 - \overline{g}_1)T_I\,\dot{Y_1} + \dot{Y_1}\left\{\frac{T_2 - T_1}{\left(\frac{\Gamma_1}{\alpha_1} + \frac{\Gamma_2}{\alpha_2}\right)}\left[\frac{\frac{\rho_1 c_1^2}{\alpha_1} + \frac{\rho_2 c_2^2}{\alpha_2}}{\rho_I} - \left(\frac{c_1^2}{\alpha_1} + \frac{c_2^2}{\alpha_2}\right)\right]\right\} \geq 0 \tag{5.9}$$

where an interface temperature appears: $T_I = \left(\frac{\Gamma_1 T_1}{\alpha_1} + \frac{\Gamma_2 T_2}{\alpha_2}\right) / \left(\frac{\Gamma_1}{\alpha_1} + \frac{\Gamma_2}{\alpha_2}\right)$.
Extended chemical potentials also appear:

$$\forall k = 1,2 \qquad \overline{g}_k = h_k - \overline{T} s_k$$

with h_k the enthalpy of the phase k and $\overline{T} = T_1 T_2 / T_I$.
It appears clearly that the first term relative to heat exchanges is necessarily positive. The second term will be positive if we assume that:

$$\dot{Y}_1 = \nu(\overline{g}_2 - \overline{g}_1)$$

where ν is a positive relaxation parameter that controls the rate at which the mixture tends to thermodynamic equilibrium. This corresponds to the form of mass transfer terms we were seeking for.

The sign of the last term of (5.9) depends on the sign of $\dot{Y}_1$ and of the sign of the temperature difference. A solution in order to guaranty positive entropy production is to take it equal to zero. This assumption provides the last unknown of the system:

$$\rho_I = \frac{\frac{\rho_1 c_1^2}{\alpha_1} + \frac{\rho_2 c_2^2}{\alpha_2}}{\frac{c_1^2}{\alpha_1} + \frac{c_2^2}{\alpha_2}}$$

We now have a symmetric hyperbolic non equilibrium compressible two-phase flow model with heat exchange and mass exchanges:

$$\begin{cases} \dfrac{\partial \alpha_1}{\partial t} + \vec{u} \cdot \vec{\nabla} \alpha_1 = K div(\vec{u}) + \dfrac{\alpha_1 \alpha_2}{\alpha_2 \rho_1 c_1^2 + \alpha_1 \rho_2 c_2^2} \left(\dfrac{\Gamma_1}{\alpha_1} + \dfrac{\Gamma_2}{\alpha_2}\right) Q_1 + \dfrac{\frac{\rho_1 c_1^2}{\alpha_1} + \frac{\rho_2 c_2^2}{\alpha_2}}{\frac{c_1^2}{\alpha_1} + \frac{c_2^2}{\alpha_2}} \dot{m}_1 \\ \dfrac{\partial \alpha_1 \rho_1}{\partial t} + div(\alpha_1 \rho_1 \vec{u}) = \dot{m}_1 = \rho \dot{Y}_1 \\ \dfrac{\partial \alpha_2 \rho_2}{\partial t} + div(\alpha_2 \rho_2 \vec{u}) = -\dot{m}_1 \\ \dfrac{\partial \rho \vec{u}}{\partial t} + div(\rho \vec{u} \otimes \vec{u}) + \vec{\nabla} p = 0 \\ \dfrac{\partial \rho E}{\partial t} + div \vec{u}(\rho E + p) = 0 \end{cases} \tag{5.10}$$

Where:
$K = \alpha_1 \alpha_2 (\rho_2 c_2^2 - \rho_1 c_1^2) / (\alpha_2 \rho_1 c_1^2 + \alpha_1 \rho_2 c_2^2)$
$\dot{Y}_1 = \nu(\overline{g}_2 - \overline{g}_1)$
And the mixture pressure is given by (4.4):

$$p(\rho, e, \alpha_1, \alpha_2, Y_1, Y_2) = \frac{\rho(e - Y_1 q_1 - Y_2 q_2) - \left(\frac{\alpha_1 \gamma_1 p_{\infty,1}}{\gamma_1 - 1} + \frac{\alpha_2 \gamma_2 p_{\infty,2}}{\gamma_2 - 1}\right)}{\frac{\alpha_1}{\gamma_1 - 1} + \frac{\alpha_2}{\gamma_2 - 1}}$$

The determination of the temperature relaxation parameter H for a two-phase mixture with arbitrary interfacial area is still an issue. The same remark is done for the phase transition kinetics parameter ν that does not depend only of interfacial area but also of local chemical relaxation. To circumvent these difficulties we will use a solution procedure based on infinite relaxation parameters, but at selected locations only. More precisely, in order to retain metastable states, the relaxation parameters H and ν will be set to zero for locations far from the interfaces. At the interfaces, they will be taken infinite in order to fulfill equilibrium interface conditions with mass transfer. When dealing with interfaces of simple contact, they will be set to zero everywhere. Such procedure is described in the next section.

6 Numerical method

The numerical method to solve the compressible two-phase flow system (5.10) with heat and mass transfer proceeds in two steps. At each time step, the hyperbolic system in absence of heat and mass transfer is solved. This provides the non equilibrium hydrodynamic field. Stiff thermal and chemical relaxations are then solved at the interfaces only. The interfaces are detected from the knowledge of volume fraction and mass fraction fields.

The basic ingredients of the hyperbolic solver are summarized hereafter. This solver is not conventional as the hyperbolic system is not conservative. Conventional Godunov type schemes or other existing methods are not suitable for its resolution. The present hyperbolic solver is described in details in the case of the Euler equations in Saurel et al. (2007a).

We second present a new stiff differential solver specifically derived for the present model. It is used for the integration of stiff relaxation terms that express heat and mass transfers.

6.1 Hyperbolic solver

The hyperbolic system (5.10) without heat and mass transfer reads:

$$\begin{cases} \dfrac{\partial \alpha_1}{\partial t} + \vec{u} \cdot \vec{\nabla} \alpha_1 = K div(\vec{u}) \\[2ex] \dfrac{\partial \alpha_1 \rho_1}{\partial t} + div(\alpha_1 \rho_1 \vec{u}) = 0 \\[2ex] \dfrac{\partial \alpha_2 \rho_2}{\partial t} + div(\alpha_2 \rho_2 \vec{u}) = 0 \\[2ex] \dfrac{\partial \rho \vec{u}}{\partial t} + div(\rho \vec{u} \otimes \vec{u}) + \vec{\nabla} p = 0 \\[2ex] \dfrac{\partial \rho E}{\partial t} + div \vec{u}(\rho E + p) = 0 \end{cases} \tag{6.1}$$

The volume fraction equation of system (6.1) is not conservative. This poses two important difficulties regarding numerical resolution:

- Conventional shock relations are not available. In absence of these relations, the Riemann problem cannot be solved. The Riemann problem is the cornerstone of modern numerical methods to solve hyperbolic systems.

- The average of the volume fraction variable within a computational cell has no physical sense. Cell averages have sense only for conservative variables.

Non conventional shock relations for system (6.1) are proposed in Saurel et al. (2007b). These relations:
- Conserve mass, momentum and energy of the mixture.
- Tend to the single-phase shock relations when one of the phases disappears. They thus guarantee a correct behavior in the single-phase limit. This feature is important for interface problems.
- Preserve volume fractions positivities. This is also an important feature for numerical resolution in the presence of material interfaces.
- Involves a symmetric treatment for all phases. This symmetry allows a possible extension of the model to an arbitrary number of components.
- Is in perfect agreement with experimental shock measurements. It has been validated for more than 100 experimental tests involving a wide range of shock strength, very different acoustic impedance ratios between phases, different initial volume fractions of the phases, different EOS of pure components.
- Last, the mixture Hugoniot curve is tangent to the mixture isentrope. It means that multiphase weak shock waves behave like simple compression waves. This feature is also important for the Riemann problem resolution.

These relations are summarized hereafter (6.2):

$$\begin{aligned}
&\forall k = 1,2 \quad Y_k = Y_k^0 \\
&\rho(u-\sigma) = \rho_0(u_0-\sigma) = m \\
&p - p_0 + m^2(v - v_0) = 0 \\
&\forall k = 1,2 \quad e_k - e_k^0 + \tfrac{p+p_0}{2}(v_k - v_k^0) = 0
\end{aligned} \tag{6.2}$$

With the help of these relations and Riemann invariants, exact or approximate Riemann solvers can be derived (Petitpas et al. (2007)). Once the Riemann problem is solved, the second difficulty is to average or project the Riemann problem solution to the computational cell. The presence of the volume fraction variable prevents from the use of conventional projection methods. Thus a specific relaxation-projection method has been derived in the case of the Euler equations in Saurel et al. (2007a). The basic ideas are recalled hereafter in the multiphase case.

Relaxation-projection method This method consists in the resolution of a relaxation problem between the various sub-volumes present in a computational cell. These sub-volumes correspond to the ones traveled by the various waves that produce states with different pressures, velocities, densities and temperatures. For the sake of simplicity and efficiency the method is developed in a Lagrange - Projection method context. This method uses an approximate Riemann solver where each wave is considered as a shock. The resolution of the Riemann problem at each cell boundary of an Eulerian cell defines a Lagrangian cell where only two waves propagate. A schematic view of the wave's propagation into a Lagrangian cell is shown in the Figure 7.

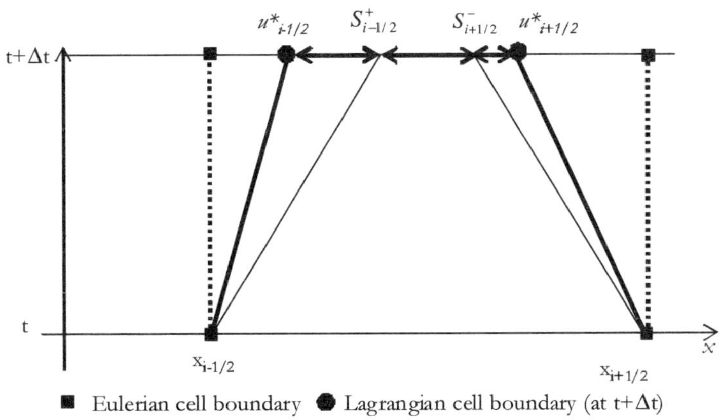

Figure 7. Schematic view of the waves incoming a Lagrangian cell.

Only two waves enter the Lagrangian cell: $S^+_{i-\frac{1}{2}}$ is a right-facing shock or rarefaction wave and $S^-_{i+\frac{1}{2}}$ is a left-facing shock or rarefaction wave.
At the end of the time step, the Lagrangian cell is divided into three sub-volumes. Each sub-volume $j = 1,.,3$ contains several phases k in a state different of those in the direct neighboring sub-volume. The goal of the relaxation-projection method is to determine from these non equilibrium states the mechanical equilibrium state where the three sub-volumes have the same pressure and velocity. The mechanical equilibrium state is obtained by the asymptotic solution of the relaxation system (6.3) when the pseudo time τ tends to infinity:

$$\forall j = 1,2,3 \quad \forall k = 1,2 \quad \left| \begin{aligned} \frac{\partial \alpha_{jk}\beta_j}{\partial \tau} &= \frac{\delta}{Z_{jk}}\left(p_j - p_I\right) \\ \frac{\partial \alpha_{jk}\beta_j\rho_{jk}}{\partial \tau} &= 0 \\ \frac{\partial \alpha_{jk}\beta_j\rho_{jk}u_j}{\partial \tau} &= \delta Z_{jk}\left(u_I - u_j\right) \\ \frac{\partial \alpha_{jk}\beta_j\rho_{jk}E_{jk}}{\partial \tau} &= \delta\left[u_I Z_{jk}\left(u_I - u_j\right) - \frac{p_I}{Z_j}\left(p_j - p_I\right)\right] \end{aligned} \right. \tag{6.3}$$

with

$$u_I = \frac{\sum_{jk} Z_{jk} u_j}{\sum_{jk} Z_{jk}} \quad , \quad p_I = \frac{\sum_{jk} \frac{p_j}{Z_{jk}}}{\sum_{jk} \frac{1}{Z_{jk}}} \quad \text{and} \quad \delta = 2/\Delta x$$

where β_j, u_j, p_j are respectively the sub-cell volume fraction, velocity and pressure of the sub-volume j in the Lagrangian computational cell. α_{jk}, ρ_{jk}, E_{jk} and Z_{jk} represents respectively the volume fraction, the density, the total energy and the acoustic impedance of the phase k in the sub-volume j.

The next difficulty is to determine the asymptotic solution of the relaxation system. The steady state solution of system (6.3) only is needed. Let us denote the variables coming from the resolution of the Riemann problem with the superscript "0" and the variables at the end of the relaxation process by the superscript "*" (mechanical equilibrium state). The mixture mass conservation implies:

$$\rho^* = \sum_{jk} \alpha_{jk}^* \beta_j^* \rho_{jk}^* = \sum_{jk} \alpha_{jk}^0 \beta_j^0 \rho_{jk}^0 = \rho^0 = \rho$$

Using the phase's mass equations (6.3), we obtain:

$$\forall j = 1,2,3 \; and \; k = 1,2 \qquad Y_{jk}^* = \frac{\alpha_{jk}^* \beta_j^* \rho_{jk}^*}{\rho^*} = \frac{\alpha_{jk}^0 \beta_j^0 \rho_{jk}^0}{\rho^0} = Y_{jk}^0 = Y_{jk}$$

The equilibrium velocity is determined by using the mixture momentum equation that reads:

$$\sum_{jk} \alpha_{jk}^* \beta_j^* \rho_{jk}^* u_j^* = \sum_{jk} \alpha_{jk}^0 \beta_j^0 \rho_{jk}^0 u_j^0$$

Reminding that the mechanical equilibrium state is characterized by $u_j^* = u^*$, we have:

$$u^* = \frac{\sum_{jk} \alpha_{jk}^0 \beta_j^0 \rho_{jk}^0 u_j^0}{\sum_{jk} \alpha_{jk}^* \beta_j^* \rho_{jk}^*} = \sum_{jk} Y_{jk} u_j^0$$

This results is not different to the one obtained with the Godunov method. The goal is now to determine the lasts seven variables p^*, and v_{jk}^* ($j = 1,2,3 \; and \; k = 1,2$). To do this, we write the energy equations under the following form:

$$\frac{\partial \alpha_{jk} \beta_j \rho_{jk} E_{jk}}{\partial \tau} = -p_I \frac{\partial \alpha_{jk} \beta_j}{\partial \tau} + u_I \frac{\partial \alpha_{jk} \beta_j \rho_{jk} u_j}{\partial \tau}$$

By using the mass and momentum equations, we obtain:

$$\frac{\partial e_{jk}}{\partial \tau} + p_I \frac{\partial v_{jk}}{\partial \tau} = (u_I - u_j) \frac{\partial u_j}{\partial \tau}$$

These six equations can be integrated between the initial and final state:

$$\int_0^\infty \frac{\partial e_{jk}}{\partial \tau} d\tau + \hat{p}_{Ijk} \int_0^\infty \frac{\partial v_{jk}}{\partial \tau} = q_{jk}$$

where $\hat{p}_{Ijk}$ and q_{jk} are functions that have to be defined.

The integration leads to:

$$e^*_{jk}\left(v^*_{jk}, p*\right) - e^0_{jk} + \hat{p}_{Ijk}\left(v^*_{jk} - v^0_{jk}\right) = q_{jk} \tag{6.4}$$

The six equations system (6.4) is closed by a seventh equation, the saturation condition that reads: $\sum_{jk} \alpha_{jk}\beta_j = 1$. It can be written as:

$$\sum_{jk} m_{jk} v_{jk} = 1 \tag{6.5}$$

where the mass $m_{jk} = \alpha_{jk}\beta_j\rho_{jk}$ is constant during the relaxation process.
The previous system requires expressions for the preceding averages and functions ($\hat{p}_{Ijk}$ and q_{jk}). Admissible forms of these functions are:

$$\begin{cases} \hat{p}_{Ijk} = p* \\ q_{jk} = \frac{1}{2Z_{jk}} \sum_{m=1,6} Y_m Z_{m,jk} \left(u^0_m - u^0_j\right)^2 \end{cases} \tag{6.6}$$

with $\frac{1}{Z_{m,jk}} = \frac{1}{Z_m} + \frac{1}{Z_{jk}}$. These relations guarantee energy conservation and entropy inequality.
The six equations (6.4) and the saturation condition (6.5) form now a non linear system of 7 equations with 7 unknowns. It can be reduced to a single non-linear equation function of p^*. It is then solved by the Newton-Raphson method. All details are available in Saurel et al. (2007a).
Once the equilibrium solution is reached in the Lagrangian cell, the projection onto the Eulerian grid is obtained by applying the same method with new definitions of the sub-volumes. The sub-volumes now correspond to the volumes occupied by the Lagrangian cells in the Eulerian one.
This method has been tested for interface computations in the presence of shocks and rarefaction waves and produced excellent results. It is also able to deal with wave propagation into multiphase mixtures.

6.2 Stiff thermo-chemical solver

The cell now contains a multiphase mixture in mechanical equilibrium but remains out of thermal and chemical equilibrium. Indeed, each phase in the cell has its own temperature and chemical potential. In order to fulfill interface conditions in the presence of heat and mass transfer (equal temperatures and chemical potentials) a relaxation method is used. The interface is located by the following procedure:

- A cell is considered filled by a pure fluid when its volume and mass fractions are closed to 1. The interface corresponds to mixture cells when volumes and mass fractions range between ϵ and $1-\epsilon$ (typically, $\epsilon = 10^{-6}$).

- Mass transfer is allowed if one of the fluid in the mixture cell is metastable ($T_k > T_{sat}(P)$)). Then, the following thermo-chemical solver is used.

The hydrodynamics evolution has been achieved by the numerical approximation of system (6.1) and the goal is now to solve the following system of ordinary differential equations at the interface only:

$$\left\{\begin{array}{l} \dfrac{\partial \alpha_1}{\partial t} = \dfrac{\alpha_1 \alpha_2}{\alpha_2 \rho_1 c_1^2 + \alpha_1 \rho_2 c_2^2}\left(\dfrac{\Gamma_1}{\alpha_1} + \dfrac{\Gamma_2}{\alpha_2}\right) Q_1 + \dfrac{\frac{\rho_1 c_1^2}{\alpha_1} + \frac{\rho_1 c_1^2}{\alpha_1}}{\frac{c_1^2}{\alpha_1} + \frac{c_2^2}{\alpha_1}} \dot{m}_1 = S_{\alpha_1} \\ \dfrac{\partial \alpha_1 \rho_1}{\partial t} = \dot{m}_1 = S_{Y_1} \\ \dfrac{\partial \alpha_2 \rho_2}{\partial t} = -\dot{m}_1 \\ \dfrac{\partial \rho \vec{u}}{\partial t} = 0 \\ \dfrac{\partial \rho E}{\partial t} = 0 \end{array}\right. \tag{6.7}$$

The integration of this system necessitates closure relations for Q_1 and $\dot{Y}_1$, in particular regarding the relaxation parameters h and ν. In order that the model be free of parameter we assume that thermodynamic local equilibrium is reached at the interface at any time. It means that h and ν tend to infinity locally. This assumption is conventional at equilibrium interfaces when mass transfer occurs. It does not mean that the entire flow evolves at thermodynamic equilibrium. The states remain out of equilibrium far from interfaces.
In the context of numerical integration, taking infinite relaxation parameters means that the equilibrium has to be reached during each time step. The time step is imposed by CFL restriction of the hydrodynamic system. Using this time step, we determine Q_1 and $\dot{Y}_1$ in order that thermodynamic equilibrium be reached at the end of each time step. To determine these source terms, the equations for the temperatures and chemical potentials differences are necessary. They read:

$$\left\{\begin{array}{l} \dfrac{\partial \Delta T}{\partial t} = A Q_1 + B \dot{m}_1 \\ \dfrac{\partial \Delta g}{\partial t} = A' Q_1 + B' \dot{m}_1 \end{array}\right. \tag{6.8}$$

where A, B, A', B' are functions of all flow variables.
The simplest numerical approximation of these equations is used. Let "n" and "n+1" denote two successive time steps. The variables at time t^n are taken equal to those resulting of the numerical integration of system (6.1). The variables at time t^{n+1} denote

the end of the integration process, including both hydrodynamic effects and source terms (6.7). The simplest numerical approximation of this ODE system reads:

$$\left\{\begin{array}{l} \frac{(\Delta T)^{n+1}-(\Delta T)^{n}}{\Delta t} = A^{n}Q_1^n + B^n\dot{m}_1^n \\ \\ \frac{(\Delta g)^{n+1}-(\Delta g)^{n}}{\Delta t} = A'^{n}Q_1^n + B'^n\dot{m}_1^n \end{array}\right. \tag{6.9}$$

Assuming that at the end of the time step we have: $(\Delta T)^{n+1} = 0$ and $(\Delta g)^{n+1} = 0$, the corresponding heat and mass transfer terms are given by:

$$\left\{\begin{array}{l} Q_1 = -\frac{B'}{AB'-A'B}\frac{(\Delta T)^n}{\Delta t} + \frac{B}{AB'-A'B}\frac{(\Delta g)^n}{\Delta t} \\ \\ \dot{m}_1 = \frac{A'}{AB'-A'B}\frac{(\Delta T)^n}{\Delta t} - \frac{A}{AB'-A'B}\frac{(\Delta g)^n}{\Delta t} \end{array}\right. \tag{6.10}$$

These approximations of heat and mass transfer terms allow the calculation of the source terms S_{α_1} and S_{Y_1} of system (6.7), in order that thermodynamic equilibrium be reached at the end of the time step. Nevertheless, there is no guarantee that positivity of the solution be preserved, in particular regarding mass and volume fractions. A limitation of these source terms has to be done. Let us express the maximum admissible source term for the volume fraction evolution in order to preserve positivity of this variable:

$$S_{max,\alpha_1} = \left\{\begin{array}{lll} \frac{1-\alpha_1}{\Delta t} & if & S_{\alpha_1} > 0 \\ \\ \frac{-\alpha_1}{\Delta t} & otherwise & \end{array}\right. \tag{6.11}$$

The same maximum source term is computed for mass fraction positivity. Thus, if $|S_{max,\alpha_1}| > |S_{\alpha_1}|$ and $|S_{max,Y_1}| > |S_{Y_1}|$, the numerical integration can be done with the hydrodynamics time step. Otherwise, equations are stiff and the integration time step has to be reduced. The ratio $R_{\alpha_1} = \frac{S_{max,\alpha_1}}{S_{\alpha_1}}$ is computed and the system (6.7) is integrated on a fraction of the time step, typically: $\Delta t_{chem} = \frac{1}{2}R_{\alpha_1}\Delta t$. Successive integrations are done in order that the complete hydrodynamic step be covered.
With this algorithm, thermodynamic equilibrium is reached very fast and positivity of the solution is preserved. Using this relaxation method corresponds to the local resolution (in fact at the interface only) of the following limit system, corresponding to the mixture Euler equations:

$$\left\{\begin{array}{l} \frac{\partial \rho}{\partial t} + div\,(\rho\vec{u}) = 0 \\ \\ \frac{\partial \rho\vec{u}}{\partial t} + div\,(\rho\vec{u}\otimes\vec{u} + p) = 0 \\ \\ \frac{\partial \rho E}{\partial t} + div\,[(\rho E + p)\,\vec{u}] = 0 \end{array}\right. \tag{6.12}$$

Where the mixture density reads $\rho = \alpha_1\rho_1 + \alpha_2\rho_2$ and the mixture internal energy reads $\rho e = \alpha_1\rho_1 e_1 + \alpha_2\rho_2 e_2$. The total mixture energy is still defined by $E = e + \frac{1}{2}u^2$.

This system is closed by the three thermodynamic equilibrium conditions:

$$p_1 = p_2 = p \tag{6.13}$$
$$T_1 = T_2 = T \tag{6.14}$$
$$g_1 = g_2 \tag{6.15}$$

With the help of the SG EOS (4.1) with conditions (6.13) and (6.14), each phase variable can be expressed as a function of pressure and temperature. The definitions of mixture density and internal energy now read:

$$\left\{\begin{array}{l} \rho = \alpha_2 \left[\rho_2(p,T) - \rho_1(p,T)\right] + \rho_1(p,T) \\ e = \frac{1}{\rho}\left[\alpha_2 \left[\rho_2(p,T)e_2(p,T) - \rho_1(p,T)e_1(p,T)\right] + \rho_1(p,T)e_1(p,T)\right] \end{array}\right. \tag{6.16}$$

The condition (6.15) now reduces to the mixture EOS (4.3) and system (6.16) becomes:

$$\alpha_2(T) = \frac{\rho - \rho_1(T)}{\rho_2(T) - \rho_1(T)} \tag{6.17}$$
$$e(T) = \frac{1}{\rho}\left[\alpha_2(T)\left[\rho_2(T)e_2(T) - \rho_1(T)e_1(T)\right] + \rho_1(T)e_1(T)\right] \tag{6.18}$$

Equation (6.18) can be solved numerically. It allows the calculation of α_2 with (6.17) and p with (4.3). Thus, the system is closed.
By using the sound speed definition $c^2 = \left(\frac{\partial p}{\partial \rho}\right)_s$ we obtain the following expression:

$$\frac{1}{\rho c_{eq}^2} = \frac{\alpha_1}{\rho_1 c_1^2} + \frac{\alpha_2}{\rho_2 c_2^2} + T\left[\frac{\alpha_1 \rho_1}{C_{p,v}}\left(\frac{ds_1}{dp}\right)^2 + \frac{\alpha_2 \rho_2}{C_{p,l}}\left(\frac{ds_2}{dp}\right)^2\right] \tag{6.19}$$

corresponding to the equilibrium mixture speed of sound. The Wood formula (4.5) is recovered with the first two terms of (6.19). The extra terms result in mixture sound speed decrease due to heat and mass transfers. This model is again hyperbolic with the characteristic waves speeds: $u + c_{eq}$, $u - c_{eq}$ and u.
However, with the present algorithm based on system (6.1) coupled to the relaxation system (6.7) with the preceding stiff integrator, there is no need to solve explicitly (6.12).

7 Numerical Results

The aim of this section is to highlight the model capabilities for the numerical resolution of interface problems with or without mass transfer. One dimensional tests are first considered for validation issues and two-dimensional tests are secondly achieved for the dynamic creation of supercavitating flow pockets.

7.1 One-dimensional tests

Two-phase shock tube. In this example, the left part of a shock tube is filled with liquid dodecane at high pressure $p_l = 10^8\ Pa$ with density $\rho_l = 500\ kg/m^3$. The

right chamber is set at atmospheric pressure and filled with vapor dodecane at density $\rho_v = 2\ kg/m^3$. The initial discontinuity is located at $x = 0.75\ m$ in a 1-m length tube. For numerical reasons, each chamber of the tube contains a weak volume fraction of the other fluid (typically 10^{-8}).
In the first example, the liquid - gas interface is solved as a contact discontinuity: heat and mass transfer are removed. The results are shown at time $t = 473\ \mu s$ (Figure 8) and consist in three conventional waves. From left to right, a left-facing rarefaction wave propagates through the liquid, the contact discontinuity is moving from left to right and a right-facing shock propagates through dodecane vapor. The numerical solution is compared to the exact one and shows a perfect agreement.
We now rerun this test case and consider heat and mass transfer at the interface. Indeed,

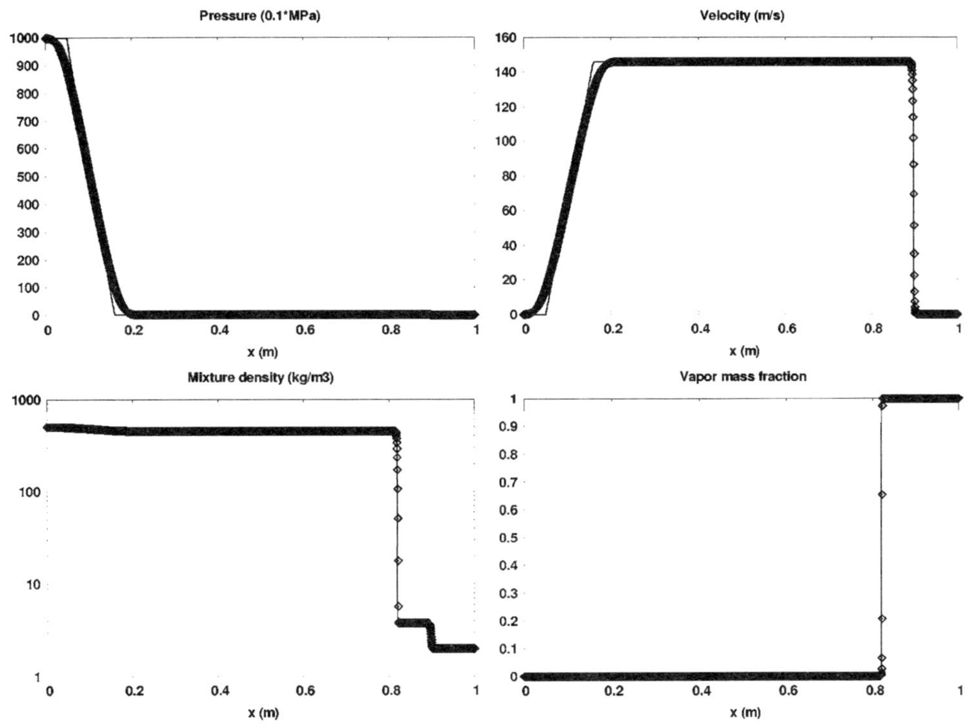

Figure 8. Dodecane liquid-vapor shock tube without evaporation. The numerical solution (symbols) is compared to the exact one (lines)

the rarefaction wave propagation transforms the stable high pressure liquid dodecane to a superheated liquid and evaporation has to be considered (Figure 9). An additional left-facing wave (evaporation front), which looks like an expansion wave, appears between the rarefaction wave and the contact discontinuity. It propagates through the superheated liquid and produces a liquid-vapor mixture at thermodynamic equilibrium and high velocity. Four waves are now present: the left facing expansion wave, the

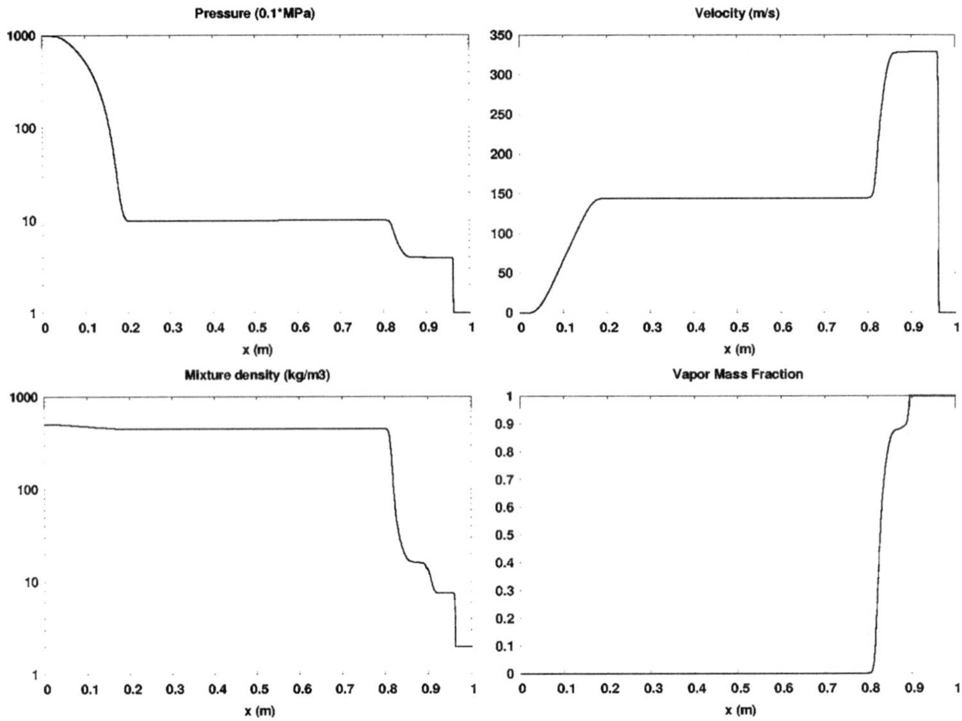

Figure 9. Stiff evaporation in a rarefaction tube. Thermo-chemical solution

evaporation front, the contact discontinuity and the right-facing shock. These waves are clearly visible on the mixture density graph. The evaporation front makes the vapor mass fraction increase, but no totally. The second jump in mass fraction is related to the contact discontinuity.

As mentioned in the preceding section, the interface is solved as a relaxation zone where the multiphase flow model tends to the equilibrium Euler equations. This last system admits very different waves speeds from the ones of the temperature non equilibrium model. This results in the appearance of a transition front that corresponds to an expansion wave of the equilibrium model. This expansion wave propagates to the left at a velocity approximately equal to $u - c_{eq}$. This observation gives an interpretation to the experimental results of Simões-Moreira and Shepherd (1999) where the Chapmann Jouguet kinetic relation ($\sigma = u \pm c$ where σ represents the front velocity) was proposed as closure relation for evaporation fronts in metastable liquids. The present results show that this kinetic relation is appropriate as the evaporation front is an expansion wave. This interpretation is confirmed by the following numerical experiment.

Two-phase expansion tube In this example, a 1-m length tube is filled with liquid water at atmospheric pressure and with density $\rho = 1150\ kg/m^3$. An initial velocity

discontinuity is located at $x = 0.5\ m$. On the left, the velocity is set to $u = -2\ m/s$ and on the right, $u = 2\ m/s$. An initial weak volume fraction of vapor water ($\alpha_v = 10^{-2}$) is added to the liquid.
First, the numerical solution without mass transfer is compared to the exact one. The solution is represented in Figure 10 at time $t = 3200\ \mu s$. It results in left- and right-facing rarefaction waves. The vapor volume fraction increase at the center of the domain. This is due to the gas mechanical expansion present in small proportions. This effect tends to create new interfaces in the domain. Dynamic appearance of these interfaces is considered in the numerical resolution as well as in the exact solution, provided in Petitpas et al. (2007). The gas pocket that appears is not due to mass transfer, but to mechanical relaxation only.
However, these rarefaction waves make the liquid metastable and phase transition ap-

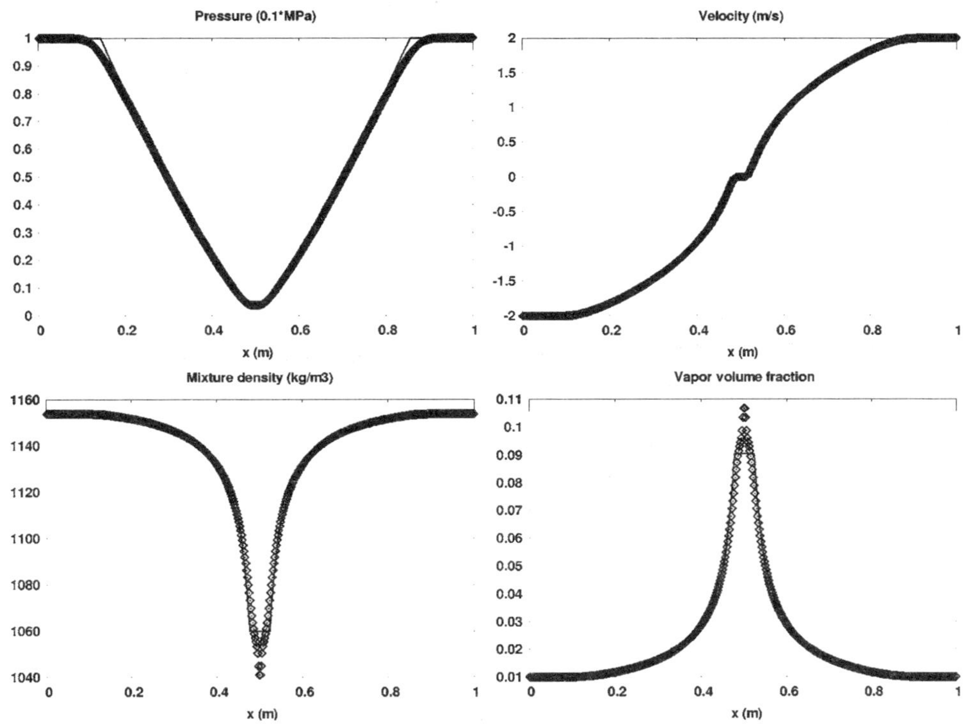

Figure 10. The multiphase model computed solution (symbols) without mass transfer is compared to the exact one (lines) on the symmetric expansion tube.

pears in order to prevent vacuum creation. Figure 11 represents the solution when mass transfer is involved. The solution is shown at the same instant as previously and compared to the previous exact solution without evaporation. Liquid water is expanded until the saturation pressure is reached (see the pressure graph) then evaporation appears and a quite small amount of vapor is created (see the mass fraction graph). This small amount of vapor has nevertheless strong consequences on the vapor volume fraction that increases significantly.

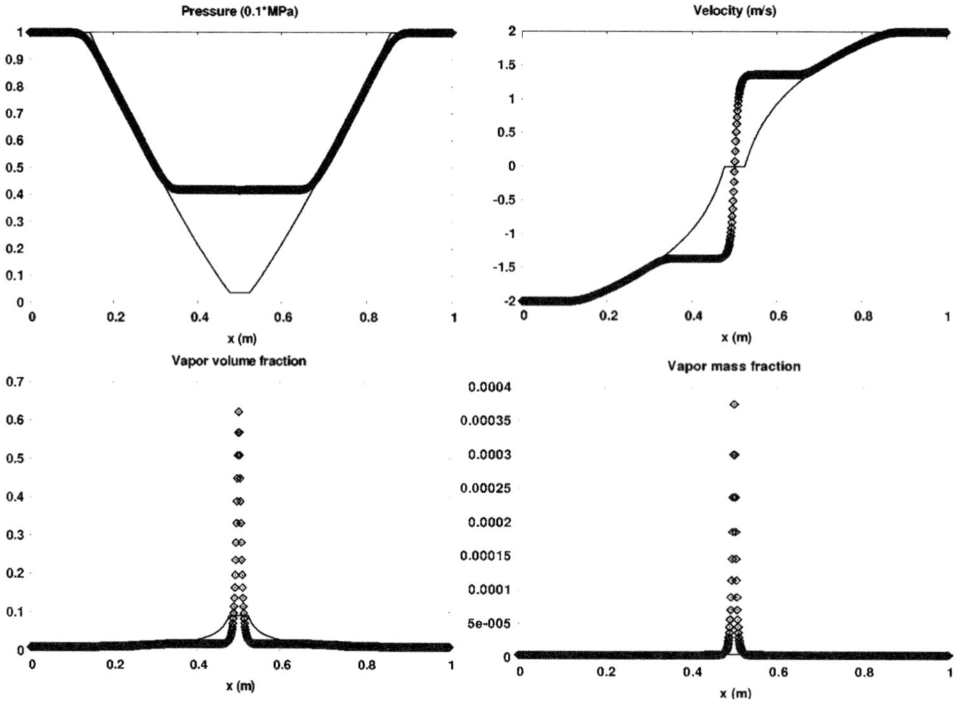

Figure 11. The multiphase model computed solution (symbols) with mass transfer is compared to the exact one (lines) without mass transfer on the symmetric expansion tube. Mass transfer appears and results in important differences for all flow variables.

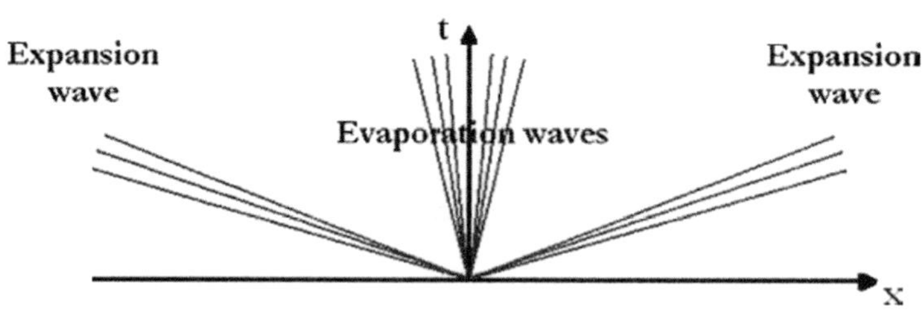

Figure 12. Wave propagation diagram of the symmetric expansion tube.

The solution with mass transfer is composed of 4 expansion waves, as shown in Figure 12. In order to highlight the two slow expansion waves, the solution is now shown at time $t = 59\ ms$ in Figure 13. The two leading fast expansion waves have exited the tube. The two slow evaporation fronts have characteristic profiles of expansion waves. This observation confirms the preceding interpretation of evaporation fronts as expansion waves of the equilibrium system.

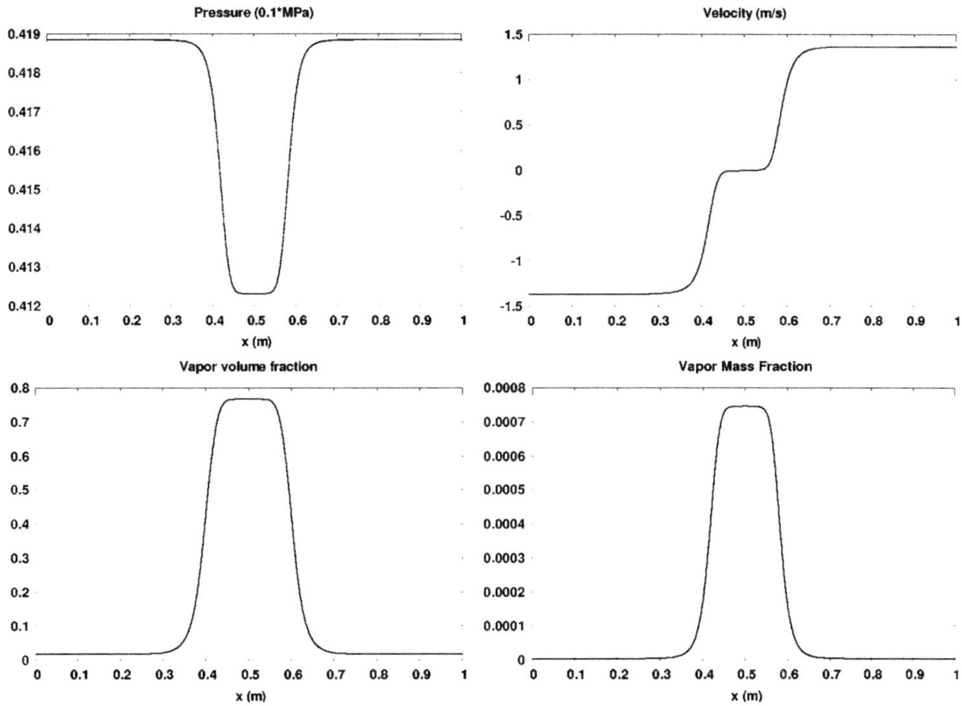

Figure 13. Results of Figure 11 at long time

When rarefaction effects become stronger, it is possible to see the four waves present on a single graph. In Figure 14 the same conditions are used except for velocities which are set to $u = -500\ m/s$ on the left, and $u = 500\ m/s$ on the right. In this case, evaporation is much intense resulting in a large cavitation pocket where the gas volume fraction is closed to 1. However, this pocket contains a mixture at thermodynamic equilibrium as shown on the mass fraction profile. The various expansion waves have now comparable velocities, as shown in Figure 15.

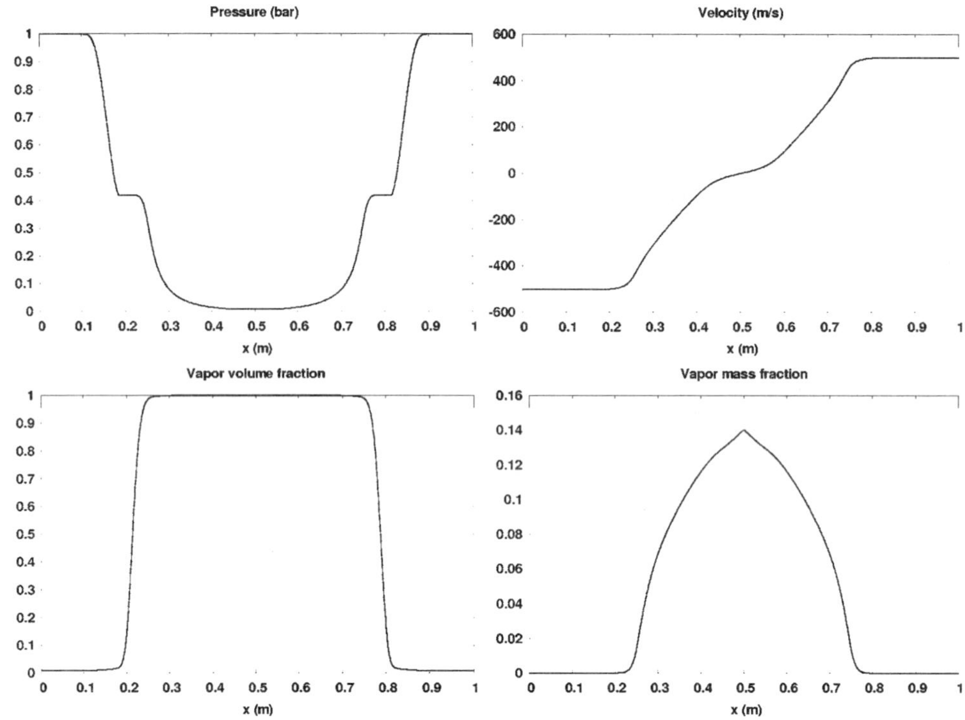

Figure 14. Numerical solution of the multiphase flow model with mass transfer and strong velocity difference in the expansion tube. The four expansion waves are clearly visible.

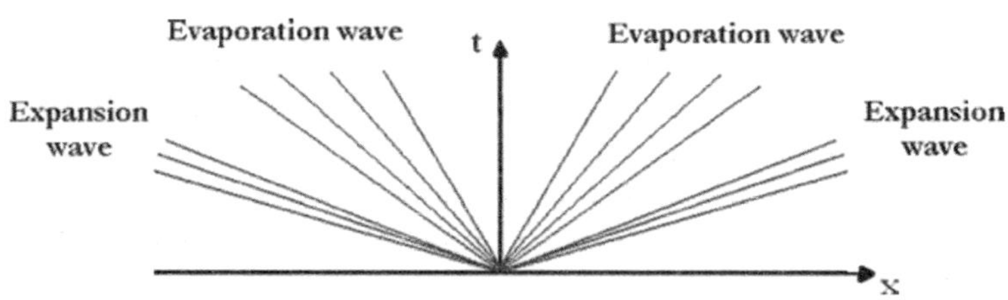

Figure 15. Schematic (x,t) diagram of the strong symmetric expansion tube.

7.2 Two-dimensional illustrations

The model capabilities are now illustrated on severe test problems involving cavitation pockets with or without evaporation. The first example is related to supercavitation around a high velocity underwater projectile. The second test is related to cavitation pockets in fuel injector nozzles.

High velocity underwater projectile A liquid water flow at velocity 600 m/s is considered around an immersed obstacle. Liquid watter is initially at atmospheric pressure with a density of 1150 kg/m^3. We remind that thermodynamic parameters of water are given in Table 1 . A weak volume fraction of vapor ($\alpha_v = 10^{-3}$) is initially present in water.

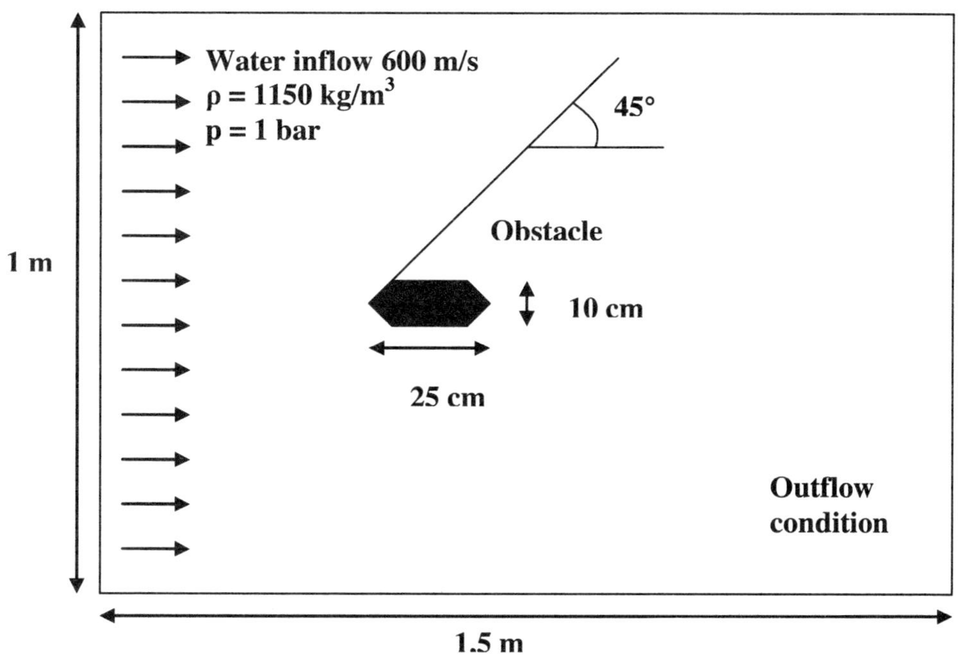

Figure 16. Initial configuration of an high velocity underwater projectile

At the leading edge of the obstacle a strong compression zone is present and the pressure exceeds 2000 atm. Then the high velocity liquid flow undergoes strong rarefaction waves at each geometrical singularity and the pressure decreases. On the left column of Figure 17 the computational results are shown at steady state without mass transfer. On the right of the same figure, mass transfer is involved. Qualitatively, both results are very similar: a cavitation wake appears. But the mechanisms responsible for these cavitation wakes are very different. On the left, only mechanical effects are responsible for bubble growth while on the right mechanical and chemical relaxation effects are responsible of the pocket growth. It means that the results without mass transfer can be used for a qualitative prediction of the locations where cavitation appears. For a quantitative prediction, results with mass transfer are more realistic. It is interesting to note that the maximum vapor mass fraction is not very large (less than 0.13). The pressure level in the cavitation pocket is quite different: 0.4 atm with mass transfer, corresponding to the saturation pressure, to be compared with 10^{-3} atm in absence of mass transfer.

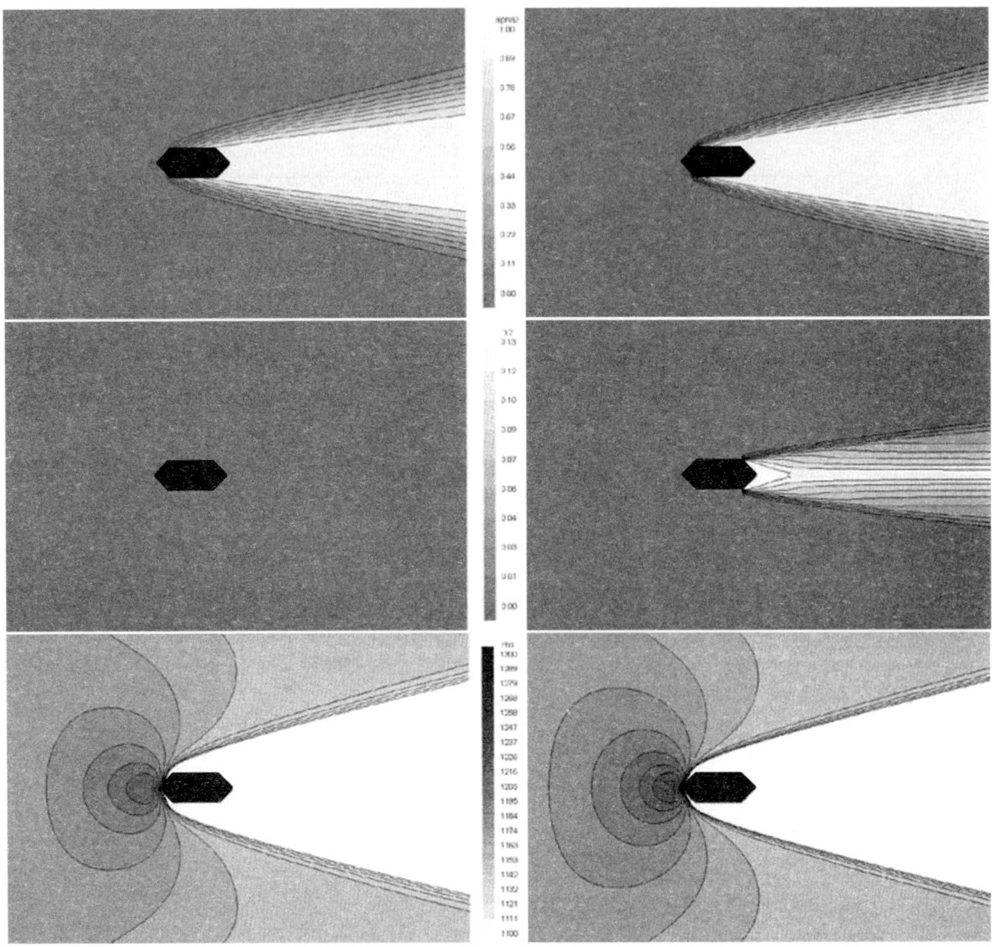

Figure 17. High velocity underwater projectile. On the left, the solution obtained with the inert solver and on the right the solution with the new thermo-chemical solver. Solutions are shown at steady state.

High pressure fuel injector The second situation under study consists in a nozzle where a liquid fuel, coming from a high pressure chamber is injected into another chamber at atmospheric pressure. The high pressure tank (1000 atm) is filled with liquid dodecane at density 570 kg/m^3, corresponding to the temperature $T = 640\ K$ (lower than the critical temperature). As previously a weak volume fraction of vapor ($\alpha_v = 10^{-4}$) is present in the liquid.

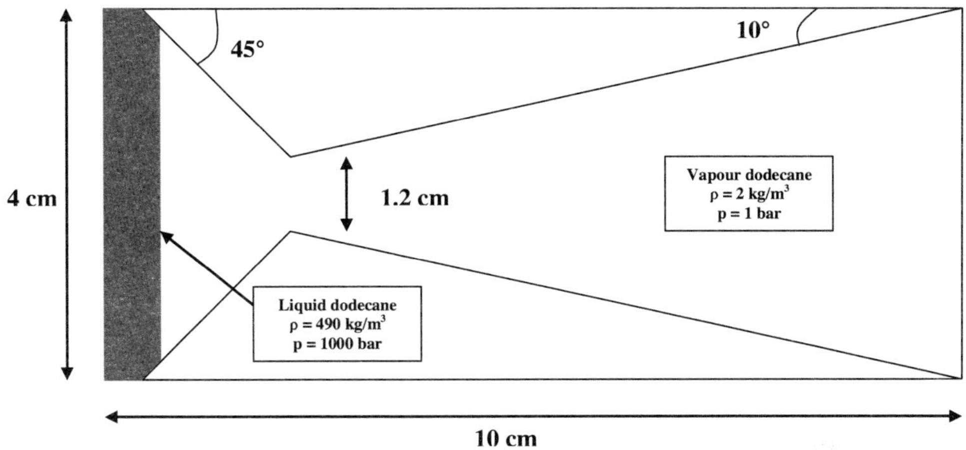

Figure 18. Initial configuration of an high pressure fuel injector

The initial conditions consist in an initial contact discontinuity between liquid dodecane at 1000 bar and 570 kg/m^3 and its vapour at atmospheric pressure. Results are shown in Figure 19 at times $t = 24\ \mu s$, $t = 60\ \mu s$, $t = 120\ \mu s$, $t = 600\ \mu s$. On the left, computed results without mass transfer are presented, and on the right those including evaporation. As in the preceding example, the results are qualitatively equals. But quantitative differences are present regarding flow variables.
It is interesting to note that the cavitation pocket sizes in both simulations are different but these differences are not very large. Two-phase nozzle flows have been used in the past to determine mass transfer rate in non-equilibrium two-phase flow models. Mass transfer rate was adjusted in order to match the cavitation pocket size. The present computations show that such method is inappropriate. Indeed, the size of the cavitation pocket has a too weak dependence regarding mass transfer. Mechanical effects alone are able to produce significant cavitation pockets.

8 Conclusion

A new hyperbolic compressible flow model has been built for the computation of cavitating flows. Relaxation effects are modeled and are able to connect the non equilibrium flow

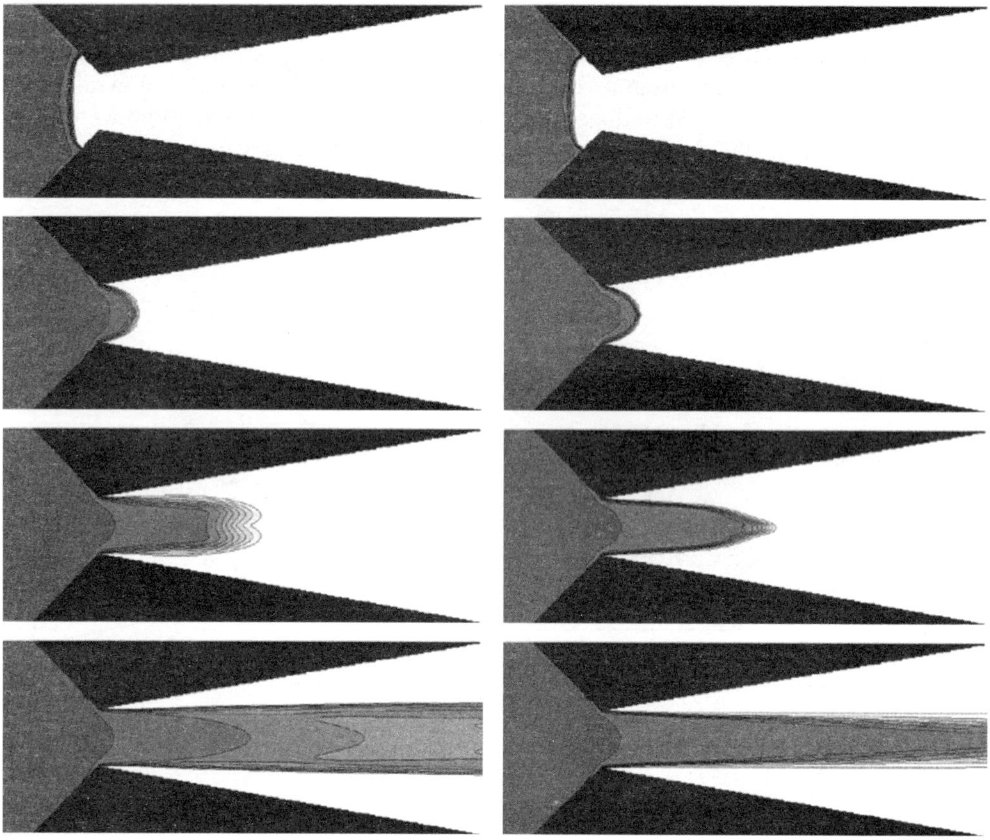

Figure 19. Flow in a high pressure fuel injector. Mixture density contours are shown at times $t = 24\ \mu s$, $t = 60\ \mu s$, $t = 120\ \mu s$ and $t = 600\ \mu s$. On the left, the solutions are obtained with the inert solver and on the right the solution are obtained with the thermo-chemical solver.

model to the mixture Euler model, valid at thermodynamic equilibrium. The connection occurs through rarefaction waves where mass transfer occurs. The overall model is able to predict dynamic evaporation waves as well as interfaces of simple contact separating non-miscible phases.

The model analysis shows that the Chapmann Jouguet kinetic relation for the dynamics of evaporation fronts (Simões-Moreira and Shepherd (1999)) is recovered. It corresponds to an expansion wave of the present model in the limit of stiff thermal and chemical relaxation.

Numerical examples show that the model is able to predict complex flow pattern in one and two dimensions.

Acknowledgements. The first author is grateful to Prof. M.V. Salvetti and Prof. L. d'Agostino for their kind invitation to the Udine summer school. He also addresses special thanks to Dr. F. Caro for fruitful discussions during his stay at Onera.

Bibliography

R. Abgrall and R. Saurel. Discrete equations for physical and numerical compressible multiphase mixtures. *Journal of Computational Physics*, 186(2):361–396, 2003.

M.R. Baer and J.W. Nunziato. A two-phase mixture theory for the deflagration-to-detonation transition (DDT) in reactive granular materials. *Int. J. of Multiphase Flows*, 12:861–889, 1986.

P.B. Butler, M.F. Lambeck, and H. Krier. Modeling of shock development and transition to detonation initiated by burning in porous propellant beds. *Comb. Flame*, 46:75–93, 1982.

J.W. Cahn and J.E. Hilliard. Free energy of a nonuniform system i: Interfacial free energy. *J. Chem. Physics*, 28(2):258, 1958.

A. Chinnayya, E. Daniel, and R. Saurel. Computation od detonation waves in heterogeneous energetic materials. *Journal of Computational Physics*, 196:490–538, 2004.

J.M. Delhaye and J.A. Boure. *General Equations and Two-Phase Flow Modeling. In: Handbook of Multiphase Systems.* In: Handbook of Multiphase Systems, edited by G. Hestroni, Hemisphere Publishing Corp., New-York, 1982.

E. Faucher, J.M. Herard, M. Barret, and C. Toulemonde. Computation of flashing flows in variable cross section ducts. *Int. J. Comp. Fluid Dynamics*, 13(4):365–391, 2000.

D.L. Frost, J.H.S. Lee, and G. Ciccarelli. The use of hugoniot analysis for the propagation of vapor explosion waves. *Shock Waves*, 1:99–110, 1991.

M. Ishii. *Thermo-Fluid Dynamic Theory of Two-Phase Flow.* Eyrolles, Paris, 1975.

D. Jamet, O. Lebaigue, N. Coutris, and J.M. Delhaye. The second gradient method for the direct numerical simulation of liquid-vapor flows with phase change. *J. Comp. Physics*, 169:624–651, 2001.

A. Kapila, R. Menikoff, J. Bdzil, S. Son, and D. Stewart. Two-phase modeling of DDT in granular materials: reduced equations. *Physics of Fluids*, 13:3002–3024, 2001.

Th. Kurschat, H. Chaves, and G.E.A. Meier. Complete adiabatic evaporation of highly superheated liquid jets. *J. Fluid Mech.*, 236:43–59, 1992.

M.S. Liou and J.R. Edwards. *AUSM Schemes and Extensions for Low Mach and Multiphase Flows.* 30th Computational Fluid Dynamics, March 8-12, Lect. Series 1999-03, Von Karman Institute for Fluid Dynamics, 1999.

O. Le Métayer, J. Massoni, and R. Saurel. Elaboration equations of state of a liquid and its vapor for two-phase flow models. *Int. J. of Thermal Sciences*, 43:265–276, 2004.

O. Le Métayer, J. Massoni, and R. Saurel. Modeling evaporation fronts with reactive riemann solvers. *Journal of Computational Physics*, 205:567–610, 2005.

R. Oldenbourg. Properties of water and steam in si-units. *Springer*, 1989.

G. Perigaud and R. Saurel. A compressible flow model with capillary effects. *J. Comp. Phys.*, 209:139–178, 2005.

F. Petitpas, E. Franquet, R. Saurel, and O. Le Métayer. A relaxation-projection method for compressible single phase and multiphase flows. Part II: Artificial heat exchanges for multiphase shocks. *Journal of Computational Physics*, 2007. In press.

P. Reinke and G. Yadigaroglu. Explosive vaporization of superheated liquids by boiling fronts. *Int. J. of Multiphase Flows*, 27:1487–1516, 2001.

R. Saurel and R. Abgrall. A multiphase Godunov method for compressible multifluid and multiphase flows. *Journal of Computational Physics*, 150:425–467, 1999.

R. Saurel, J.P. Cocchi, and P.B. Butler. A numerical study of cavitation in the wake of a hypervelocity underwater projectile. *J. of Propulsion and Power*, 15(4):513–522, 1999.

R. Saurel, E. Franquet, E. Daniel, and O. Le Métayer. A relaxation-projection method for compressible single phase and multiphase flows. Part I: The numerical equation of state for the Euler equations. *Journal of Computational Physics*, 223:822–845, 2007a.

R. Saurel, S. Gavrilyuk, and F. Renaud. A multiphase model with internal degrees of freedom: application to shock-bubble interaction. *Journal of Fluid Mechanics*, 495: 283–321, 2003.

R. Saurel and O. Le Métayer. A multiphase model for compressible flows with interfaces, shocks, detonation waves and cavitation. *Journal of Fluid Mechanics*, 431:239–271, 2001.

R. Saurel, O. Le Métayer, J. Massoni, and S. Gavrilyuk. Shock jump relations for multiphase mixtures with stiff mechanical relaxation. *Shock Waves*, 16:209–232, 2007b.

J.R. Simões-Moreira and J.E. Shepherd. Evaporation waves in superheated dodecane. *J. Fluid Mech.*, 382:63–86, 1999.

P.A. Thompson, H. Chaves, G.E.A Meier, Y.G. Kim, and H.D. Speckmann. Wave splitting in a fluid of large heat capacity. *J. Fluid Mech.*, 185:385–414, 1987.

Towards the simulation of cavitating flows in inducers through a homogeneous barotropic flow model

Maria-Vittoria Salvetti [*], E. Sinibaldi [†] and F. Beux [†]

[*] Department of Aerospace Engineering, University of Pisa, Italy

[†] Scuola Normale Superiore di Pisa, Italy

Abstract The present paper describes the starting efforts made for constructing a numerical frame aimed at simulating propellant flows occurring in the feed turbo-pumps of modern liquid propellant rocket engines. A homogeneous-flow cavitation model, accounting for thermal effects and active nuclei concentration, is considered, which leads to a barotropic state law. The 3D continuity and momentum equations for compressible inviscid flows are discretized by a finite-volume approach, applicable to unstructured grids. The numerical fluxes are computed through a shock-capturing Roe-type upwind scheme, defined for barotropic flows. The accuracy of the proposed method at low Mach numbers is ensured by *ad-hoc* preconditioning, which only modifies the upwind part of the numerical flux; thus, the time consistency is maintained and the proposed method can also be used for unsteady problems. Time advancing is carried out by an implicit linearized scheme, in which the linearization only exploits the properties of the Roe matrix. Examples of applications are provided for flow configurations of increasing geometric complexity, viz. some 1D validation benchmarks, the flow around a hydrofoil mounted in a tunnel and the flow in a turbo-pump inducer.

1 Introduction

The final goal of the present work is to set-up a numerical frame for the simulation of propellant flows occurring in the feed turbo-pumps of modern liquid propellant rocket engines, and, more precisely, the three-dimensional unsteady flows around the axial inducers which are part of the aforementioned turbo-machines.

Several factors make it challenging to accurately and efficiently simulate the considered liquid flows. Firstly, the severe weight and size constraints to which the considered engines are subjected impose high rotor speeds which, in turn, systematically entail the occurrence of cavitation phenomena. The simultaneous presence of the pure liquid (which is almost incompressible) and of the liquid-vapor mixture (which behaves like a highly compressible fluid) dramatically changes the local flow properties at the unknown interface, thus rendering common numerical methods hardly applicable. Furthermore, the very complex geometry of the axial inducers adds to the difficulty of the problem. Indeed, it imposes to adopt huge computational grids which, if unstructured, generally do no permit to straightforwardly extend some well-known numerical techniques and, in any case, require efficient, hopefully parallel, algorithms to be defined.

1.1 Overview and choice of the cavitation model

Current models for the description of cavitating flows can be classified as follows:

- *free streamline models*, where the cavity region is separated by a sharp interface from the region occupied by the pure liquid (e.g. Wu (1972));
- *equivalent fluid models*, where volume, time or ensemble averaging is used to account for the presence of two phases (e.g. Ishii (1975));
- *direct simulation models*, where the coupled Navier-Stokes equations of the two phases are solved simultaneously.

None of these models is free from inherent limitations. Free streamline models, where a well-defined interface separates the pure liquid from the cavity region occupied by the non-condensed phase, introduce prohibitive complications in 3D configurations and are not realistic in the thermal cavitation conditions typical of cryogenic propellants of rocket motors, where traveling bubble cavitation prevails (Brennen (1994)). On the other hand, direct simulation methods are extremely demanding in terms of computational resources and their superior accuracy is eluded in practice by the uncertain knowledge of the initial state of the system, especially the nature, concentration and susceptibility of cavitation nuclei.

This brief overview indicates that the successful choice of a model for simulating cavitation in technical applications must be based on careful consideration of the final objectives and implementation constraints, in order to exploit all opportunities to simplify the formulation of the problem by including only the essential physical phenomena. The typical requirements of space propulsion applications put especial prize on the suction and dynamic performance of the turbo-pump rather than on its resistance to erosion and other long-term effects of cavitation, which are typically a major concern in other applications. These properties are essentially related to the large-scale characteristics of the flow field, while the local behavior of the cavities, on the other hand, mainly controls aspects such as erosion, high frequency vibrations and noise, which are less important in liquid propellant rocket engines in view of their limited expected life time. These considerations indicate the opportunity of choosing an equivalent fluid model where the fine details of the cavity growth and collapse are neglected, and the cavitating flow is visualized in terms of a single fluid, whose properties are obtained by introducing suitable simplifying assumptions. Indeed, almost all the formulations used for computations of industrial interest are based on *equivalent fluid models*, namely:

(BH) barotropic *homogeneous flow models*, according to which the pressure and the density are linked to each other by an invertible relation within both the pure liquid and the cavitating mixture. Examples of this kind of approach may be found, for instance, in Choi et al. (2003), Coutier-Delgosha and Astolfi (2003), Coutier-Delgosha et al. (2005a), Coutier-Delgosha et al. (2005b), Coutier-Delgosha et al. (2005c), Coutier-Delgosha et al. (2001), Hoeijmakers et al. (1998), Janssens et al. (1997), Pouffary et al. (2003), Qin et al. (2003), Reboud et al. (1998), Song and He (1998), van der Heul et al. (1999) and van der Heul et al. (2000);

(DS) *dual species* models, in which a convection equation for the volume or mass void fraction is introduced, explicitly accounting for the mass transfer at phase transition. Examples of this approach may be found, for instance, in Barberon and

Helluy (2005), Hosangadi et al. (2003), Hosangadi et al. (2004), Kunz et al. (2000), Merkle et al. (1998), Saito et al. (2003), Senocak and Shyy (2002), Shyy et al. (2004) and Wu et al. (2003).

Moreover, in cavitating liquids with relatively high vapor pressures (like most cryogenic propellants) the transfer of heat at the interface represents the most important interaction between the two phases because of its influence on the cavity pressure (thermodynamic effect) and indirectly on the other flow variables. Conversely, in these flows mass and momentum exchanges usually play a comparatively minor role. In view of these considerations, pressure and velocity differences between the two phases can be neglected and the choice of a homogeneous flow model explicitly accounting, at least approximately, for thermal cavitation effects seems to be an efficient approach to the simulation of cavitating flows for performance predictions in space propulsion applications. Based on the previous considerations, the model proposed by d'Agostino and Rapposelli (2001) has been adopted here. In this model the liquid/vapor mixture behaves isentropically, so that it is possible to use the mixture energy balance in order to evaluate the mass interaction term accounting for evaporation/condensation phenomena between the two phases and ultimately derive a monotonic constitutive relation between the density ρ and the pressure p of the cavitating mixture (i.e. a barotropic thermal state law). In addition, the model naturally accounts for the effects of thermal cavitation by exploiting the specific properties of the thermally-controlled dynamics of cavitating spherical bubbles. Finally, if required, the effects of the active nuclei concentration in the liquid phase can be readily incorporated in the model. The essential features of the model are described in sec. 2, together with the governing equations. We wish just to stress here that, in spite of the formal simplicity of the constitutive relations provided by barotropic homogeneous-flow models, they are characterized by an extremely abrupt transition between wetted (nearly-incompressible) and cavitating regimes (highly-compressible) to be simulated at the same time. This clearly leads to formidable numerical problems.

1.2 Choice of the numerical method

In order to simultaneously deal with nearly-incompressible and highly-compressible flow regions, two opposite ways can be followed: adaptation to the compressible case of numerical pressure-based methods suitable for incompressible flows (see e.g. Karki and Patankar (1989), Mc-Guirck and G.J. (1990), van der Heul et al. (2000), Coutier-Delgosha and Astolfi (2003) and Senocak and Shyy (2002)), or, conversely, adaptation to the low Mach number limit of compressible solvers. On the basis of the literature, both the pressure-based and the density-based approaches seem to possess points of strength as well as weaknesses, so that there is no clear advantage in preferring one to the other.

Our approach belongs to the second class, i.e. compressible flow solvers preconditioned for low Mach numbers. The governing equations for compressible inviscid barotropic flows are discretized by a finite-volume approach, in which the numerical fluxes are evaluated by an ad-hoc developed adaptation of the Roe scheme, suitable for a generic barotropic fluid. This type of discretization can be applied to unstructured tetrahedral grids, thus simplifying the treatment of complex geometries, as those typical of turbo-pumps.

Standard compressible solvers, while being more efficient than modified incompressible schemes at high Mach numbers, exhibit efficiency problems as well as accuracy problems when dealing with nearly-incompressible flows. Indeed, at the low Mach number limit, the numerical difficulties are of two types: first, due to the stiffness of the equations with a very large disparity between acoustic and convective time-scales, the numerical schemes result inefficient, i.e. severe limitations of the allowable time step arise. Secondly, the accuracy of the solution is lost, as explained in Guillard and Viozat (1999) through an asymptotic analysis in power of the Mach number for both continuous and discrete equations. The derivation of a time-preconditioning was proposed in Turkel (1987) to overcome the efficiency problems. Moreover, as shown in Turkel et al. (1994) and Viozat (1997), the use of a preconditioning also solves the difficulties related to the accuracy in the low Mach number limit.

Following Guillard and Viozat (1999), the low Mach number asymptotic behavior of both the continuous and the discrete problems here considered was investigated in Sinibaldi et al. (2003) and Sinibaldi (2007). Similar difficulties as for the case of a perfect-gas state law in Guillard and Viozat (1999) were found in our case, i.e. it was observed that the discrete solution admits pressure fluctuations in space much larger than those of the exact one. Thus, the same type of preconditioning procedure as in Guillard and Viozat (1999) and Viozat (1997) was formulated for the barotropic case and it was verified that the obtained discrete formulation has an asymptotic behavior in agreement with the continuous case (see again Sinibaldi et al. (2003) and Sinibaldi (2007)). Moreover, since the adopted preconditioning only modifies the upwind part of the flux function, the scheme remains consistent with the time-dependent equations, and thus, the resulting method can be also used for unsteady flows (see Guillard and Viozat (1999); Viozat (1997); Sinibaldi et al. (2003); Sinibaldi (2007)).

The space-discretized equations are then advanced in time (method of lines). Since it was observed (Sinibaldi et al. (2003); Sinibaldi (2007)) that the preconditioning at hand restricts the stability region of common explicit time-advancing schemes, leading to unacceptable decrease of efficiency, a linearized implicit time-advancing strategy has been proposed, in which the linearization is based only on the algebraic properties of the Roe flux function. The implicit formulation has also been extended to the preconditioned scheme. Finally, the set-up numerical methodology was implemented within a parallel framework, using MPI.

All the previous numerical ingredients are briefly described in Sec. 3.

1.3 Examples of applications

A few examples of application of the set-up numerical methodology are presented in Sec. 4 for flow configurations of increasing geometric complexity.

First, two different 1D test-cases have been considered for validation. The first one is the quasi-1D non cavitating liquid flow in a duct and is aimed to appraise the accuracy of the preconditioning strategy and the numerical efficiency of the linearized implicit scheme for a smooth non cavitating flow. The second 1D test-case is a Riemann problem, whose exact solution has been derived in Sinibaldi (2007) for a generic barotropic law (and, thus, for the considered state law also). The chosen initial conditions lead to the presence of a cavitating region in the flow, and, thus, this provides an exact benchmark to accurately investigate the behavior of the considered numerical scheme at cavitation inception (thus addressing most of the difficulties related to the phase transition, as described by a homogeneous flow model).

As a second test-case, the cavitating flow around a NACA0015 hydrofoil mounted in a tunnel is considered. A quantitative appraisal of the accuracy of the numerical results can be obtained by comparison with experimental pressure data Cervone et al. (2006) and with the results obtained through a different numerical code Beux et al. (2005). For sake of brevity, we only report here the validation against experimental data and we refer to Beux et al. (2005) for the comparison with other numerical results.

Finally, the rotating flow in a complex 3D configuration, which realistically represents an axial turbo-pump inducer, is simulated in non-cavitating conditions.

2 Governing equations

A weakly-compressible liquid at constant temperature T_L is considered as working fluid. The liquid density ρ is allowed to locally fall below the saturation limit $\rho_{Lsat} = \rho_{Lsat}(T_L)$ thus originating cavitation phenomena. A regime-dependent (wetted/cavitating) constitutive relation is therefore adopted.

As for the wetted regime ($\rho \geq \rho_{Lsat}$), a barotropic model of the form:

$$p = p_{sat} + \frac{1}{\beta_{sL}} \ln\left(\frac{\rho}{\rho_{Lsat}}\right) \tag{2.1}$$

is adopted, $p_{sat} = p_{sat}(T_L)$ and $\beta_{sL} = \beta_{sL}(T_L)$ being the saturation pressure and the liquid isentropic compressibility, respectively. As previously discussed in the Introduction, for the cavitating regime ($\rho < \rho_{Lsat}$), a homogeneous-flow model explicitly accounting for thermal cavitation effects and for the concentration of the active cavitation nuclei in the pure liquid has been adopted (d'Agostino and Rapposelli (2001)). According to this model, pressure and velocity differences between the liquid and the vapor phases are neglected and the mixture density ρ is directly related to the void fraction α: $\rho = \rho_{Lsat}(1-\alpha)$. By assuming that the evaporation/condensation phenomena taking place at the liquid-vapor interface do not alter the global entropy of the mixture, the model exploits the energy balance to derive a (differential) barotropic relation for the cavitating mixture as well:

$$\frac{p}{\rho}\frac{d\rho}{dp} = (1-\alpha)\left[(1-\varepsilon_L)\frac{p}{\rho_{Lsat}a_{Lsat}^2} + \varepsilon_L g^\star \left(\frac{p_c}{p}\right)^\eta\right] + \alpha\left[\frac{1}{\gamma_V}\right] \tag{2.2}$$

where $g^\star$, η, γ_V and p_c are liquid parameters, a_{Lsat} is the liquid sound speed at saturation and $\varepsilon_L = \varepsilon_L(\alpha, \delta_T/R)$ is a given function representing the volume fraction ($0 \leq \varepsilon_L \leq 1$) of the liquid that is in thermal equilibrium with the bubbles, δ_T/R being a free parameter accounting for thermal effects and the concentration of the active cavitation nuclei (see d'Agostino and Rapposelli (2001) for more details). By assigning the value of $dp/d\rho(\rho = \rho_{Lsat})$ as provided by (2.1) to a_{Lsat}^2 and starting the (a priori) integration of (2.2) from $p(\rho = \rho_{Lsat}) = p_{sat}$, a C^1-continuous junction is obtained between the two branches of the constitutive relation. Thanks to the physical foundations of the pure liquid and the cavitation models, $dp/d\rho$ is always strictly positive and may be regarded to as the square of the fluid sound speed a. Despite the model simplifications leading to a unified barotropic state law (only depending on T_L and δ_T/R), the transition between wetted and cavitating regimes is extremely abrupt, as shown in the upper-right corner of Fig. 1. Indeed, the fluid sound speed falls from $O(10^3)$ m/s in the pure liquid down to $O(10^{-1} \div 10^0)$ m/s in the mixture, while the void fraction increment just below the saturation point is of the order of $10^{-6} \div 10^{-4}$. As mentioned in the Introduction, the corresponding Mach number variation renders this state law very stiff from a numerical viewpoint.

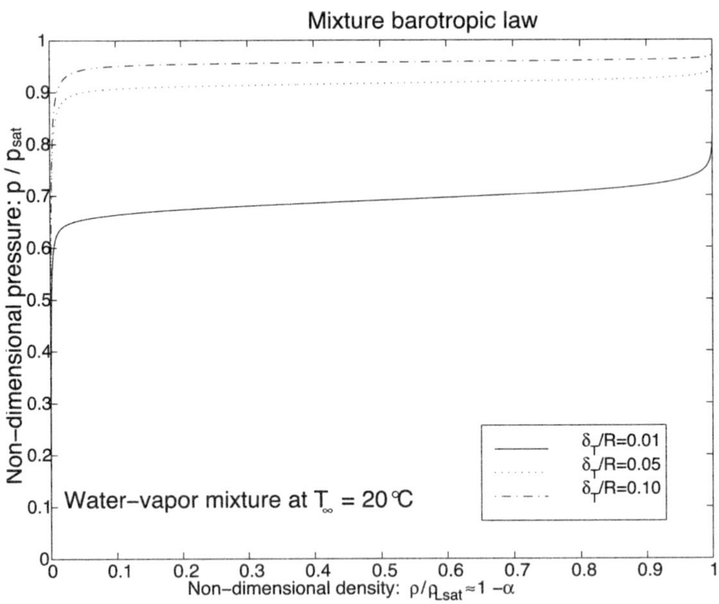

Figure 1. Pressure variation with density for a water-vapor mixture a $T_L = 20°$ C. Water parameters: $g^\star = 1.67$, $\eta = 0.73$, $\gamma_V = 1.28$ and $p_c = 2.21 \cdot 10^7$ Pa.

Consistently with the fact that viscous effects are usually negligible, at least at a preliminary stage, with respect to the huge dynamic actions typical of modern hydraulic turbo machinery, the 3D Euler equations for an inviscid fluid are considered as governing equations. By virtue of the barotropic state law, the energy balance is decoupled from the mass and momentum balances; hence, it is possible to consider the following reduced set of governing equations:

$$\partial_t W + \vec{\nabla} \cdot \vec{\mathcal{F}}(W) = S(W) \tag{2.3}$$

where $W = (\rho, \rho u, \rho v, \rho w)^T$ is the conservative variables vector (u,v and w denoting the components of the velocity vector $\vec{u}$), $\vec{\mathcal{F}}(W) = (F_x, F_y, F_z)^T$ and:

$$F_x(W) = \begin{pmatrix} \rho u \\ \rho u^2 + p \\ \rho uv \\ \rho uw \end{pmatrix} \quad F_y(W) = \begin{pmatrix} \rho v \\ \rho uv \\ \rho v^2 + p \\ \rho vw \end{pmatrix} \quad F_z(W) = \begin{pmatrix} \rho w \\ \rho uw \\ \rho vw \\ \rho w^2 + p \end{pmatrix}$$

The source term at the right-hand side of Eq. (2.3) accounts for centrifugal and Coriolis effects, when the reference frame rotates with a constant angular velocity $\vec{\omega}$, as in the simulation of flows in inducers. The source term has the following expression:

$$S(W) = \begin{pmatrix} 0 \\ \rho((\omega_y^2 + \omega_z^2)x - \omega_x(\omega_y y + \omega_z z)) - 2(\omega_y \rho v - \omega_z \rho w) \\ \rho((\omega_x^2 + \omega_z^2)y - \omega_y(\omega_x x + \omega_z z)) - 2(\omega_z \rho u - \omega_x \rho w) \\ \rho((\omega_x^2 + \omega_y^2)y - \omega_z(\omega_x x + \omega_y y)) - 2(\omega_y \rho w - \omega_z \rho v) \end{pmatrix}$$

The source term can also be rewritten in a vectorial form as: $-\rho\vec{\omega} \wedge (\vec{\omega} \wedge \vec{x}) - 2\vec{\omega} \wedge (\rho\vec{u})$.

The constitutive law $p = p(\rho)$ (Eqs. (2.1) and (2.2)) provides the closing relation.

3 Numerical Ingredients

The framework of our numerical formulation is based on a numerical solver (Farhat et al. (1999)) for the simulation of compressible, viscous, perfect-gas flows on tetrahedral unstructured grids, which has been adapted to a generic barotropic state law.

3.1 Space discretization

As far as the space discretization is concerned, the finite-volume approach already present in the original solver has been maintained. Starting from an unstructured tetrahedral grid, a dual finite-volume tessellation is obtained by the rule of medians: a cell C_i is built around each vertex i, and boundaries between cells are made of triangular interface facets. Each of these facets has a mid-edge, a face centroid, and a tetrahedron centroid as vertices.

The semi-discrete balance applied to cell C_i reads (not accounting for boundary contributions):

$$V_i \frac{dW_i}{dt} + \sum_{j \in K(i)} \Phi_{ij} = S_i \tag{3.1}$$

where $W_i = (\rho_i, \rho_i u_i, \rho_i v_i, \rho_i w_i)^T$ is the semi-discrete unknown associated with C_i, V_i is the cell volume, $K(i)$ represents the set of nodes joined to N_i through an edge and Φ_{ij} denotes the numerical flux crossing the boundary ∂C_{ij} shared by C_i and C_j (positive towards C_j). Finally S_i is semi-discrete form of the source term appearing on the right-hand-side of Eq. (2.3).

Once defined $\vec{\nu}_{ij}$ as the integral over ∂C_{ij} of the outer unit normal to the cell boundary, it is possible to approximate Φ_{ij} by exploiting a 1D flux function between W_i and W_j, along the direction $\vec{\nu}_{ij}$. By adopting the Roe numerical flux function Roe (1981), as for the original solver, Φ_{ij} reads:

$$\Phi_{ij}\left(W_i, W_j, \vec{\nu}_{ij}\right) = \frac{1}{2}\left(\vec{\mathcal{F}}(W_i) + \vec{\mathcal{F}}(W_j)\right) \cdot \vec{\nu}_{ij} - \frac{1}{2}|\tilde{A}_{ij}|(W_j - W_i) \tag{3.2}$$

where $\tilde{A}_{ij}\left(W_i, W_j, \vec{\nu}_{ij}\right)$ is a suitable *R*oe matrix, which must be diagonalizable with real eigenvalues λ_{ij}: $\tilde{A}_{ij} = T_{ij} Diag(\lambda_{ij}) T_{ij}^{-1}$; $|\tilde{A}_{ij}|$ is consequently well-defined as $|\tilde{A}_{ij}| = T_{ij} Diag(|\lambda_{ij}|) T_{ij}^{-1}$. Moreover, the Roe matrix should satisfy the following properties:

$$\lim_{W_i, W_j \to W^*} \tilde{A}_{ij} = \frac{\partial \vec{\mathcal{F}}}{\partial W}(W^*) \cdot \vec{\nu}_{ij}$$

$$\tilde{A}_{ij}\left(W_j - W_i\right) = \left(\vec{\mathcal{F}}(W_j) - \vec{\mathcal{F}}(W_i)\right) \cdot \vec{\nu_{ij}}$$

A suitable Roe matrix for a generic barotropic fluid has been defined (see Sinibaldi et al. (2003) or Sinibaldi et al. (2004)), namely:

$$\tilde{A}_{ij} = \begin{pmatrix} 0 & \nu_{ij}^x & \nu_{ij}^y & \nu_{ij}^z \\ \nu_{ij}^x \tilde{a}_{ij}^2 - \tilde{u}_{ij} X_{ij} & \nu_{ij}^x \tilde{u}_{ij} + X_{ij} & \nu_{ij}^y \tilde{u}_{ij} & \nu_{ij}^z \tilde{u}_{ij} \\ \nu_{ij}^y \tilde{a}_{ij}^2 - \tilde{v}_{ij} X_{ij} & \nu_{ij}^x \tilde{v}_{ij} & \nu_{ij}^y \tilde{v}_{ij} + X_{ij} & \nu_{ij}^z \tilde{v}_{ij} \\ \nu_{ij}^z \tilde{a}_{ij}^2 - \tilde{w}_{ij} X_{ij} & \nu_{ij}^x \tilde{w}_{ij} & \nu_{ij}^y \tilde{w}_{ij} & \nu_{ij}^z \tilde{w}_{ij} + X_{ij} \end{pmatrix}$$

where $X_{ij} = \tilde{u}_{ij}\nu_{ij}^x + \tilde{v}_{ij}\nu_{ij}^y + \tilde{w}_{ij}\nu_{ij}^z$ and $\tilde{u}_{ij}$, $\tilde{v}_{ij}$ and $\tilde{w}_{ij}$ are the following *R*oe averages (Roe (1981)):

$$\begin{cases} \tilde{u}_{lr} = \dfrac{\sqrt{\rho_l}u_l + \sqrt{\rho_r}u_r}{\sqrt{\rho_l} + \sqrt{\rho_r}} \\[2ex] \tilde{v}_{lr} = \dfrac{\sqrt{\rho_l}v_l + \sqrt{\rho_r}v_r}{\sqrt{\rho_l} + \sqrt{\rho_r}} \\[2ex] \tilde{w}_{lr} = \dfrac{\sqrt{\rho_l}w_l + \sqrt{\rho_r}w_r}{\sqrt{\rho_l} + \sqrt{\rho_r}} \end{cases} \tag{3.3}$$

Finally, $\tilde{a}_{ij}^2$ is given by (Sinibaldi et al. (2003)):

$$\tilde{a}_{ij}^2 = \begin{cases} \dfrac{p_j - p_i}{\rho_j - \rho_i} & if \mid \rho_j - \rho_i \mid > 0 \\[2ex] a^2(\rho_i) = a^2(\rho_j) & otherwise \end{cases} \tag{3.4}$$

As far as the numerical implementation is concerned, the condition in (3.4) must be replaced with: $\mid \rho_j - \rho_i \mid > \epsilon$, where ϵ is a suitable numerical threshold. If $\mid \rho_j - \rho_i \mid < \epsilon$, then $\tilde{a}_{ij}^2$ must be computed as $a^2(\bar{\rho}_{ij})$, where $\bar{\rho}_{ij}$ is an average value tending to $\rho(W^*)$ when both W_i and W_j tend to the same value W^* (for instance $\bar{\rho}_{ij} = \sqrt{\rho_i \rho_j}$).

As for the source term in the right-hand-side of Eq. (2.3), once introduced the following position:

$$\vec{\sigma}_i = -\vec{\omega} \wedge \left(\vec{\omega} \wedge \frac{1}{V_i} \int_{C_i} \vec{x} dV \right)$$

the corresponding semi-discrete source term S_i now appearing in the right-hand-side of eq. (3.1) reads:

$$S_i = V_i \begin{pmatrix} 0 \\ \rho_i \vec{\sigma}_i - 2\vec{\omega} \wedge \rho_i \vec{u}_i \end{pmatrix} \tag{3.5}$$

3.2 Preconditioning at low Mach numbers

It is well known that compressible solvers exhibit accuracy problems when dealing with nearly-incompressible flows. An explanation of this behavior, for the case of the Roe flux function associated with a perfect-gas state law, has been proposed by Guillard and Viozat (1999) by applying an asymptotic analysis in power of a Mach number, $M_\star$, characteristic of the nearly-incompressible limit of the flow. By performing the same kind of analysis for a generic barotropic fluid in a 1D context, it has been shown (Sinibaldi et al. (2003) and Sinibaldi (2007)) that the nearly-incompressible limit of the semi-discrete solution admits pressure fluctuations in space of the order of $M_\star$, which are larger than those allowed by the analytical one, of the order of $M_\star^2$.

In order to counteract this discrepancy, a Turkel-like preconditioning has been adopted, of the same kind of that proposed in Guillard and Viozat (1999). A preconditioned flux function is obtained by replacing $|\tilde{A}_{ij}|$ in (3.2) with $P_{ij}^{-1}|P_{ij}\tilde{A}_{ij}|$:

$$\Phi_{ij}^{prec} = \frac{1}{2}\left(\vec{\mathcal{F}}(W_i) + \vec{\mathcal{F}}(W_j)\right) \cdot \vec{\nu}_{ij} - \frac{1}{2} P_{ij}^{-1} \, |P_{ij}\tilde{A}_{ij}|(W_j - W_i) \tag{3.6}$$

where P_{ij} is a Turkel-type preconditioner Turkel (1987) which, as a function of the primitive variables $W_p = (p, u, v, w)^T$, reads: $P(W_p) = Diag(\beta^2, 1, 1, 1)$. Returning now to the conservative variables, the following expression of $P(W)$ is obtained:

$$P(W) = \frac{\partial W}{\partial W_p} P(W_p) \frac{\partial W_p}{\partial W} = \begin{pmatrix} \beta^2 & 0 & 0 & 0 \\ u(\beta^2-1) & 1 & 0 & 0 \\ v(\beta^2-1) & 0 & 1 & 0 \\ w(\beta^2-1) & 0 & 0 & 1 \end{pmatrix} \tag{3.7}$$

Then, using the previous expression of $P(W)$, the matrix $P(\tilde{W}_{ij})\tilde{A}_{ij} = P_{ij}\tilde{A}_{ij}$ is defined by:

$$P_{ij}\tilde{A}_{ij} = \begin{pmatrix} 0 & \beta^2 \nu^x_{ij} & \beta^2 \nu^y_{ij} & \beta^2 \nu^z_{ij} \\ \nu^x_{ij}\tilde{a}^2_{ij} - \tilde{u}_{ij}X_{ij} & \beta^2 \nu^x_{ij}\tilde{u}_{ij} + X_{ij} & \beta^2 \nu^y_{ij}\tilde{u}_{ij} & \beta^2 \nu^z_{ij}\tilde{u}_{ij} \\ \nu^y_{ij}\tilde{a}^2_{ij} - \tilde{v}_{ij}X_{ij} & \beta^2 \nu^x_{ij}\tilde{v}_{ij} & \beta^2 \nu^y_{ij}\tilde{v}_{ij} + X_{ij} & \beta^2 \nu^z_{ij}\tilde{v}_{ij} \\ \nu^z_{ij}\tilde{a}^2_{ij} - \tilde{w}_{ij}X_{ij} & \beta^2 \nu^x_{ij}\tilde{w}_{ij} & \beta^2 \nu^y_{ij}\tilde{w}_{ij} & \beta^2 \nu^z_{ij}\tilde{w}_{ij} + X_{ij} \end{pmatrix} \tag{3.8}$$

in which $X_{ij} = \vec{\tilde{U}}_{ij} \cdot \vec{\nu}_{ij}$.

In particular, $P_{ij}\tilde{A}_{ij}$ is diagonalizable with real eigenvalues:

$$\lambda^{(1)}_{ij} = \lambda^{(2)}_{ij} = X_{ij},\ \lambda^{(3)}_{ij} = \frac{1+\beta^2}{2}X_{ij} + \sqrt{\omega_{ij}} \ \text{ and } \ \lambda^{(4)}_{ij} = \frac{1+\beta^2}{2}X_{ij} - \sqrt{\omega_{ij}}$$

where:

$$\omega_{ij} = \left(\frac{1-\beta^2}{2}\right)^2 X^2_{ij} + \beta^2\, \tilde{a}^2_{ij} \|\vec{\nu}_{ij}\|^2$$

Furthermore, P_{ij} is invertible and P^{-1}_{ij} is expressed as:

$$P^{-1}_{ij} = \frac{1}{\beta^2} \begin{pmatrix} 1 & 0 & 0 & 0 \\ -(\beta^2-1)\,\tilde{u}_{ij} & \beta^2 & 0 & 0 \\ -(\beta^2-1)\,\tilde{v}_{ij} & 0 & \beta^2 & 0 \\ -(\beta^2-1)\,\tilde{w}_{ij} & 0 & 0 & \beta^2 \end{pmatrix} \tag{3.9}$$

Thus, since the matrices P^{-1}_{ij} and $|P_{ij}\tilde{A}_{ij}|$ exist, the preconditioned numerical flux function expressed in eq. (3.6) is well-defined.

It has been shown in Sinibaldi et al. (2003) and Sinibaldi (2007) that the preconditioned semi-discrete solution (i.e. that satisfying (3.1) with Φ^{prec}_{ij} in place of Φ_{ij}) recovers the same asymptotic behavior of the analytical one, provided the free parameter β^2 is of the order of $M_\star$.

It is worth nothing that the proposed preconditioning technique does not destroy the time-consistency of the discrete scheme, since it only affects the "upwind" part of the flux function (3.6) (i.e. that associated with $W_j - W_i$). As a consequence, the proposed preconditioned scheme can be exploited for unsteady computations as well.

Finally, the preconditioning should be applied only to those regions where the flow is nearly incompressible, i.e. to the pure liquid. Thus, even if the theoretical analysis has been performed for a constant β value, in some of the applications presented herein, a local preconditioning strategy has been implemented, in which β_{ij} defined locally following semi-empirical criteria that will specified in the following, is used instead of β (see Sec. 4).

3.3 Time discretization

In order to avoid the severe time-step limitations related to common explicit time-advancing techniques, a linearized implicit time-advancing was developed.

Let us start from the following fully implicit backward Euler scheme:

$$V_i \frac{\Delta^n W_i}{\Delta^n t} + \sum_{j \in K(i)} \Phi_{ij}^{n+1} = 0 \tag{3.10}$$

where $\Delta^n(\cdot) = (\cdot)^{n+1} - (\cdot)^n$, superscript n denoting the time-level of the fully-discrete solution. To avoid the solution of a non-linear system, Eq. (3.10) must be linearized. Classical linearization technique, relying on the 1-st order homogeneity of the flux function (see, e.g., Fezoui and Stoufflet (1989)), holding true for the perfect-gas state law, can not be used here, since this property is not satisfied in the barotropic case. Thus, a new linearization of the flux function (3.2) has been proposed (Sinibaldi et al. (2003)), namely:

$$\Delta^n \Phi_{ij} \cong \left(\tilde{A}_{ij}^n\right)^+ \Delta^n W_i + \left(\tilde{A}_{ij}^n\right)^- \Delta^n W_j \tag{3.11}$$

where $2\tilde{A}_{ij}^{\pm} = \tilde{A}_{ij} \pm |\tilde{A}_{ij}|$ and $\Delta^n(\cdot) = (\cdot)^{n+1} - (\cdot)^n$, superscript n denoting the time-level of the fully-discrete solution.

The above linearization is based only on the algebraic properties of the Roe flux function and it is therefore applicable to a variety of problems. It is possible to show (Sinibaldi et al. (2003); Sinibaldi (2007)) that, under some regularity assumptions, the error introduced by (3.11) is of the order of $O(\delta t^2, \delta x \delta t)$, where δt and δx are characteristic sizes of the time and space discretization, respectively. By substituting (3.11) into the fully-implicit scheme (3.10), a linear system is obtained, of the form:

$$\left[\frac{V_i}{\Delta^n t} I + \sum_{j \in K(i)} \left(\tilde{A}_{ij}^n\right)^+\right] \Delta^n W_i + \sum_{j \in K(i)} \left[\left(\tilde{A}_{ij}^n\right)^-\right] \Delta^n W_j = - \sum_{j \in K(i)} \Phi_{ij}^n \tag{3.12}$$

where I denotes the identity matrix.

When the proposed preconditioning is applied, it is possible to introduce the same kind of linearization as in (3.11), once $\tilde{A}_{ij}^{n\pm} = \tilde{A}^{\pm}(W_i^n, W_j^n)$ have been replaced with $\left(P_{ij}^n\right)^{-1} \left(P_{ij}^n \tilde{A}_{ij}^n\right)^{\pm}$ respectively. As a result, the following scheme can be obtained for the preconditioned case:

$$\left[\frac{V_i}{\Delta^n t} I + \sum_{j \in K(i)} \left(P_{ij}^n\right)^{-1} \left(P_{ij}^n \tilde{A}_{ij}^n\right)^+\right] \Delta^n W_i + \sum_{j \in K(i)} \left[\left(P_{ij}^n\right)^{-1} \left(P_{ij}^n \tilde{A}_{ij}^n\right)^-\right] \Delta^n W_j = - \sum_{j \in K(i)} \Phi_{ij}^n \tag{3.13}$$

Once solved the block tridiagonal linear system given by eq. (3.12) or eq. (3.13), the unknowns at time level $n+1$ are simply given by: $W_i^{n+1} = W_i^n + \Delta^n W_i$.

4 Applications

4.1 1D validation test-cases

A quasi-1D, inviscid, liquid flow within a duct having variable cross-sectional area $A = A(x)$ (e.g. a convergent-divergent nozzle) is considered. The effect of the cross-sectional area variation is introduced in the continuity and momentum equations through a source term having the following form:

$$\mathrm{s} = -\frac{d}{dx} \ln (A(x)) \begin{pmatrix} \rho u \\ \rho u^2 \end{pmatrix}$$

where ρ and ρu are the unknown flow variables for the considered 1D problem. A smooth solution is sought in the present context. In particular, the steady, nearly-incompressible solution described below is considered in order to define a benchmark for the proposed preconditioning strategy.

Let $M_\star \ll 1$ denote the characteristic Mach number of a flow field exhibiting a roughly constant density:

$$\rho \approx \rho_\infty \tag{4.1}$$

where the subscript ∞ hereafter refers to the inlet conditions, associated with the section located at $x = x_{min} \in [x_{min}, x_{max}]$. In consideration of (4.1), the conservation of the mass approximately reduces to the following relation:

$$u\, A \approx u_\infty\, A_\infty \tag{4.2}$$

Moreover, by invoking the well-known Bernoulli theorem (for incompressible, non-dissipativ steady flows), the momentum balance can be approximated as follows:

$$p + \frac{1}{2} \rho_\infty u^2 \approx p_\infty + \frac{1}{2} \rho_\infty u_\infty^2 \tag{4.3}$$

Then, once defined:

$$\alpha(x) := \frac{A(x)}{A_\infty} \tag{4.4}$$

it is possible to respectively recast (4.2) and (4.3) as follows:

$$\frac{u}{u_\infty} \approx \alpha^{-1} \tag{4.5}$$

$$\frac{p}{p_\infty} \approx 1 + \frac{1}{2} \frac{\rho_\infty u_\infty^2}{p_\infty} (1 - \alpha^{-2}) \tag{4.6}$$

The relations (4.5) and (4.6) provide a nearly-exact, steady solution (they tend to be exact for $M_\star \to 0$) which can be exploited for validating the discrete scheme.

Benchmark	κ	$\varkappa$	γ	ρ_∞	u_∞	x_{min}	x_1	x_2	x_{max}	σ
BN	10^6	1	0	1	1	-2000	-1000	1000	2000	$2.5 \cdot 10^{-2}$

Table 1. Main flow parameters for the quasi-1D flow in a nozzle.

As far as the variation of the cross-sectional area is concerned, the following definition is adopted, in particular, for $\alpha(x)$:

$$\alpha(x) := \begin{cases} 1 & \text{if} \quad x_{min} \leq x \leq x_1 \\ 1 - \sigma\left(1 - \cos\left(2\pi \dfrac{x - x_1}{x_2 - x_1}\right)\right) & \text{if} \quad x_1 < x < x_2 \\ 1 & \text{if} \quad x_2 \leq x \leq x_{max} \end{cases} \tag{4.7}$$

where x_1, x_2 and $0 < \sigma < 1/2$ are adjustable parameters. The function (4.7) represents a sinusoidal reduction of the cross-sectional area between x_1 and x_2. In particular, the minimum value of α is given by:

$$\alpha_{min} := 1 - 2\,\sigma \tag{4.8}$$

and it is obtained for $x = (x_1 + x_2)/2$.

The considered benchmark is summarized in Tab. 1. In this table, κ, $\varkappa$ and γ are the parameters characterizing the following barotropic state law:

$$p = \kappa \rho^{\varkappa} + \gamma$$

used for the liquid in the present case. The previous law introduces, in particular, a constant sound speed $\tilde{a} = \sqrt{\kappa} = 10^3$ for the flow. In consideration of (4.5), the maximum value of u is given by $u \approx \alpha_{min}^{-1} \approx 1.05$ and therefore $M_\star = 10^{-3}$ can be regarded to as a characteristic Mach number of the considered flow.

As far as the BCs are concerned, a Dirichlet-like inlet BC is enforced by defining the fictitious state vector $W(x_0, t)$ as follows:

$$W(x_0, t) = W_\infty \tag{4.9}$$

while the following transmissive BC is adopted at the outlet:

$$W(x_{N_c+1}, t) = W(x_{N_c}, t) \tag{4.10}$$

Two runs have been carried out; for the first one (IR-NOPREC) the proposed preconditioning strategy is not activated (indeed $\beta^2 = 1$ is used in the preconditioner (3.9)). Conversely, the second run (IR-PREC) exploits the preconditioning strategy at hand, by

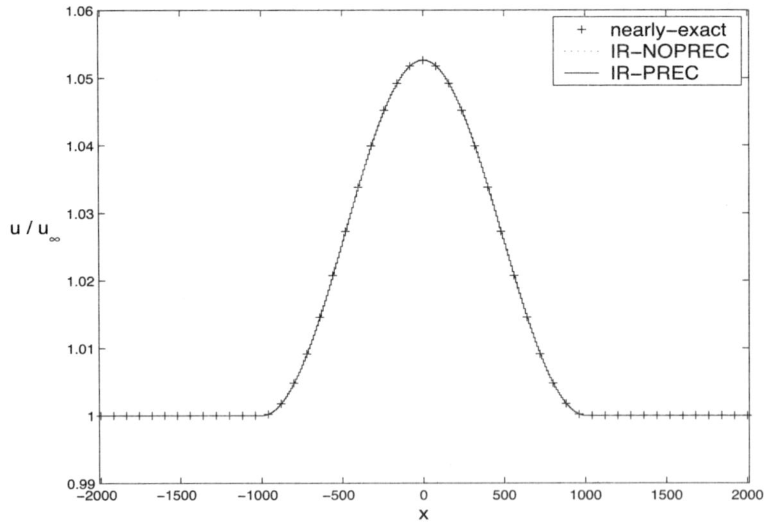

Figure 2. Approximation of u for the quasi-1D nozzle liquid flow obtained with and without local preconditioning.

choosing $\beta^2 = M_\star^2$ in (3.9). The $x-$domain $[x_{min}, x_{max}]$ (with x_{min} and x_{max} defined in the aforementioned Tab. 1) is uniformly discretized (with cells having size μ) for both cases.

The behavior of the corresponding numerical solutions is shown in Figs. 2 and 3. It should be noticed that the approximation of u does not suffer from the accuracy problems related to the low Mach number flow (its main driver being the area variation, according to (4.2)), while the approximation of p exhibits the problems discussed in Sec. 3.2. The proposed preconditioning strategy, however, seems to effectively counteract these problems, as shown in Fig. 3.

As for the computational efficiency, it turns out that, for both the considered test-cases, a practically "unbounded" time-step can be adopted for advancing the numerical solution by means of the proposed linearized implicit strategy. In the carried out numerical experiments the time step τ has been increased up to 10^5, thus reaching the steady-state solution in a very few (namely 2 to 5) iterations. Conversely, for the same test case, with a 4th-stage explicit Runge-Kutta scheme the time step was found to be limited to $5 \cdot 10^{-3}$ for the simulation without preconditioning and to $5 \cdot 10^{-6}$ with preconditioning (see Sinibaldi (2007)).

To validate the proposed numerical method in cavitating conditions, we consider the following Riemann problem (RP), whose exact solution has been derived in Sinibaldi

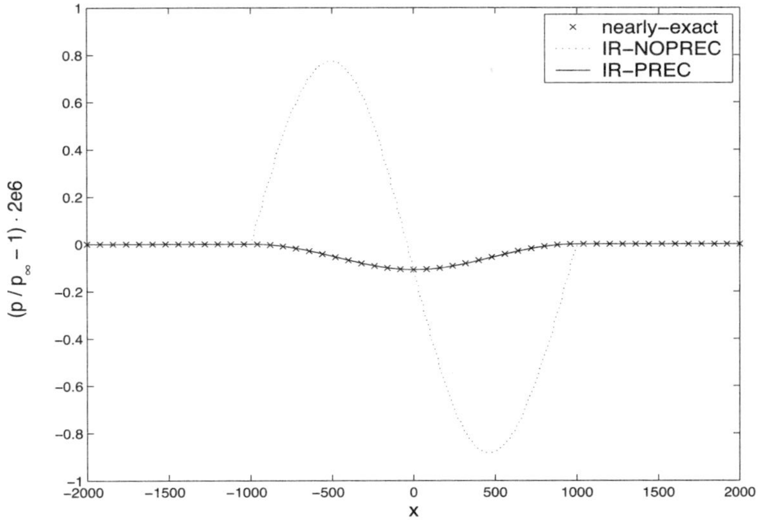

Figure 3. Approximation of p for the quasi-1D nozzle liquid flow obtained with and without local preconditioning. The pressure variation on the $y-$axis is scaled for ease of readability.

(2007) for a generic barotropic state law:

$$\begin{cases} \dfrac{\partial W(x,t)}{\partial t} + \dfrac{\partial F_x(W(x,t))}{\partial x} = 0 \\ W(x,0) = \begin{cases} W_L \text{ if } x < 0 \\ W_R \text{ if } x > 0 \end{cases} \end{cases} \tag{4.11}$$

where $W = (u, \rho u)^T$ and:

$$F_x(W) = \begin{pmatrix} \rho u \\ \rho u^2 + p \end{pmatrix}$$

The considered benchmark is defined in Tab. 2, in which δ_T/R is the free -parameter in the adopted barotropic state law for the cavitating mixture (2.2). The other relevant parameters in Eqs. (2.1) and (2.2) are set to: $1/\beta_{sL} = 8.55 \cdot 10^5$, $g^* = 1.67$, $p_c = 2.21\ 10^7$, $p_{sat} = 2086.82$, $\eta = 0.73$ and $1/\gamma_V = 0.78$. At the chosen temperature T_L, the liquid saturation density is $\rho_{Lsat} = 997.95$ and therefore the IC in Tab. 2 defines two liquid states[1]. Moreover, the speeds u_L and u_R are chosen so as to obtain two rarefactions (symmetrical with respect to the original discontinuity at $x = 0$) which lead to a cavitating star region characterized by $\rho_\star \approx 960.47 < \rho_{Lsat}$ and $u_\star = 0$ (by symmetry). The sound speed in the liquid is $a_L = a_R \approx 1415.63$ while in the cavitating region it falls down to approximately $a_{cav} \approx 0.37$; the resulting flow is entirely subsonic. In consideration of the aforementioned variation of the sound speed, it is to be expected that the star region is hardly observable as part of the solution.

[1]The SI units are tacitly understood

Benchmark	Liquid	T_L	δ_T/R	ρ_L	u_L	ρ_R	u_R	t_{eval}
B4	water	293.16	0.1	998	−0.1	998	0.1	1

Table 2. Considered benchmark for 1D cavitating flow.

Test-case	Benchmark	μ	(n_L, n_R)	τ
LdA1	B4	1	$(2,2)\cdot 10^3$	10^{-4}
LdA2	B4	1	$(2,2)\cdot 10^3$	10^{-3}
LdA3	B4	1	$(2,2)\cdot 10^3$	10^{-2}

Table 3. Considered test-cases.

For practical reasons a limited domain is considered, $x\epsilon[x_{min}, x_{max}]$, with $x_{min} = -200$ and $x_{max} = 200$. Zero gradient of variables is assumed at the boundaries. A uniform space discretization having measure $\mu = 1$ and a constant time-step τ have been adopted. The considered test-cases are reported in Tab. 3. The preconditioned linearized implicit scheme described in Sec. 3 has been used in all cases.

The numerical approximation of ρ, p and u is shown in Figs. 4-8. The density undergoes a spike-like variation close to the cavitating region. It is practically impossible to distinguish the head as well as the tail of the density waves, because the density variation close to rarefaction head is squashed by the considerable variation occurring towards the cavitating region. Furthermore, the width of the star region is not resolved by the adopted space discretization. When examining in Fig. 5 a narrower sub-domain around $x = 0$ it is evident that, as expected, the accuracy of the numerical solution improves when adopting smaller time-steps. The pressure exhibits a remarkably different trend with respect to the density, as shown in Fig. 6. Indeed, the head of the rarefactions is clearly visible; that of the left rarefaction, for example, is marked by P1 in the considered figure. Furthermore, most of the pressure variation takes place, in practice, near the head of the rarefactions. For instance, as far as the left rarefaction is concerned, the pressure abruptly reaches the saturation value p_{sat}, marked by P2 in Fig. 6, as well as the "right corner" of the relevant pressure curve in Fig. 1 (very close to the right margin of the figure), marked by P3 in Fig. 6. The transition between P2 and P3 is aligned with that between P1 and P2 (i.e. no abrupt changes occur when entering the mixture region). Indeed, the rightmost portion of the relevant cavitating curve in Fig. 1 is practically vertical near the saturation point (due to the very weak compressibility of the liquid) and therefore it behaves like a prolongation of the adopted liquid model. The cavity, whose width is $O\left(10^{-1}\right)$ (see above), is indicated in Fig. 6 by P4. When moving from P3 to P4, the pressure weakly decreases (the variation is not resolved in the figure) and therefore this arc is not part of the star region of the considered RP (where the solution is constant). Indeed, most of this arc corresponds to the practically horizontal portion of the relevant pressure curve in Fig. 1 and the pressure decrease occurring near the left extreme of the aforementioned curve originates the variation shown in Fig. 7. As shown in Fig. 8, the approximation of u is reasonably good, even near the cavity. In consideration of the previous points, it is clear that an accurate description of the rarefaction's tail and, more in general, of the cavity is only possible at the cost of a very fine space discretization.

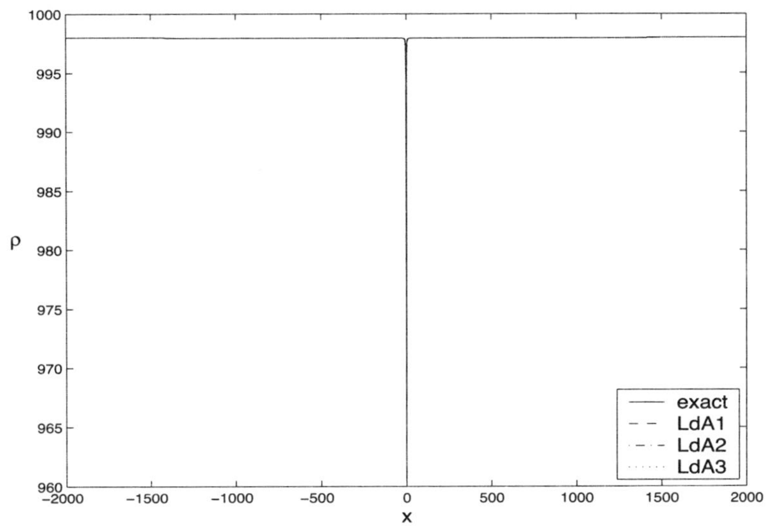

Figure 4. Approximation of ρ for the test-cases reported in Tab. 3.

As for computational efficiency, it appears from Tab. 3 that the maximum allowable time step is severely reduced when considering cavitating flows. Following the results of the investigations in Sinibaldi (2007) this is probably due to the occurrence of noticeable discontinuities -especially as far as the Mach number and the density are concerned- which are associated with the inception of cavitation. However, this problem does not seem to be specifically introduced by the proposed linearization technique (see Sinibaldi (2007)). Furthermore, it has been also observed by performing a rather extensive number of numerical simulations Bilanceri (2005), based on the proposed linearization and involving a different homogeneous flow cavitation model (namely the instance of the well-known barotropic cavitation model of Delannoy which is reported in Coutier-Delgosha et al. (2005b)).

4.2 Flow around a NACA0015 hydrofoil

The present benchmark problem concerns the numerical simulation of the liquid flow around a NACA0015 hydrofoil in cavitating conditions and was one of the test-cases considered at the conference "Mathematical and Numerical aspects of Low Mach Number Flows", Porquerolles, France, June 21-25 2004 (Beux et al. (2005)).

The hydrofoil has a chord $c = 115$ mm and mounted within a water tunnel at 4° angle of attack. The geometry of the test-chamber is sketched in Fig. 10 while the test-section, which is obtained by cutting the chamber along its symmetry plane, is sketched in Fig. 11. The considered liquid is water at a temperature $T_L = 293.16$ K. Let the subscript ∞ denote the free-stream (unperturbed) flow, the following reference experimental conditions are considered, for which experimental data are available in Cervone et al. (2006): $p_\infty = 12000$ Pa, $\|\mathbf{u}_\infty\| = 3.46$ m/s, $M_\infty = 2.3 \cdot 10^{-3}$ and $\sigma_\infty = 1.5$.The velocity $\mathbf{u}_\infty$ is orthogonal to the inlet section. The symbols M_∞ and σ_∞ respectively denote the free-stream Mach number, defined as follows:

$$M_\infty := \frac{\|\mathbf{u}_\infty\|}{a_\infty} \tag{4.12}$$

and the cavitation number, defined as follows (Brennen (1994)):

$$\sigma_\infty := \frac{p_\infty - p_{sat}}{\frac{1}{2}\,\rho_\infty\,\|\mathbf{u}_\infty\|^2} \tag{4.13}$$

In view of the definition (4.13), it is clear that cavitation phenomena are likely to take place in correspondence of low cavitation numbers. The considered conditions are associated with a cavitating flow and, at the given liquid temperature, an extremely abrupt transition between non-cavitating and cavitating flow regions occurs (see Fig. 1), whose numerical treatment is particularly tough.

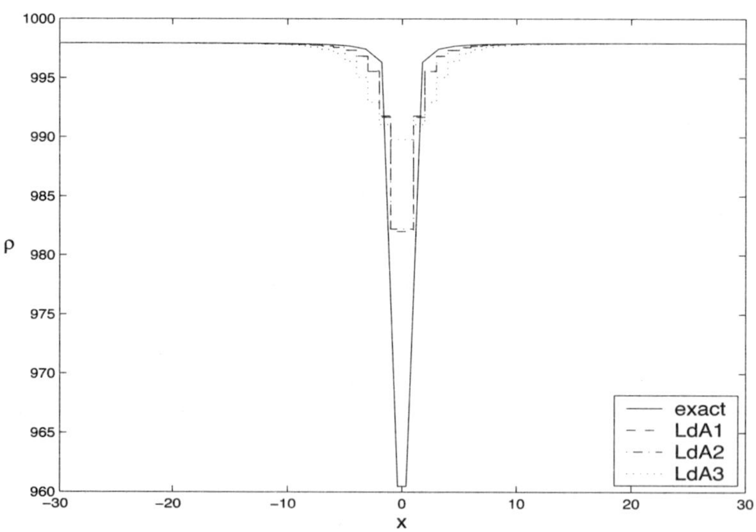

Figure 5. Detail of Fig. 4.

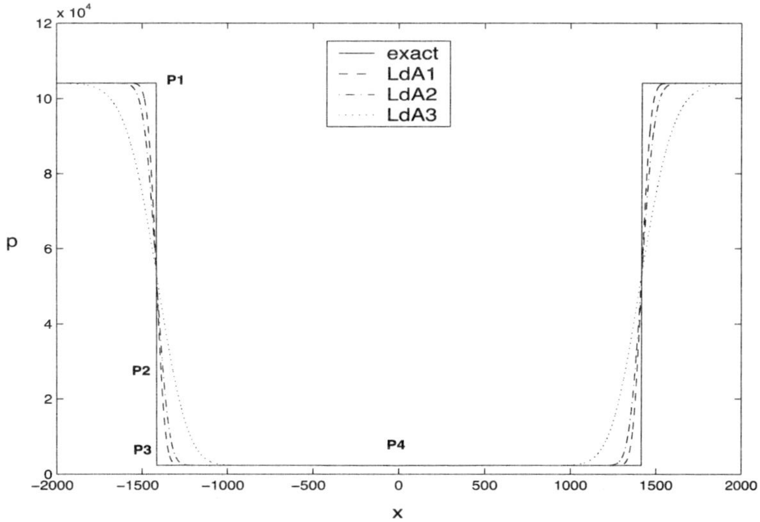

Figure 6. Approximation of p for the test-cases reported in Tab. 3. The labels P1-P4 are added for ease of discussion.

The domain sketched in Fig. 10 is discretized by means of a 3D tetrahedral unstructured grid. The considered grids are reported in Tab. 4, in which the symbols N_c and N_t respectively denote the number of cells (i.e. nodes) and elements (i.e. tetrahedrons).

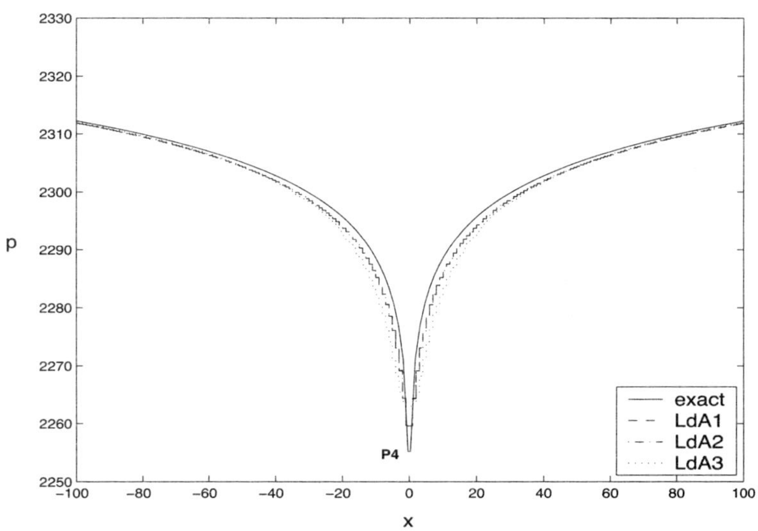

Figure 7. Detail of Fig. 6.

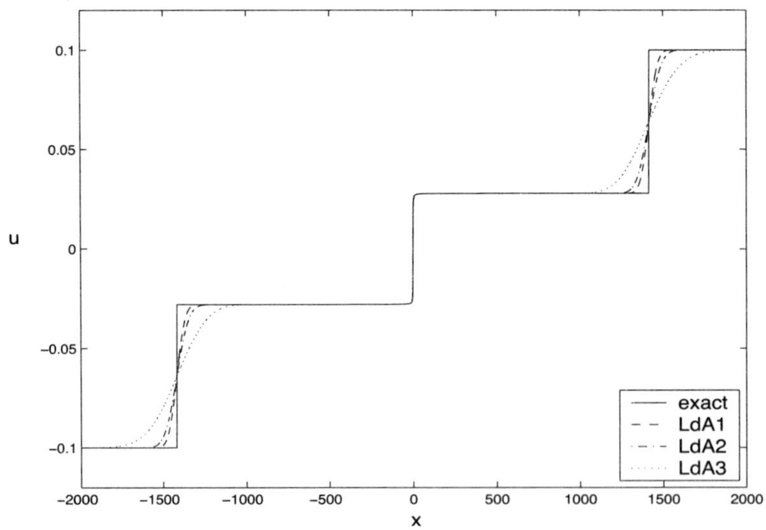

Figure 8. Approximation of u for the test-cases reported in Tab. 3.

Neither GR1 nor GR2 is highly refined in order to contain the computational cost of the simulations while validating/developing the considered numerical schemes (examples of finer grids discretizing the domain under consideration can be found in Beux et al. (2005)). While GR1 represents the whole test-chamber, GR2 only discretizes a "slice"

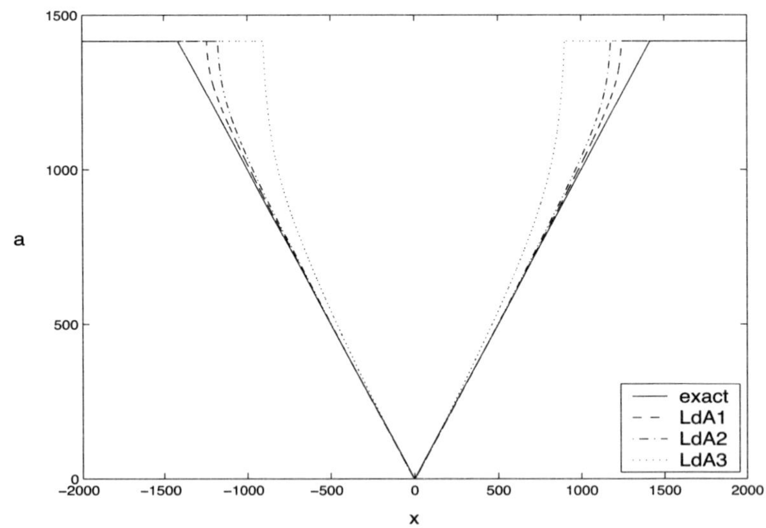

Figure 9. Approximation of a for the test-cases reported in Tab. 3.

of it (its span-wise width being 0.1 c instead of 0.7 c) and it is used for reducing the computational cost while validating/developing the considered numerical schemes. A variable time-step is adopted, defined as follows:

$$\delta^n t = c^{(CFL)\,n} \min_{h\in\mathcal{H}} \left(\frac{\lambda_h}{\tilde{s}_h^n} \right) \tag{4.14}$$

where:

- λ_h denotes the minimum among the four heights which are associated with the $h-$th tetrahedron T_h $(h \in \mathcal{H} := \{1, \ldots, N_t\})$;
- $\tilde{s}_h^n$ denotes the value at time-level n of an estimate of the maximum wave speed $\tilde{s}_h$ associated with T_h. More precisely, $\tilde{s}_h$ is chosen as the maximum among the wave speeds arising in the Roe-linearized RPs associated with the four vertices of T_h;
- $c^{(CFL)\,n}$ denotes the value at time-level n of a CFL-like coefficient, which can be modulated during the simulation.

Since, as shown by the experiments, the flow associated with the considered free-stream conditions, although cavitating, turns out to be substantially steady due to the low angle of attack, the numerical simulations are advanced in time up to a steady-state.

The free-stream Mach number, M_∞, is assumed to be the characteristic Mach number of the liquid region. However, in consideration of the fact that no preconditioning is required within the cavitating region (where the flow can be easily hypersonic), a local preconditioning strategy is heuristically adopted. More precisely, a local preconditioning parameter β_{ij}^2, defined as follows:

$$\beta_{ij}^2 := \begin{cases} M_\infty^2 & \text{if} \quad \min(\rho_i, \rho_j) \geq \rho_{Lsat} \\ 1 & \text{otherwise} \end{cases}$$

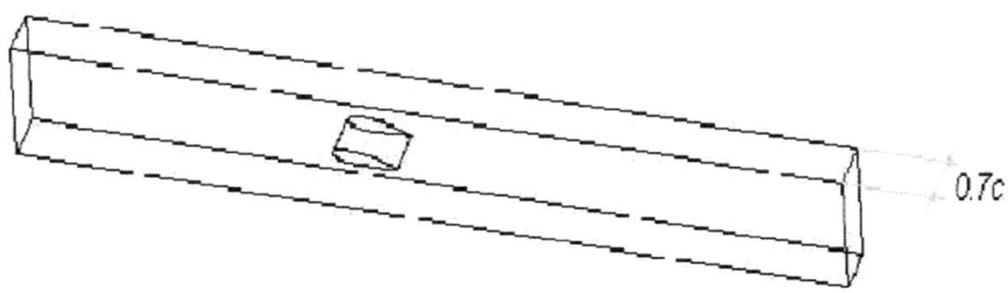

Figure 10. Sketch of the 3D test-chamber.

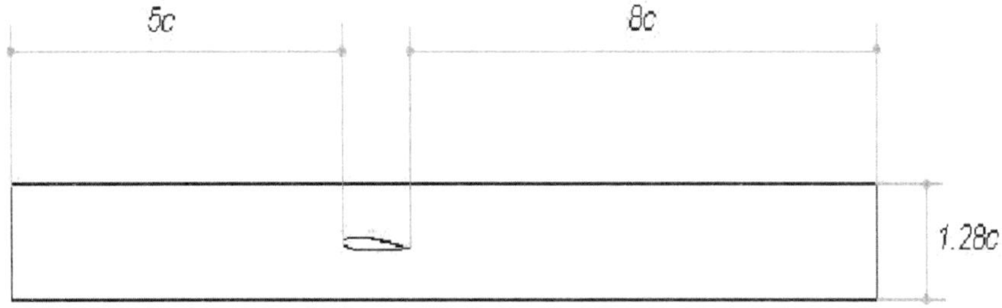

Figure 11. Sketch of the test-section.

Grid	N_c	N_t
GR1	27220	137756
GR2	19322	88400

Table 4. Computational grids used for the simulation of the flow around a NACA0015 hydrofoil.

Test-case	δ_T/R	Grid
CAV1	0.1	GR1
CAV2	0.1	GR2
CAV3	0.01	GR2

Table 5. Different simulations of the cavitating flow around a NACA0015 hydrofoil.

is introduced into the preconditioning matrix (3.7) in place of the original parameter β^2.

Different simulations have been carried out characterized by the parameters reported in Tab. 5. In In Figs. 12 and 13, the distribution of the pressure coefficient, obtained from the considered simulations on the suction side of the hydrofoil, is compared with the aforementioned experimental measurements. The pressure coefficient is defined as:

$$C_p := \frac{p - p_\infty}{\frac{1}{2}\,\rho_\infty \,\|\mathbf{u}_\infty\|^2}$$

First, it is worth noticing that only small differences, located near the leading edge of the hydrofoil (i.e. where cavitation occurs), appear in the C_p distribution when adopting different grids (see Fig. 12). Second, the agreement between the numerical results and

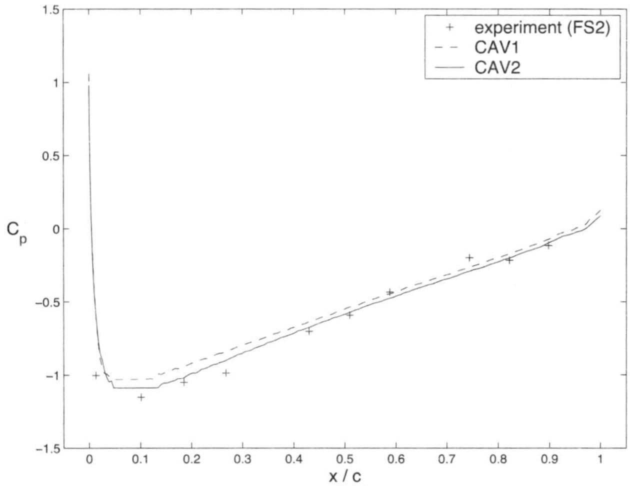

Figure 12. C_p distribution (suction side) for the test-cases CAV1 and CAV2 reported in Tab. 5.

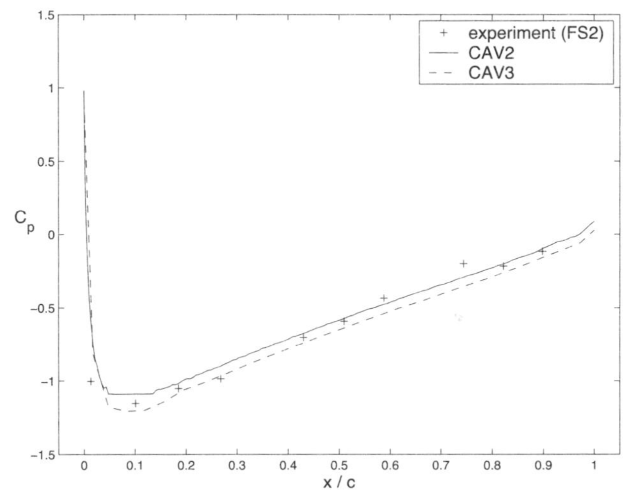

Figure 13. C_p distribution (suction side) for the test-cases CAV2 and CAV3 reported in Tab. 5.

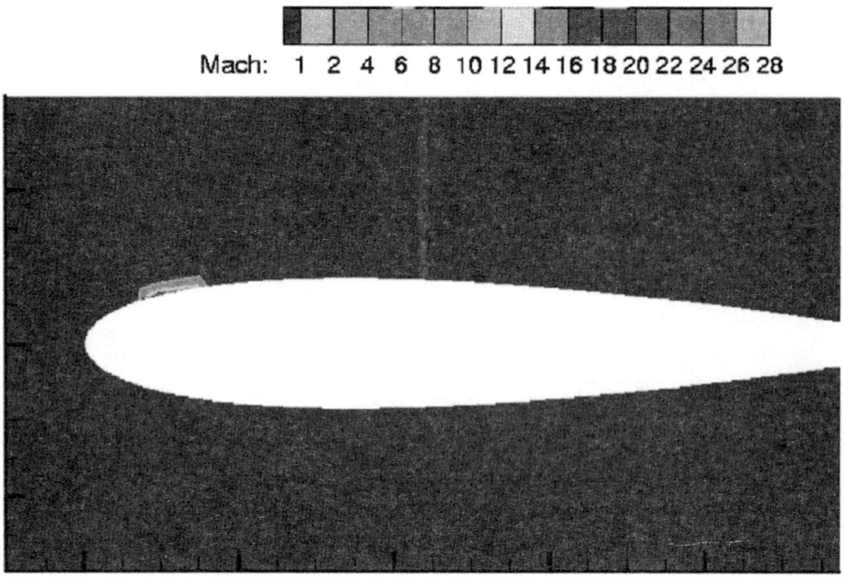

Figure 14. Contour plot of the local Mach number on the test-section (detail), for the test-case CAV2 reported in Tab. 5.

the experimental data can be considered reasonably good, in view of the fact that it is very challenging to accurately simulate the cavitation phenomena at hand. On the basis of the numerical results in Fig. 13, it seems that the C_p distribution smoothly varies with respect to δ_T/R. Moreover, a lower value of δ_T/R correctly leads to a less pronounced Mach number variation as shown in Figs. 14 and 15. Indeed, the minimum sound speed -in the cavitating region- increases when reducing δ_T/R.

Furthermore, once defined a local cavitation number as follows (compare with (4.13)):

$$\sigma := \frac{p - p_{sat}}{\frac{1}{2}\,\rho_\infty \left\|\mathbf{u}_\infty\right\|^2}$$

it is possible to identify the cavity with the fluid sub-domain within which $\sigma < 0$. Then, as shown in Figs. 16 and 17, it is possible to see that a lower value of δ_T/R results in a more extended cavity. Also this result seems to be correct, since the nearly constant pressure value in Fig. 1, which roughly provides a characteristic value of the cavity pressure, decreases when decreasing δ_T/R.

As for the computational efficiency, before the inception of cavitation, the coefficient $c^{(CFL)}$ introduced in (4.14) can be increased during the simulation up to $\mathrm{O}\left(10^2\right)$ for all the considered cavitating test-cases However, as soon as cavitation occurs, it must be reduced to $\mathrm{O}\left(10^{-2}\right)$ for all the considered simulations to remain stable. This observed stability restriction is probably caused by the presence/onset of discontinuities in the

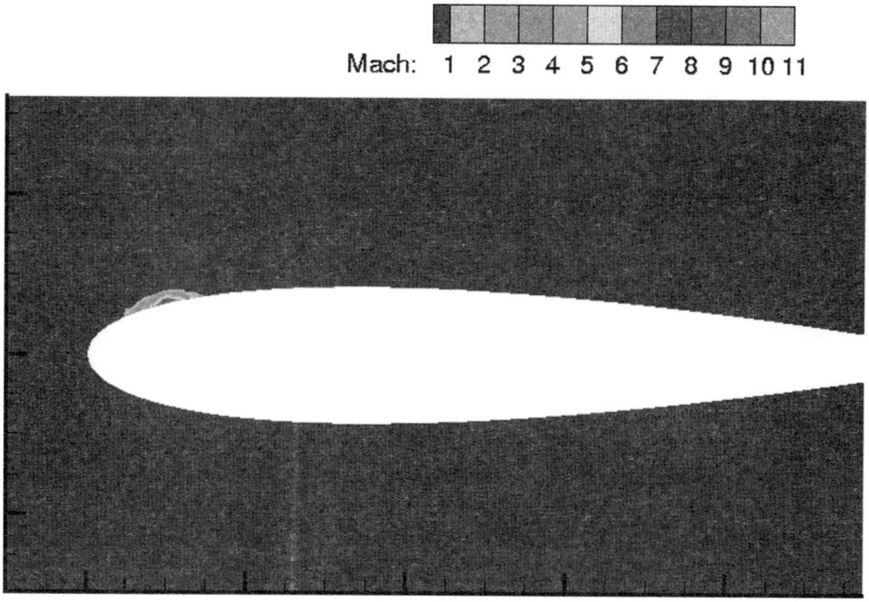

Figure 15. Contour plot of the local Mach number on the test-section (detail) for the test-case CAV3 reported in Tab. 5.

flow field (caused by the cavitation inception in the present case), which render it more difficult to exploit the proposed linearized scheme (see also sec. 4.1). However, this point definitely deserves further investigation in a future stage of the present work.

4.3 3D rotating flow in a turbo-pump inducer

The water flow in a turbo-pump inducer is considered in the present section, as a qualitative validation benchmark for the set-up numerical methodology.

In consideration of the efficiency problems already discussed in Secs. 4.1 and 4.2, no cavitating simulations are considered for the inducer flow at hand. Indeed, the huge increase in computational cost, which is here amplified by the complexity of the considered geometry (see below), makes it practically impossible to advance the simulation unless exploiting huge computational resources. Indeed, a cavitating simulation was stopped at the inception stage, due to the aforementioned efficiency problems.

The considered geometry is sketched in Fig. 18, where the inducer is denoted by "I". A nose "N" as well as an after-body "A" smoothly join "I"; in particular, the nose is part of an axisymmetrical ellipsoid while the after-body is a circular cylinder having a diameter equal to the base diameter of the inducer. The flow domain is bounded by a cylindrical case, whose diameter is equal to the maximum blade tip diameter D; hence, there is no tip clearance and a shrouded inducer is considered. The length L_{out} of the

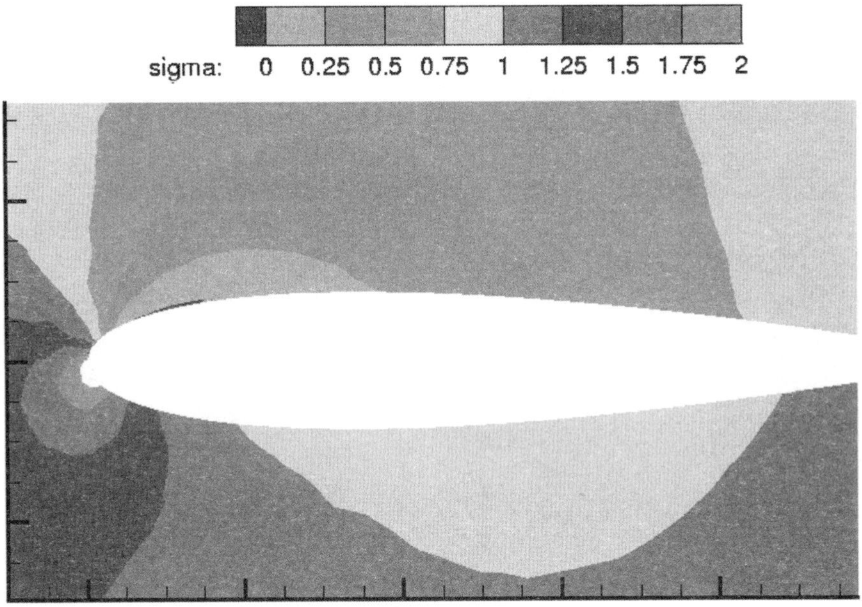

Figure 16. Contour plot of the local cavitation number (sigma) on the test-section (detail), for the test-case CAV2 reported in Tab. 5.

after-body, as well as the length L_{in} of the inflow section, are chosen equal to $1.5\,D$. The inducer angular speed is equal to 2000 rpm.

The chosen temperature of the water is $T_L = 296.16$ K. The considered free-stream conditions are the following: $p_\infty = 115000$ Pa, $\|\mathbf{u}_\infty\| = 0.476$ m/s, $M_\infty = 3.4 \cdot 10^{-4}$ and $\sigma_\infty = 990$. Both M_∞ and σ_∞, respectively defined in (4.12) and (4.13), are computed by exploiting the absolute velocity $\mathbf{u}_\infty$.

The domain sketched in Fig. 18 is discretized by means of a 3D tetrahedral unstructured grid having 549139 nodes and 2588501 tetrahedrons. It is worth mentioning that the size of the grid elements smoothly transitions between different regions on the body surface (e.g. the nose-inducer junction shown in Fig. 19) and accurately follows the solid walls even within high-curvature regions (e.g. the hub-blade intersection shown in Fig. 20).

As far as the time-advancing is concerned, the variable time-step (4.14) is adopted.

In consideration of the fact that, with respect to the rotating frame, the local Mach number can undergo substantial variations along the radial direction due to the dragging velocity, a local preconditioning strategy is particularly suitable for this test case. As previously, a local preconditioning parameter β_{ij}^2 is used in the preconditioning matrix

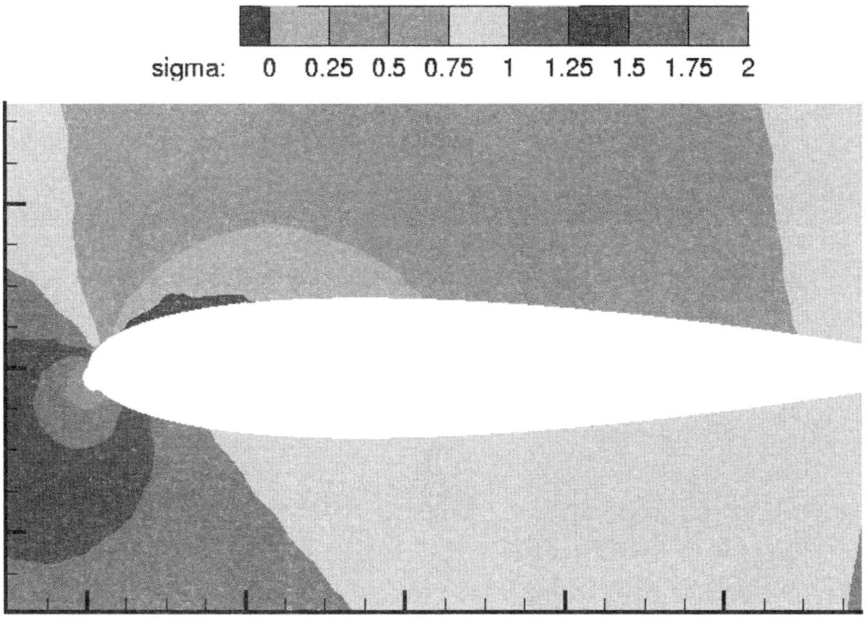

Figure 17. Contour plot of the local cavitation number (sigma) on the test-section (detail), for the test-case CAV3 reported in Tab. 5.

(3.7) in place of the original parameter β^2. This is heuristically defined as follows:

$$\beta_{ij}^2 := 1 - \exp\left(-\left(\hat{M}_{ij}\right)^2\right)$$

where:

$$\hat{M}_{ij} := \frac{\|\mathbf{u}_{ij}\|}{a_{ij}}$$

with $\mathbf{u}_{ij}$ and a_{ij} respectively defined in (3.3) and (3.4).

The pressure contours on the inducer surface, obtained after 27000 time iterations, are reported in Fig. 21. It is worth noticing that the behavior of the flow field, as described by the numerical simulation, is in a good qualitative agreement with that one observed in a number of experimental works. Indeed, the working fluid gradually undergoes a pressure rise while flowing within the vanes between the rotating blades, as shown in Fig. 21. Moreover, according to this figure, the flow region which is most prone to cavitation is located near that portion of the blades where the volutes, detaching from the hub, firstly reach the external tip diameter D. This is in agreement with the experiments which, for similar flow conditions, observe the cavitation inception exactly in the flow region under consideration (see e.g. Cervone et al. (2005)). Furthermore, the considerable axial back-flow occurring near the blade tip where the diameter is less

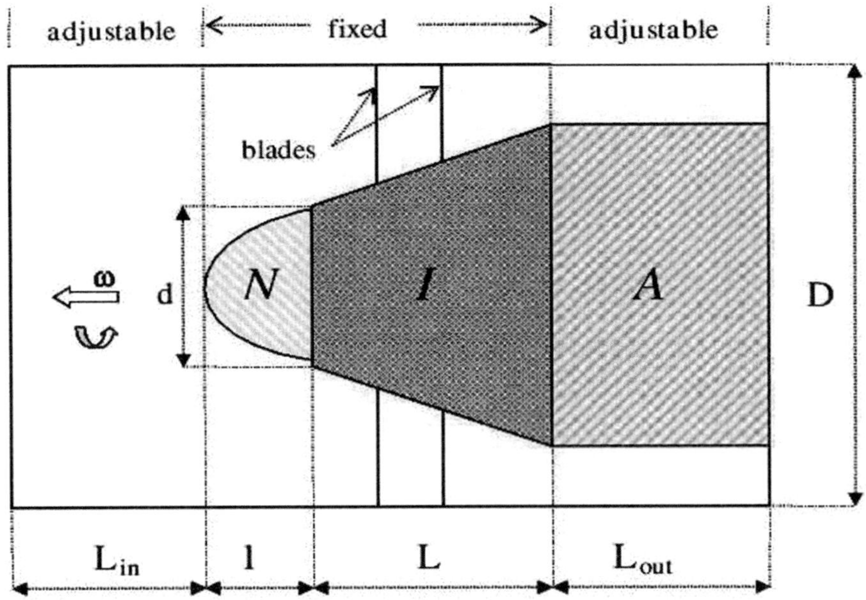

Figure 18. Schematic representation of the considered inducer geometry.

than D (i.e. where the volutes are not completely shrouded), which is well documented in a number of experimental works (e.g. Yamada et al. (2002)), is described by the numerical simulation as well, as shown in Fig. 22. As for the computational efficiency, the coefficient $c^{(CFL)}$ introduced in (4.14) has been increased during the simulation for smoothly abandoning the initial flow field (which is, in general, a crude approximation of the final one). In particular, it has been increased up to 350, confirming that the proposed linearized implicit time advancing is efficient for smooth non cavitating flows (see also Sec. 4.1).

5 Concluding remarks and perspectives

The final goal of the research activity here described was to develop a numerical tool for simulating propellant flows around 3D rotating axial inducers belonging to the feed turbo-pump system of a liquid propellant rocket engine.

In view of the fact that, under typical operational conditions, cavitation phenomena can take place within the aforementioned turbo-machines, the choice of a suitable cavitation model was initially addressed. A barotropic homogeneous flow model was adopted, which can take into account thermal cavitation effects and, possibly, the concentration of the active cavitation nuclei. This model was coupled with the mass and momentum balances for compressible flows; the effects of viscosity were neglected.

Thus, a numerical method for simulating 3D barotropic flows in complex, possibly ro-

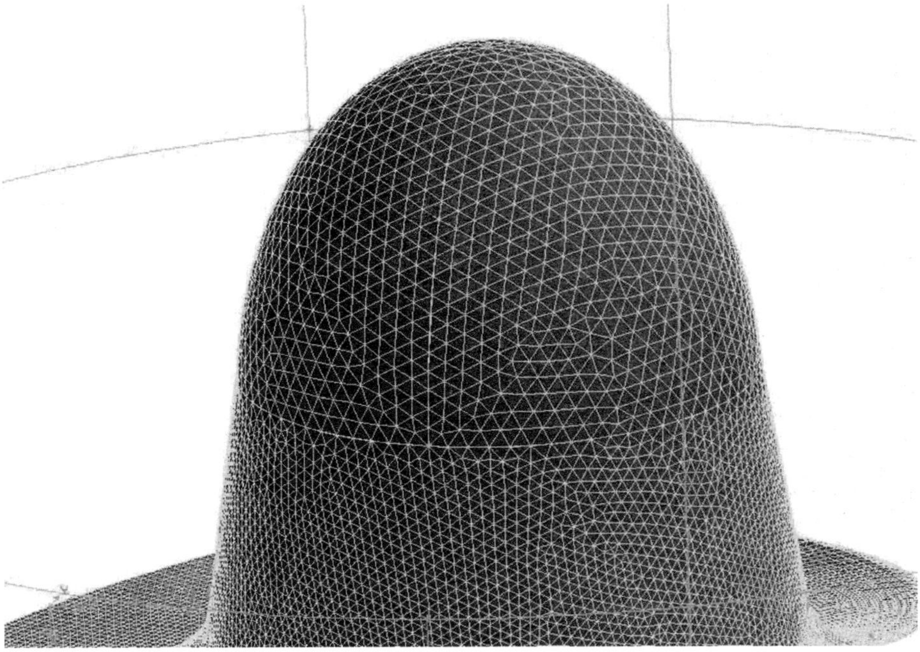

Figure 19. Detail of the grid at the nose-inducer junction.

tating, geometries was developed and has been described herein. The space discretization is based on a finite-volume approach, applicable to unstructured tetrahedral grids. The numerical fluxes at cell interfaces are computed through an adaptation of the Roe flux function to a generic barotropic case. Preconditioning is required by the compressible solver when dealing with nearly-incompressible flows. A technique already proposed for perfect gases, which can be exploited for unsteady computations as well, was adapted to the barotropic case. However, the introduction of the preconditioning narrows the stability region of common explicit time-advancing schemes. To counteract this problem, a linearized implicit time-advancing strategy was proposed, only relying on the algebraic properties of the Roe flux function and therefore applicable to a variety of problems. In particular, differently from classical linearization techniques, the proposed one does not rely on the first-order homogeneity of the analytical flux function (since this properties, satisfied by the ideal gas state law, does not hold for the barotropic one). The implicit scheme was further extended so as to incorporate the aforementioned preconditioning strategy.

A first 1D validation of the proposed numerical methodology has been presented for the water flow in a convergent-divergent nozzle in non-cavitating conditions. The proposed preconditioning technique turns out to effectively counteract the accuracy problem at low Mach numbers. Furthermore, the proposed linearized implicit scheme allows for an efficient time-advancing to be performed when considering non-cavitating flows. A 1D Riemann problem, whose exact solution was derived in Sinibaldi (2007), has been

Figure 20. Detail of the hub-blade intersection.

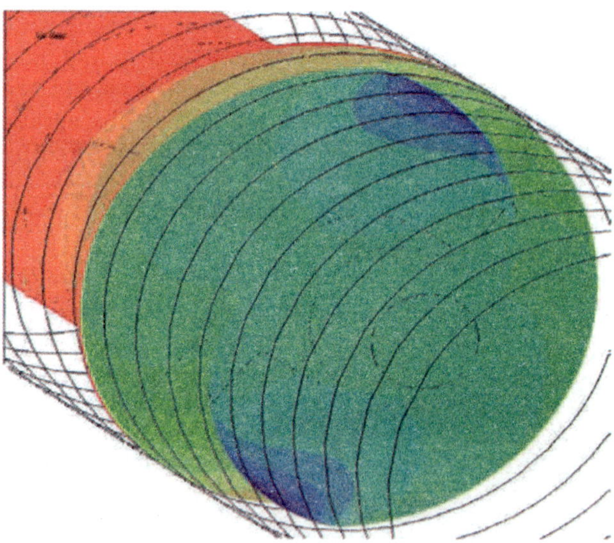

Figure 21. Pressure contours on the inducer surface: max [red] 177700 (Pa), min [blue] 79700 (Pa), spacing 5000 (Pa).

considered as an exact benchmark for the validation of the numerical set-up in cavitating conditions. Efficiency problems have been observed also for the implicit linearized scheme as soon as cavitation occurs, in that the time-step has to be dramatically reduced to obtain stable simulations.

A few simulations of the water flow around a NACA0015 hydrofoil, for which experimental data concerning the pressure coefficient distribution were available, have also been presented. In particular, water at 20° C was considered, leading to the occurrence of "cold cavitation" phenomena whose numerical treatment is extremely challenging. All the issues highlighted in the 1D numerical experiments appears in the 3D case as well; in particular, the proposed scheme proves out to efficiently compute non-cavitating flows but, as soon as cavitation takes place, its efficiency is significantly reduced. As far as the accuracy is concerned, the obtained results seem reasonably good, in view of the fact that the considered 3D numerical scheme was only first-order accurate and the used grids were relatively coarse.

A few thousand iterations of a non-cavitating simulation of the water flow around an axial turbo-pump inducer were carried out as well. The behavior of the flow field, as described by the considered numerical simulation, turns out to be in a good qualitative agreement with that one observed in a number of experimental works. In particular, the numerical solution correctly describes the pressure contours on the surface of the inducer blades as well as the considerable axial back-flow occurring where the inducer volutes are not completely shrouded. Moreover, also for the non-cavitating case under consideration,

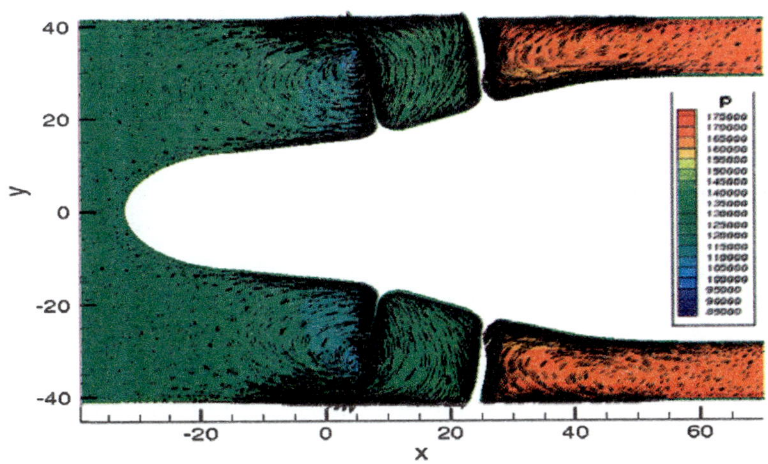

Figure 22. Velocity field (x: axial component, y: radial component) in a longitudinal cut plane of the flow domain. Pressure contours are drawn in the background.

an efficient time-advancing can be performed.

Clearly, the efficiency problem emerging when considering non-smooth flow fields, like those originating from cavitation inception when adopting realistic homogeneous flow models, deserves special attention. This efficiency issue seems to be imputable to the approximate linearization of the Roe numerical flux in the implicit time-advancing (as briefly mentioned, the specifically adopted linearization does not seem to play a crucial role in this problem). Consequently, it could be of interest to also consider different (i.e. more robust, even if less refined) numerical flux functions as, for instance, the Rusanov flux, the HLL/HLLC flux, etc... (Toro (1997)). The aforementioned point could be supported by the fact that the considered Roe flux function (as it stands, without fixes) may provide entropy-violating solutions within the transonic regime associated with cavitation inception. It is therefore evident that there is room for improvement while keeping the adopted numerical frame (i.e. homogeneous flow cavitation model, compressible -generally preconditioned- algorithms, finite volume space discretization, linearized implicit time-advancing); further investigation in this direction is definitely recommended. Simultaneously, it would be of interest to increase the order of accuracy of the proposed method as well as to investigate additional/different numerical ingredients as, for instance, relaxation techniques (see e.g. Barberon and Helluy (2005), Coutier-Delgosha et al. (2005b) and Coutier-Delgosha et al. (2005a)) and dual time-stepping strategies (see e.g. Coutier-Delgosha et al. (2005b), Coutier-Delgosha et al. (2005a) and Kunz et al. (2000)), which seem to improve the convergence properties of the considered algorithms. Finally, additional physical phenomena should also be taken into account, such as viscosity and turbulence.

Bibliography

T. Barberon and P. Helluy. Finite volume simulation of cavitating flows. *Computers & Fluids*, 34:832–858, 2005.

F. Beux, M.-V. Salvetti, A. Ignatyev, D. Li, C. Merkle, and E. Sinibaldi. A numerical study of non-cavitating and cavitating liquid flow around a hydrofoil. *Mathematical Modelling and Numerical Analysis*, 39(3):577–590, 2005.

M. Bilanceri. Studio dell'effetto della legge di stato nella simulazione di un flusso cavitante barotropico. Bsc Thesis in Aerospace Engineering, University of Pisa, Pisa (Italy), a.a. 2005, 2005.

C. E. Brennen. *Hydrodynamics of Pumps.* Concepts ETI Inc. and Oxford University Press, 1994.

A. Cervone, L. Torre, C. Bramanti, E. Rapposelli, and L. d'Agostino. Experimental characterization of the cavitation instabilities in the AVIO FAST2 inducer. In *Proc. 41st AIAA/ASME/SAE/ASEE Joint Propulsion Conference*, Tucson (Arizona, USA), July 2005.

A. Cervone, C. Bramanti, E. Rapposelli, and L. d'Agostino. Thermal cavitation experiments on a naca 0015 hydrofoil. *Journal of Fluids Engineering, Transactions of the ASME*, 128(2):326–331, 2006.

C.H. Choi, S.-S. Hong, B.J. Cha, and S. Yang. Study on the hydraulic performance of a turbopump inducer. In *Proc. ASME FEDSM'03 - 4th ASME/JSME Joint Fluids Engineering Conference*, Honolulu (Hawaii, USA), July 2003.

O. Coutier-Delgosha and J.A. Astolfi. Numerical prediction of the cavitating flow on a two-dimensional symmetrical hydrofoil with a single fluid model. In *Proc. CAV2003 - Fifth International Symposium on Cavitation*, Osaka (Japan), November 2003.

O. Coutier-Delgosha, J-L. Reboud, and R. Fortes-Patella. Numerical study of the effect of the leading edge shape on cavitation around inducer blade sections. In *Proc. CAV2001 - Fourth International Symposium on Cavitation*, Pasadena (California, USA), June 2001.

O. Coutier-Delgosha, R. Fortes-Patella, J-L. Reboud, N. Hakimi, and C. Hirsch. Stability of preconditioned Navier-Stokes equations associated with a cavitation model. *Computers & Fluids*, 34:319–349, 2005a.

O. Coutier-Delgosha, R. Fortes-Patella, J. L. Reboud, N. Hakimi, and C. Hirsch. Numerical simulation of cavitating flow in 2D and 3D inducer geometries. *International Journal for Numerical Methods in Fluids*, 48:135–167, 2005b.

O. Coutier-Delgosha, P. Morel, R. Fortes-Patella, and J. L. Reboud. Numerical simulation of turbopump inducer cavitating behavior. *International Journal of Rotating Machinery*, 2:135–142, 2005c.

L. d'Agostino and E. Rapposelli. A modified bubbly isenthalpic model for numerical simulation of cavitating flows. *AIAA paper 2001-3402*, 2001.

C. Farhat, B. Koobus, and H. Tran. Simulation of vortex shedding dominated flows past rigid and flexible structures. In *Computational Methods for Fluid-Structure Interaction*, pages 1–30. Tapir, 1999.

L. Fezoui and B. Stoufflet. A class of implicit upwind schemes for Euler simulations with unstructured meshes. *Journal of Computational Physics*, 84:174–206, 1989.

H. Guillard and C. Viozat. On the behaviour of upwind schemes in the low Mach number limit. *Computers & Fluids*, 28:63–86, 1999.

H.W.M. Hoeijmakers, M.E. Janssens, and W. Kwan. Numerical simulation of sheet cavitation. In *Proc. Third International Symposium on Cavitation*, Grenoble (France), April 1998.

A. Hosangadi, V. Ahuja, and R.J. Ungewitter. Simulations of cavitating flows in turbopumps. *AIAA paper 2003-1261*, 2003.

A. Hosangadi, V. Ahuja, and R.J. Ungewitter. Simulations of cavitating cryogenic inducers. *AIAA paper 2004-4023*, 2004.

M. Ishii. *Thermo-fluid Dynamic Theory of Two-Phase Flow.* Eyrolles, 1975.

M.E. Janssens, S.J. Hulshoff, and H.W.M. Hoeijmakers. Calculation of unsteady attached cavitation. *AIAA paper 97-1936*, 1997.

K.C. Karki and S.V. Patankar. Pressure based calculation procedure for viscous flows at all speed in arbitrary configurations. *AIAA Journal*, 27(9):1167–1174, 1989.

R. F. Kunz, D. A. Boger, D. R. Stinebring, T. S. Chyczewski, J. W. Lindau, H. J. Gibeling, S. Venkateswaran, and T. R. Govindan. A preconditioned Navier-Stokes method for two-phase flows application to cavitation prediction. *Computers & Fluids*, 29:849–875, 2000.

J.J. Mc-Guirck and Page G.J. Shock-capturing using a pressure correction method. *AIAA Journal*, 28(10):1751–1757, 1990.

C. L. Merkle, J. Feng, and P. E. O. Buelow. Computational modeling of the dynamics of sheet cavitation. In *Proc. Third International Symposium on Cavitation*, Grenoble (France), April 1998.

B. Pouffary, R. Fortes-Patella, and J-L Reboud. Numerical simulation of cavitating flow around a 2D hydrofoil: a barotropic approach. In *Proc. CAV2003 - Fifth International Symposium on Cavitation*, Osaka (Japan), November 2003.

Q. Qin, C. C. S. Song, and R.E.A. Arndt. A virtual single-phase natural cavitation model and its application to CAV2003 hydrofoil. In *Proc. CAV2003 - Fifth International Symposium on Cavitation*, Osaka (Japan), November 2003.

J-L. Reboud, B. Stutz, and O. Coutier. Two-phase flow structure of cavitation: experiment and modelling of unsteady effects. In *Proc. Third International Symposium on Cavitation*, Grenoble (France), April 1998.

P. L. Roe. Approximate Riemann solvers, parameter vectors, and difference schemes. *Journal of Computational Physics*, 43:357–372, 1981.

Y. Saito, I. Nakamori, and T. Ikohagi. Numerical analysis of unsteady vaporous cavitating flow around a hydrofoil. In *Proc. CAV2003 - Fifth International Symposium on Cavitation*, Osaka (Japan), November 2003.

I. Senocak and W. Shyy. A pressure-based method for turbolent cavitating flow computations. *Journal of Computational Physics*, 176:363–383, 2002.

W. Shyy, J. Wu, and Y. Utturkar. Computational modeling of cavitation for liquid rocket applications. *AIAA paper 2004-3985*, 2004.

E Sinibaldi. *Implicit preconditioned numerical schemes for the simulation of three-dimensional barotropic flows.* Edizioni della Normale, 2007. In Press.

E. Sinibaldi, F. Beux, and M.V. Salvetti. A preconditioned implicit Roe scheme for barotropic flows: towards simulation of cavitation phenomena. Rapport de recherche 4891, INRIA, 2003.

E. Sinibaldi, F. Beux, and M.V. Salvetti. A preconditioned compressible flow solver for numerical simulation of 3D cavitation phenomena. In *Proc. ECCOMAS2004*, Jyväskylä (Finland), July 2004.

C. C. S. Song and J. He. Numerical simulation of cavitating flows by single-phase flow approach. In *Proc. Third International Symposium on Cavitation*, Grenoble (France), April 1998.

E. F. Toro. *Riemann solvers and numerical methods for fluid dynamics.* Springer, 1997.

E. Turkel. Preconditioned methods for solving the incompressible and low speed compressible equations. *Journal of Computational Physics*, 72:277–298, 1987.

E. Turkel, A. Fiterman, and B. van Leer. *Frontiers of Computational Fluid Dynamics*, chapter Preconditioning and the limit of the compressible to the incompressible flow equations for finite difference schemes., pages 215–234. Chichester, Wiley, 1994.

D. R. van der Heul, C. Vuik, and P. Wesseling. A staggered scheme for hyperbolic conservation laws applied to unsteady sheet cavitation. *Computing and Visualization in Science*, 2:63–68, 1999.

D. R. van der Heul, C. Vuik, and P. Wesseling. Efficient computation of flow with cavitation by compressible pressure correction. In *Proc. ECCOMAS 2000*, Barcelona (Spain), September 2000.

C. Viozat. Implicit upwind schemes for low mach number compressible flows. Technical Report RR-3084, INRIA, 1997.

J. Wu, Y. Utturkar, and W. Shyy. Assessment of modeling strategies for cavitating flow around a hydrofoil. In *Proc. CAV2003 - Fifth International Symposium on Cavitation*, Osaka (Japan), November 2003.

T. Y. Wu. Cavity and wake flows. *Annual Review of Fluid Mechanics*, 3:243–284, 1972.

H. Yamada, S. Hasegawa, M. Watanabe, T. Hashimoto, T. Kimura, J. Takita, and I. Kubota. Observation of the inner flow in the inducer. In *Proc. 9th International Symposium on Transport Phenomena and Dynamics of Rotating Machinery*, Honolulu (Hawaii, USA), February 2002.